Recent Improvements in the Theory of Chaotic Attractors

This book presents some exceptional developments in chaotic attractor theory encompassing several new directions of research such as three-dimensional axiom A-diffeomorphisms, Shilnikov attractors, dendrites, and finite graphs.

The theory of chaotic attractors has experienced exceptional development over the last fifty years since the revelation of chaos in mathematics (invented by James Yorke) and symbolized by the "butterfly effect". Relevant new results have been collected in this book, including the following:

- Some remarks on minimal sets on dendrites and finite graphs and the study of recurrence and nonwandering sets of local dendrite maps.
- Ramified continua as global attractors of C^1-smooth self-maps of a cylinder close to skew products.
- Chaotic behaviour of countable products of homeomorphism groups and dynamics of three-dimensional axiom A-diffeomorphisms with two-dimensional attractors and repellers.
- The search for invariant sets of the generalized tent map and quasi-hyperbolic regime in a certain family of 2-D piecewise linear map.
- Shilnikov attractors of three-dimensional flows and maps, right fractional calculus to inverse-time chaotic maps and asymptotic stability analysis and diffeomorphisms with infinitely many Smale horseshoes.

The theory of chaotic attractor is also used as a core for evolutionary algorithms and metaheuristic optimizers in this volume.

This book will be of great value to students and researchers in mathematics, physics, engineering, and related disciplines seeking to deepen their understanding of chaotic dynamical systems and their applications. The chapters in this book were originally published in *Journal of Difference Equations and Applications.*

René Lozi is Emeritus Professor at University Côte d'Azur, Nice, France, and Vice-President of the International Society of Difference Equations. His research areas include complexity and emergence theory, dynamical systems, bifurcations, control of chaos, cryptography based on chaos, and memristors.

Lyudmila Efremova is Professor at Nizhny Novgorod State University and Moscow Institute of Physics and Technology, Russia. Her scientific interests include regular and chaotic properties of low-dimensional discrete dynamical systems.

Michal Pluhacek is Associate Professor at Tomas Bata University in Zlin, Czech Republic. His research focus includes theory and applications of evolutionary computation, swarm intelligence, swarm robotics, and artificial intelligence in general.

Recent Improvements in the Theory of Chaotic Attractors

Edited by
René Lozi, Lyudmila Efremova and
Michal Pluhacek

CRC Press
Taylor & Francis Group
Boca Raton London New York

CRC Press is an imprint of the
Taylor & Francis Group, an **informa** business
AN AUERBACH BOOK

First edition published 2025
by CRC Press
4 Park Square, Milton Park, Abingdon, Oxon, OX14 4RN

and by CRC Press
2385 NW Executive Center Drive, Suite 320, Boca Raton FL 33431

CRC Press is an imprint of Informa UK Limited

British Library Cataloguing-in-Publication Data
A catalogue record for this book is available from the British Library

ISBN13: 978-1-032-94001-4 (hbk)
ISBN13: 978-1-032-94002-1 (pbk)
ISBN13: 978-1-003-56851-3 (ebk)

DOI: 10.1201/9781003568513

Typeset in Minion Pro
by codeMantra

Publisher's Note
The publisher accepts responsibility for any inconsistencies that may have arisen during the conversion of this book from journal articles to book chapters, namely the inclusion of journal terminology.

Disclaimer
Every effort has been made to contact copyright holders for their permission to reprint material in this book. The publishers would be grateful to hear from any copyright holder who is not here acknowledged and will undertake to rectify any errors or omissions in future editions of this book.

Contents

Citation Information

The following chapters in this book were originally published in the *Journal of Difference Equations and Applications*, volume 29, issue 9-12 (2023). When citing this material, please use the original page numbering for each article, as follows:

Chapter 1
On the quasi-hyperbolic regime in a certain family of 2-D piecewise linear maps
Asma Ladjeroud and Elhadj Zeraoulia
Journal of Difference Equations and Applications, volume 29, issue 9–12 (2023), pp. 1127–1139

Chapter 2
Right fractional calculus to inverse-time chaotic maps and asymptotic stability analysis
Guo-Cheng Wu, Jia-Li Wei and Maokang Luo
Journal of Difference Equations and Applications, volume 29, issue 9–12 (2023), pp. 1140–1155

Chapter 3
Search for invariant sets of the generalized tent map
Kimberly Ayers, Dmitriy Dmitrishin, Ami Radunskaya, Alexander Stokolos and Kostyantyn Stokolos
Journal of Difference Equations and Applications, volume 29, issue 9–12 (2023), pp. 1156–1183

Chapter 4
On Shilnikov attractors of three-dimensional flows and maps
Yulia V. Bakhanova, Sergei V. Gonchenko, Alexander S. Gonchenko, Alexei O. Kazakov and Evgeniya A. Samylina
Journal of Difference Equations and Applications, volume 29, issue 9–12 (2023), pp. 1184–1201

Chapter 5
Chaotic attractors of discrete dynamical systems used in the core of evolutionary algorithms: state of art and perspectives
Ivan Zelinka and Roman Senkerik
Journal of Difference Equations and Applications, volume 29, issue 9–12 (2023), pp. 1202–1227

Chapter 6

Chaos in popular metaheuristic optimizers – a bibliographic analysis
Michal Pluhacek, Anezka Kazikova, Adam Viktorin, Tomas Kadavy and Roman Senkerik
Journal of Difference Equations and Applications, volume 29, issue 9–12 (2023), pp. 1228–1243

Chapter 7

Ramified continua as global attractors of C^1-smooth self-maps of a cylinder close to skew products
Lyudmila S. Efremova
Journal of Difference Equations and Applications, volume 29, issue 9–12 (2023), pp. 1244–1274

Chapter 8

Dynamics of three-dimensional A-diffeomorphisms with two-dimensional attractors and repellers
Marina Barinova, Vyacheslav Grines and Olga Pochinka
Journal of Difference Equations and Applications, volume 29, issue 9–12 (2023), pp. 1275–1286

Chapter 9

Chaotic behaviour of countable products of homeomorphism groups
Nina I. Zhukova and Alexander G. Korotkov
Journal of Difference Equations and Applications, volume 29, issue 9–12 (2023), pp. 1287–1312

Chapter 10

Remarks on minimal sets on dendrites and finite graphs
Elena N. Makhrova
Journal of Difference Equations and Applications, volume 29, issue 9–12 (2023), pp. 1313–1322

Chapter 11

Recurrence and nonwandering sets of local dendrite maps
Hafedh Abdelli, Habib Marzougui and Amira Mchaalia
Journal of Difference Equations and Applications, volume 29, issue 9–12 (2023), pp. 1323–1340

Chapter 12

Diffeomorphisms with infinitely many Smale horseshoes
Xu Zhang and Guanrong Chen
Journal of Difference Equations and Applications, volume 30, issue 12, pp. 1866–1884
https://doi.org/10.1080/10236198.2024.2368170

Notes on Contributors

Hafedh Abdelli, Monastir Preparatory Engineering Institute, University of Monastir, Tunisia; Faculty of Science of Bizerte, "Dynamical Systems and their Applications", University of Carthage, Tunisia; Department of Mathematics, Prince Sattam Bin Abdulaziz University, College of Science and Humanities in Al-Kharj, Saudi Arabia.

Kimberly Ayers, Department of Mathematics, Cal State San Marcos, USA.

Yulia V. Bakhanova, International Laboratory of Dynamical Systems and Applications, Higher School of Economics, National Research University, Nizhny Novgorod, Russia.

Marina Barinova, HSE University, Nizhny Novgorod, Russian Federation.

Guanrong Chen, Department of Electrical Engineering, City University of Hong Kong, People's Republic of China.

Dmitriy Dmitrishin, Department of Mathematical Sciences, Odessa National Polytechnic University, Ukraine.

Lyudmila Efremova, Department of Differential Equations, Mathematical and Numerical Analysis, National Research University, Nizhny Novgorod, Russia; Department of Fundamental Mathematics, Moscow Institute of Physics and Technology, Russia.

Alexander S. Gonchenko, Scientific and Educational Mathematical Center "Mathematics of Future Technologies", Lobachevsky State University of Nizhny Novgorod, Russia.

Sergei V. Gonchenko, International Laboratory of Dynamical Systems and Applications, Higher School of Economics, National Research University, Nizhny Novgorod, Russia; Scientific and Educational Mathematical Center "Mathematics of Future Technologies", Lobachevsky State University of Nizhny Novgorod, Russia.

Vyacheslav Grines, HSE University, Nizhny Novgorod, Russian Federation.

Tomas Kadavy, Faculty of Applied Informatics, Tomas Bata University in Zlin, Czech Republic.

Alexei O. Kazakov, International Laboratory of Dynamical Systems and Applications, Higher School of Economics, National Research University, Nizhny Novgorod, Russia; Scientific and Educational Mathematical Center "Mathematics of Future Technologies", Lobachevsky State University of Nizhny Novgorod, Russia.

Anezka Kazikova, Faculty of Applied Informatics, Tomas Bata University in Zlin, Czech Republic.

Alexander G. Korotkov, Lobachevsky State University of Nizhny Novgorod, Russia.

Asma Ladjeroud, Laboratory of Mathematics Informatics and Systems (LAMIS), University of Larbi Tebessi, Tebessa, Algeria.

René Lozi, Laboratoire J. A. Dieudonn'e, C.N.R.S., Université Côte d'Azur, Nice, France.

Maokang Luo, School of Mathematics, Sichuan University, Chengdu, People's Republic of China.

Elena N. Makhrova, Institute of Information Technologies, Mathematics and Mechanics, Lobachevsky State University of Nizhny Novgorod, Russia.

Habib Marzougui, Faculty of Science of Bizerte, "Dynamical Systems and their Applications", University of Carthage, Tunisia.

Amira Mchaalia, Faculty of Science of Bizerte, "Dynamical Systems and their Applications", University of Carthage, Tunisia.

Michal Pluhacek, Faculty of Applied Informatics, Tomas Bata University in Zlin, Czech Republic.

Olga Pochinka, HSE University, Nizhny Novgorod, Russian Federation.

Ami Radunskaya, Department of Mathematics and Statistics, Pomona College, Claremont, USA.

Elena A. Samylina, International Laboratory of Dynamical Systems and Applications, Higher School of Economics, National Research University, Nizhny Novgorod, Russia.

Roman Senkerik, A.I.Lab, Faculty of Applied Informatics, Tomas Bata University in Zlin, Czech Republic.

Alexander Stokolos, Department of Mathematics, Georgia Southern University, Statesboro, USA.

Kostyantyn Stokolos, Department of Mathematical Sciences, Odessa National Polytechnic University, Ukraine.

Adam Viktorin, Faculty of Applied Informatics, Tomas Bata University in Zlin, Czech Republic.

Jia-Li Wei, Data Recovery Key Laboratory of Sichuan Province, College of Mathematics and Information Science, Neijiang Normal University, People's Republic of China.

Guo-Cheng Wu, Data Recovery Key Laboratory of Sichuan Province, College of Mathematics and Information Science, Neijiang Normal University, People's Republic of China.

Ivan Zelinka, Department of Computer Science, VSB Technical University of Ostrava, Czech Republic; IT4Innovations, VSB-Technical University of Ostrava, Czech Republic.

Elhadj Zeraoulia, Department of Mathematics, University of Larbi Tebessi, Tebessa, Algeria.

Xu Zhang, Department of Mathematics, Shandong University, Weihai, People's Republic of China.

Nina I. Zhukova, HSE University, Nizhny Novgorod, Russia.

Introduction: recent improvements in the theory of chaotic attractors

René Lozi, Lyudmila Efremova and Michal Pluhacek

Introduction

The Lorenz attractor published in 1963 [8] is the first example of a continuous chaotic attractor. This publication was followed in 1976 by the first example of discrete chaotic attractor, the Hénon attractor [6]. Since then, the study of chaotic dynamical systems has been developed along several directions in the second half of twentieth century. In the early 90' unexpected applications of these pure mathematical objects to many fields, including biology, cryptography, optimization, economics, and finance, were discovered. This book collects original research papers addressing such themes, in particular the theory of chaotic attractors. For two years (from Spring 2021 to Spring 2023), 33 authors from 7 countries (Algeria, China, Czech Republic, Russia, Saudi Arabia, Tunisia, and the USA) worked jointly to publish new results on this theory.

The first chapter of this book concerns the topological properties of 2-D maps.

In Chapter 1, Ladjeroud and Zeraoulia [7] proved the quasi-hyperbolicity of a certain family of 2-D piecewise linear maps by applying the theorem of Shilnikov and Gavrilov concerning the hyperbolicity of some types of smooth maps.

The new paradigm of fractional maps is explored in the second chapter of this book in which Wu, Wei, and Luo [11] introduced in their study a right fractional calculus method for inverse problem modeling and proposed a concept of inverse-time fractional chaotic maps. First, a simple right fractional linear differential equation's terminal value problem and the solutions were investigated. Then, some basics of the right discrete fractional calculus were introduced and the idea was extended to the discrete case. Right fractional sum equations were derived and numerical schemes were provided for dynamical analysis. Discrete chaos does exist in an inverse-time fractional logistic map and Hénon map. Their local stability conditions were given. They concluded that this right fractional calculus method is simpler but more efficient than the left one.

In parallel to the control of Hénon and Lozi maps, the control of the generalized tent map is studied by Ayers et al. [2] in Chapter 3, who described a predictive control method to search for its unstable periodic orbits. The invariant set containing periodic orbits is a repelling set with a complicated Cantor-like structure. They showed that for certain values of the control parameter, both the local behaviour and the global behaviour of solutions change in the controlled system; in particular, the invariant set enlarges to become an interval or the entire real axis. The computational particularities of using the control system were considered, and necessary conditions for the orbit to be periodic were given. Moreover, they fully investigated the local asymptotic stability of subcycles of the controlled system's stable cycles, and some statistical properties of the subset of the classical Cantor middle-third set that was determined by the periodic points of the generalized tent map were described.

In Chapter 4 Bakhanova et al. [3] described scenarios for the emergence of Shilnikov attractors, i.e. strange attractors containing a saddle-focus with two-dimensional unstable manifold, in the case of three-dimensional flows and maps. The presented results were illustrated with various specific examples.

Chaos, an inherent phenomenon manifested by both continuous and discrete dynamic nonlinear systems, has become an essential component of engineering design. The field of evolutionary computing has been no exception. These heuristic optimization techniques (also known as metaheuristics) have seen researchers endeavouring to incorporate elements of chaos to enhance optimization, and, simultaneously, seeking to employ metaheuristics to control and harness chaotic systems.

Chapter 5 by Zelinka and Senkerik [12] delved into the intersection between discrete dynamical systems and bio-inspired metaheuristic algorithms. They examined the relationship between chaos and bio-inspired metaheuristic algorithms from two distinct angles: the presence of chaos within the realm of algorithms and the application of bio-inspired algorithms for the identification, control, or synthesis of complex systems. Furthermore, their paper offered an in-depth exploration of the interplay between metaheuristics and complex nonlinear dynamics, including potential avenues for future research. Their presented simulation studies, along with the design of objective functions for optimization and the implementation of metaheuristics, serve as a valuable foundation for subsequent experiments involving discrete nonlinear systems and complex systems with time delays. Such systems may benefit from multiparametric or multicriteria optimization approaches, paving the way for novel advancements in the field.

Pluhacek et al. [10] in Chapter 6 presented an overview of the history and recent efforts in combining chaos theory and evolutionary computation techniques. Various algorithms from the evolutionary computation domain, also known as metaheuristic algorithms, have been successfully enhanced with chaotic components in the past. Numerous ways to incorporate chaos have been examined, and many impressive results have been reported. Implementations of discrete chaotic maps such as Lozi, Hénon, and logistic map as generators of chaotic pseudo-random sequences for controlling evolution operators in metaheuristics have achieved significant popularity. In their survey, they focused on the research field itself and performed a bibliographical analysis to show how broad and active is nowadays the research field of chaos-enhanced metaheuristics and what are some of the most recent works published.

In this second part of this book, dynamics of maps are considered from a more general point of view.

In Chapter 7, Efremova [5] considered C^1-smooth self-maps of a cylinder close to C^1-smooth skew products (and satisfying some additional conditions) and studied a geometric property of these maps, as existence of C^1-smooth invariant local lamination, and applied this geometric property to the proof of their geometric integrability. The obtained results allowed her to construct an example of such family of maps, so that each map of this family admits a global attractor, which is a one-dimensional ramified continuum with a complicated topological structure. It consists of arcs of two types. On the unique circle (which is the arc of first type) the map is mixing; on arcs of second type of different lengths homeomorphic to a closed interval (the family of such arcs has continuum cardinality) the map is not mixing. The topological structure of the global attractor and dynamical properties of trajectories on the attractor lead to the property of dense intermittency (in the complement to the attractor) of attraction sets of different ω-limit sets, the union of which coincides with the global attractor.

Barinova, Grines et Pochinka, [4] in Chapter 8 considered a class of A-diffeomorphisms given on a 3-manifold, assuming that all the basic sets of the diffeomorphisms are two dimensional. It is known that such basic sets are either attractors or repellers and they are two types only: surface or locally direct product of 2-disc by Cantor set. They proved that different types of two-dimensional basic sets do not coexist in the non-wandering set of the same 3-diffeomorphism. They also constructively proved the existence of an energy function for systems of this class. This result was illustrated by examples that the two-dimensionality of the basic sets is essential in this matter and a decrease in the dimension can lead to the absence of the energy function for a diffeomorphism.

In Chapter 9 [14] Zhukova and Korotkov highlighted some remarks on the chaotic behaviour of countable products of homeomorphism groups. They investigated relationships between chaotic behaviour and closely related properties of topological transitivity, sensitivity to initial conditions, density of closed orbits of homeomorphism groups, and their countable products. They provided a large number of new examples of chaotic groups of homeomorphisms of countable products of various metrizable topological spaces, including infinite-dimensional topological manifolds, whose factors can be as non-compact surfaces, like triangulable closed manifolds of an arbitrary dimension.

Makhrova [9] in Chapter 10, considering X a dendrite and $f : X \to X$ a continuous map that has a minimal set M in X, obtained conditions on the structure of a dendrite X and the set M, under which f has an arc horseshoe and hence a positive topological entropy. She showed that this result is not correct both for continuous maps with an empty set of periodic points and for continuous maps with a non-empty set of periodic points on finite graphs.

In Chapter 11, Abdelli, Marzougui, and Mchaalia [1] considered recurrence and non-wandering sets of local dendrite maps. Given a local dendrite X with set *End(X)* of endpoints of X countable and $f : X \to X$ be a continuous map, let $P(f)$, $AP(f)$, $R(f)$, $\Omega(f)$, $EP(f)$, and $EAP(f)$ the sets of periodic points, almost periodic points, recurrent points, non-wandering points, eventually periodic points, and eventually almost periodic points of f, respectively. They showed that $R(f) \subset \overline{EAP(f)}$ and that $\Omega(f) \subset \overline{EAP(f)}$, whenever $P(f) = \emptyset$. As a consequence, they proved that if the set $B(X)$ of branch points is finite, then $\overline{R(f)} = \overline{AP(f)}$. They gave a counter-example showing that these results do not hold whenever $B(X)$ is infinite or $End(X)$ is uncountable.

In Chapter 12 [13], a class of planar diffeomorphims is formulated by Zhang and Chen, with infinitely many coexisting Smale horseshoes, where the Lebesgue measure of the parameters with such strange dynamics is infinite. On each horseshoe, there exists a uniformly hyperbolic invariant set, on which the map is topologically conjugate to the two-sided full–shift on two symbols. Moreover, the topological entropy is infinite in certain parameter regions.

Conclusion

Since the first example of a chaotic attractor, sixty years ago, research on such objects is still relevant. This book is only one step in this process which will continue for a long time.

References

[1] H. Abdelli, H. Marzougui & A. Mchaalia: Recurrence and nonwandering sets of local dendrite maps, *J. Differ. Equ. Appl.* 29(9–12) (2023), pp. 1323–1340. https://doi.org/10.1080/10236198.2022.2038143

[2] K. Ayers, D. Dmitrishin, A. Radunskaya, A. Stokolos & K. Stokolos: Search for invariant sets of the generalized tent map, *J. Differ. Equ. Appl.* 29(9–12) (2023), pp. 1156–1183. https://doi.org/10.1080/10236198.2022.2119851

[3] Y. Bakhanova, S. Gonchenko, A. Gonchenko, A. Kazakov & E. Samylina: On Shilnikov attractors of three-dimensional flows and maps, *J. Differ. Equ. Appl.* 29(9–12) (2023), pp. 1184–1201. https://doi.org/10.1080/10236198.2022.2063051

[4] M. Barinova, V. Grines & O. Pochinka: Dynamics of three-dimensional A-diffeomorphisms with two-dimensional attractors and repellers, *J. Differ. Equ. Appl.* 29(9–12) (2023), pp. 1275–1286. https://doi.org/10.1080/10236198.2022.2088287

[5] L. S. Efremova: Ramified continua as global attractors of C^1-smooth self-maps of a cylinder close to skew products, *J. Differ. Equ. Appl.* 29(9–12) (2023), pp. 1244–1274. https://doi.org/10.1080/10236198.2023.2204144

[6] M. Hénon: A two-dimensional mapping with a strange attractor. *Commun. Math. Phys.* 50 (1976), pp. 69–77. https://doi.org/10.1007/BF01608556.

[7] A. Ladjeroud & E. Zeraoulia: On the quasi-hyperbolic regime in a certain family of 2-D piecewise linear maps, *J. Differ. Equ. Appl.* 29(9-12) (2023), pp. 1127–1139. https://doi.org/10.1080/10236198.2022.2145890

[8] E. N. Lorenz: Deterministic nonperiodic flow. *J. Atmos. Sci.* 20 (1963), pp. 130–141. https://doi.org/10.1175/1520-0469(1963)020⟨0130:DNF⟩2.0.CO;2.

[9] E. N. Makhrova: Remarks on minimal sets on dendrites and finite graphs, *J. Differ. Equ. Appl.* 29(9–12) (2023), pp. 1313–1322. https://doi.org/10.1080/10236198.2023.2220417

[10] M. Pluhacek, A. Kazikova, A. Viktorin, T. Kadavy & R. Senkerik: Chaos in popular metaheuristic optimizers – a bibliographic analysis, *J. Differ. Equ. Appl.* 29(9–12) (2023), pp. 1228–1243. https://doi.org/10.1080/10236198.2023.2203779

[11] G.-C. Wu, J.-L. Wei & M. Luo: Right fractional calculus to inverse-time chaotic maps and asymptotic stability analysis, *J. Differ. Equ. Appl.* 29(9–12) (2023), pp. 1140–1155. https://doi.org/10.1080/10236198.2023.2198043

[12] I. Zelinka & R. Senkerik: Chaotic attractors of discrete dynamical systems used in the core of evolutionary algorithms: state of art and perspectives, *J. Differ. Equ. Appl.* 29(9–12) (2023), pp. 1202–1227. https://doi.org/10.1080/10236198.2023.2220416

[13] X. Zhang & G. Chen: Diffeomorphisms with infinitely many Smale horseshoes, *J. Differ. Equ. Appl.* (2024), Published online: 13 Jun 2024, 30(12) (2024), pp. 1866–1884. https://doi.org/10.1080/10236198.2024.2368170

[14] N. I. Zhukova & A. G. Korotkov: Chaotic behaviour of countable products of homeomorphism groups, *J. Differ. Equ. Appl.* 29(9–12) (2023), pp. 1287–1312. https://doi.org/10.1080/10236198.2022.2090838

On the quasi-hyperbolic regime in a certain family of 2-D piecewise linear maps

Asma Ladjeroud and Elhadj Zeraoulia

ABSTRACT
The main goal of this paper is to offer a proof of quasi-hyperbolicity in a certain family of 2-D piecewise linear maps by applying the theorem of Shilnikov and Gavrilov concerning the hyperbolicity of some types of smooth maps.

1. Introduction and preliminaries

In this work, we will prove the existence of a quasi-hyperbolic attractor for a particular family of 2-D piecewise linear maps. The main analysis is based on finding an absorbing region and then proving that it contains a quasi-hyperbolic attractor under some specific conditions. To do this, we need first some preliminaries about the topic of chaotic attractors of discrete mappings. Indeed, an attractor is the asymptotic limit of the solutions starting from any initial condition located in a *basin of attraction* which is a domain of non-zero volume. It is known in the current literature that several dynamical systems are chaotic deterministic and have a strange attractor [6]. According to Anishchenko & Strelkova [3], Plykin [14] and Zeraoulia [6], the strange attractors are *ideal concepts* occurring in chaotic systems satisfying a number of rigorous mathematical properties. They are classified into three principal classes which are [2]:

1.1. Hyperbolic attractors

The hyperbolic chaos is called *true chaos* because of its homogeneous and topological stable structure.

Consider $f : \Omega \longrightarrow \mathbb{R}^n$, with $\Omega \in \mathbb{R}^n$, a C^r real function defining a discrete map and Ω is a manifold. Then, we have the following definitions:

Definition 1.1: (1) If for every neighbourhood U of such a point x there is $k \geq 1$ such that $f^k(U) \cap U$ is nonempty then the point x is a non-wandering point for the map f.
(2) We call non-wandering set of f the set of all non-wandering points.
(3) An f-invariant subset Λ of $\mathbb{R}^n$ satisfies $f(\Lambda) \subset \Lambda$.
(4) If f is a diffeomorphism defined on some compact smooth manifold $\Omega \subset \mathbb{R}^n$ then, an f-invariant subset Λ of $\mathbb{R}^n$ is said to be hyperbolic if there exists $0 < \lambda_1 < 1$ and $c > 0$ such that
   (a) $T_\Lambda\Omega = \Sigma^s \bigoplus \Sigma^u$, where $\bigoplus$ means the algebraic direct sum and where $T_\Lambda\Omega$ is the tangent space of Ω.
   (b) $Df(x)\Sigma_x^s = \Sigma_{f(x)}^s$, and $Df(x)\Sigma_x^u = \Sigma_{f(x)}^u$ for each $x \in \Lambda$.
   (c) $\|Df^k V\| \leq c\lambda_1^k\|v\|$, for each $v \in \Sigma^s$ and $k > 0$.
   (d) $\|Df^{-k} =\leq c\lambda_1^k\|v\|$, for each $v \in \Sigma^u$ and $k > 0$.

where Σ^s, Σ^u are the two Df-invariant sub-manifolds (respectively, the stable and unstable sub-manifolds of the map f) and Σ_x^s, Σ_x^u are the two $Df(x)$-invariant sub-manifolds.

Definition 1.2: A hyperbolic set Λ is locally maximal (or isolated) if there exists an open set U such that $\Lambda = \bigcap_{n \in \mathbb{Z}} f^n(U)$.

The Smale horseshoe and the Plykin attractor are examples of locally maximal sets. Actually, if Λ is a hyperbolic set with nonempty interior for a map f, then f is Anosov if it is transitive, locally maximal and Ω is a surface. This type of attractors has the properties including: Shadowing property, structural stability, Markov partitions and the SRB measures. When $\Lambda = \Omega$, then the diffeomorphism f is called an *Anosov diffeomorphism or uniformly hyperbolic map*. An important example of the hyperbolic maps is the *Arnold cat map* (see [4]) given by

$$\begin{cases} x_{n+1} = x_n + y_n, & \text{mod } 1 \\ y_{n+1} = x_n + 2y_n, & \text{mod } 1 \end{cases}$$

and its small perturbation map is given by

$$\begin{cases} x_{n+1} = x_n + y_n + \delta \sin 2\pi y_n, & \text{mod } 1 \\ y_{n+1} = x_n + 2y_n, & \text{mod } 1 \end{cases}$$

where δ is a real parameter.

1.2. Lorenz-type attractors

Also called *pseudo-hyperbolic attractors*. Almost all known examples of this type of attractors does not shown to be structurally stable, although their homoclinic and heteroclinic orbits are structurally stable, and no stable periodic orbits appear under small parameter variations. One of the most used examples is the Lorenz system (see [11]) given by

$$\begin{cases} x' = \sigma(x - y) \\ y' = rx - y - xz \\ z' = -bz + \end{cases}$$

Lorenz-type attractors were considered as examples of truly strange attractors and there are a finite number of these attractors in the literature. We can summarize the properties of

Lorenz-like attractors as follow. Let Λ be an attractor of a Lorenz-like system in a compact boundaryless 3-manifold M. Let $\mathcal{C}^1(M)$ be the set of $\mathcal{C}^1$ vector fields on M endowed with the $\mathcal{C}^1$ topology. Then these systems share the following properties: (1) There is an invariant foliation whose leaves are forward contracted by the flow. (2) There is a positive Lyapunov exponent at every orbit. (3) They are expansive and so sensitive with respect to initial data. (4) They have zero volume if the flow is $\mathcal{C}^2$. (5) There is a unique physical measure whose support is the whole attractor and which is the equilibrium state with respect to the centre unstable Jacobian.

1.3. Quasi-hyperbolic attractors

This type of attractors is the main subject of our study in this paper. First of all, we should note that the most of known observed chaotic attractors are *quasi-attractors* [1,10,12,15,16], which means that they are neither *robust hyperbolic attractors* (referring to the first case defined previously) nor *pseudo-hyperbolic attractors* (referring to the second case defined previously). The fact of showing a complex structure of embedded basins of attraction from quasi-attractors in terms of initial conditions and a set of bifurcation parameters of non-zero measure does not restrict the complexity. The homoclinic tangency of stable and unstable manifolds of saddle points in the Poincaré section is the principal cause of this complexity. These attractors have the following properties: (1) Their separatrix loops of saddle-foci or homoclinic orbits of saddle cycles in the moment of tangency of their stable and unstable manifolds, because they enclose non-robust singular trajectories that are dangerous in the following sense: A *dangerous border collision bifurcation* has been defined as the dynamical instability that occurs when the basins of attraction of stable fixed points shrink to a set of zero measure as the parameter approaches the bifurcation value from either side [9]. (2) A map of Smale's horseshoe-type appears in the neighbourhood of their trajectories. This map contains both non-trivial hyperbolic subset of trajectories and a denumerable subset of stable periodic orbits [17] and Newhouse's theorem [13]. (3) The quasi-attractor is the unified limit set of the whole attracting set of trajectories including a subset of both chaotic and stable periodic trajectories which have long periods, weak and narrow basins of attraction and stability regions. This is a result of the effect that this attractor is holed by a set of basins of attraction of different periodic orbits. (4) The basins of attraction of stable cycles are very narrow. (5) Some orbits do not reveal themselves ordinarily in numeric simulations except some quite large stability windows where they are clearly visible.

2. On the quasi-hyperbolic regime in 2-D piecewise linear smooth mappings

In the present paper, we will give a specific example of *quasi-hyperbolic attractors*.

It is known that the Lozi map is the piecewise-affine version of the Hénon map, which took a large part of attention. Lozi map took the place of very important studies because of its suitable form for analysis and the much of the new chaotic phenomena that occur. As well, recently, there was a lot of work considering the analysis of a smooth version of the piecewise linear Lozi map [7,8].

We are going to present and analyse in the following lines a new result of the Lozi map generalization which is called the *border collision normal form* ($BCNF_{RC}$). We will prove that a similar form to the ($BCNF_{RC}$) has a quasi-hyperbolic attractor for some set of values of the parameters.

Let us consider the following 2-D piecewise map given by:

$$g(x,y,\mu)=\begin{pmatrix} g_1=\begin{pmatrix} f_1(x,y,\mu)\\ f_2(x,y,\mu)\end{pmatrix}, & \text{if } x<S(y,\mu)\\ g_2=\begin{pmatrix} f_1(x,y,\mu)\\ f_2(x,y,\mu)\end{pmatrix}, & \text{if } x\geq S(y,\mu)\end{pmatrix}$$

where μ is the bifurcation parameter, the smooth curve $x=S(y,\mu)$ divided the phase plane into two regions R_L and R_R given by:

$$\begin{cases} R_L=\left\{(x,y)\in\mathbb{R}^2, x<S(y,\mu)\right\}\\ R_R=\left\{(x,y)\in\mathbb{R}^2, x\geq S(y,\mu)\right\}\end{cases}$$

and the boundary between them as:

$$\Sigma=\left\{(x,y)\in\mathbb{R}^2, x=S(y,\mu)\right\}$$

We say that $g(x,y,\mu)$ is a *2-D piecewise smooth map* if the following conditions hold:

- The map g is continuous, but its derivative is discontinuous at the borderline $x=S(y,\mu)$.
- The functions g_1 and g_2 are both continuous and have continuous derivatives.
- The one-sided partial derivatives at the border are finite and in each subregion R_L and R_R.
- The map $g(x,y,\mu)$ has one fixed point in R_L and one fixed point in R_R for a value μ_* of the parameter μ.

It is possible to consider $S(y,\mu)=0$ after some change of variables as shown in [5]. Let $z=(x,y)^t$, the $BCNF_{RC}$ in two-dimensional piecewise mapping given in [5] is the continuous map

$$z_{n+1}=(x_{n+1},y_{n+1})^t=\begin{cases} J_L z_n+m, & \text{if } x_n<0\\ J_R z_n+m, & \text{if } x_n\geq 0\end{cases} \tag{1}$$

where

$$J_k=\begin{pmatrix}\tau_k & 1\\ -\delta_k & 0\end{pmatrix},\quad k=L \text{ or } R,\quad \text{and}\quad m=\begin{pmatrix}\mu\\ 0\end{pmatrix}.$$

The parameters τ_k and δ_k are taken to be fixed, and μ is the bifurcation parameter.

The fixed points of the system on the two sides, right and left, are given by:

$$X=\left(\frac{-1}{\tau_L-1-\delta_L},\frac{\delta_L}{\tau_L-1-\delta_L}\right),\quad Y=\left(\frac{1}{1+\delta_R-\tau_R},\frac{-\delta_R}{1+\delta_R-\tau_R}\right).$$

Their stability is determined by the eigenvalues $\lambda_{1,2}=\frac{1}{2}(\tau\pm\sqrt{\tau^2-4\delta})$.

2.1. Assumptions and notations

Let us consider the new mapping system which has some resemblance with the system defined in Equation (1), and we will show later in this page why we would better take this system rather than the ($BCNF_{RC}$) system, given in (1). Let F be the map defined by:

$$F(x,y) = \begin{pmatrix} f(x,y) \\ g(x,y) \end{pmatrix} = \begin{cases} \begin{pmatrix} \alpha & \tau_L \\ 0 & -\delta_L \end{pmatrix} \begin{pmatrix} x \\ y \end{pmatrix} + \begin{pmatrix} \mu \\ 0 \end{pmatrix}, & x < 0 \\ \begin{pmatrix} \alpha & \tau_R \\ 0 & -\delta_R \end{pmatrix} \begin{pmatrix} x \\ y \end{pmatrix} + \begin{pmatrix} \mu \\ 0 \end{pmatrix}, & x \geq 0 \end{cases} \tag{2}$$

where,

$$f(x,y) = \begin{cases} \tau_L y + \alpha x + \mu, & x < 0, \\ \tau_R y + \alpha x + \mu, & x \geq 0, \end{cases}$$

and,

$$g(x,y) = \begin{cases} -\delta_L y, & x < 0, \\ -\delta_R y, & x \geq 0. \end{cases}$$

The system has a single fixed point which is :

$$Y = \left(\frac{\mu}{1-\alpha}, 0 \right).$$

Its stability is determined by the eigenvalues $\lambda_1 = \alpha$ and $\lambda_2 = -\delta_k$, $k = L, R$. We shall use some assumptions on the parameters α and δ_k to find the necessary conditions for the hyperbolicity theorem given in [1, Chap 4].

Theorem 2.1 (Sufficient Conditions for Hyperbolicity): *Consider a C^1-map $F : U \longrightarrow \mathbb{R}^{m+n}$, where U is an open convex subset of $\mathbb{R}^{m+n}$, such that $F(x,y) = (\bar{x}, \bar{y})$, $x \in \mathbb{R}^m$, $y \in \mathbb{R}^n$,*

$$\bar{x} = f(x,y), \quad \bar{y} = g(x,y).$$

If,

$$\|f_x\| < 1, \tag{3}$$

$$\|g_y^{-1}\| < 1, \tag{4}$$

$$1 - \|f_x\| \|g_y^{-1}\| > 2\sqrt{\|f_y \cdot g_y^{-1}\| \|g_x\| \|g_y^{-1}\|}, \tag{5}$$

$$(1 - \|f_x\|)\left(1 - \|g_y^{-1}\|\right) > \|f_y \cdot g_y^{-1}\| \cdot \|g_x\|, \tag{6}$$

where $\| \cdot \| = \sup_{(x,y)\in U} | \cdot |$, and subscripts mean differentiation with respect to the corresponding coordinates, then:

(1) *any compact invariant set Λ in U is hyperbolic*

(2) *the stable subspace* $E^s_{p_0}$, $p_0 = (x_0, y_0) \in \Lambda$, *can be represented as the graph of an affine map*

$$y = L_{p_0}(x - x_0) + y_0 \quad \text{and} \quad \|L_{p_0}\| \leq L.$$

The unstable subspace $E^u_{p_0}$ *can be represented as the graph of an affine map*

$$x = M_{p_0}(y - y_0) + x_0,$$

and $\|M_{p_0}\| \leq M$, *where* L *and* M *are constants and*

$$LM < 1.$$

By applying this theorem on the new mappings system (2), we have

$$\left\|\frac{\partial f}{\partial x}\right\| = \alpha, \quad \left\|\frac{\partial f}{\partial y}\right\| = \tau_k, \quad \left\|\frac{\partial g}{\partial x}\right\| = 0, \quad \left\|\left(\frac{\partial g}{\partial y}\right)^{-1}\right\| \leq \frac{1}{\delta_k}. \tag{7}$$

The inequality (3) holds if $\alpha < 1$, and the inequality (4) holds if $\delta_k > 1$.
Since $\|g_x\| = 0$ then, the inequality (5) is equivalent to

$$1 - \|f_x\| \|g_y^{-1}\| > 0,$$

by using the inequalities (3) and (4), we have

$$1 - \frac{\alpha}{\delta_k} > 0,$$

then

$$\delta_k - \alpha > 0,$$

which means that the inequality (5) holds if $\delta_k > \alpha$.
Since $\|g_x\| = 0$ then, the inequality (6) is equivalent to :

$$(1 - \|f_x\|)\left(1 - \|g_y^{-1}\|\right) > 0,$$

by using the inequalities (3) and (4), we have:

$$(1 - \alpha)\left(1 - \frac{1}{\delta_k}\right) > 0,$$

which means that,

$$1 - \alpha > \frac{1 - \alpha}{\delta_k},$$

Since $\alpha < 1$ then $1 - \alpha > 0$ and by using this in the last inequality we find that the inequality (6) holds if $\delta_k > 1$. So, by collecting all of these results, we conclude that every orbit of the new normal form (2) is hyperbolic if

$$\delta_k > 1 > \alpha \quad \text{with } 0 < \alpha < 1.$$

Since $0 < \alpha < 1$, we can have a condition that is more suitable and which is:

$$1 < \delta_k < \frac{1}{\alpha} \quad \text{with } 0 < \alpha < 1. \tag{8}$$

Due to this condition, we can say that our fixed point is a saddle in the two sides where $det(J) < 0$ and $tr(J) < 0$.

The eigenvectors are $V_1 = (1, 0)$ and $V_2 = (1, \frac{-(\alpha+\delta_k)}{\tau_k})$.

Remark 2.1: The application of Theorem 2.1 on the $BCNF_{RC}$ gives us:

$$\left\| \frac{\partial f}{\partial x} \right\| = \tau_k, \quad \left\| \frac{\partial f}{\partial y} \right\| = 1, \quad \left\| \frac{\partial g}{\partial x} \right\| = -\delta_k, \quad \left\| \frac{\partial g}{\partial y} \right\| = 0.$$

So, we note that the inequality (4) in the Theorem 2.1 will not be reachable in that case, and then we conclude that we can not apply the Theorem 2.1 on the normal form $BCNF_{RC}$, which means the conditions of the last theorem are not applicable and maybe the normal form with that structure does not contain the quasi-hyperbolic regime.

2.2. *Hyperbolicity and trapping region*

In this section, we want to find an absorbing region and determine the necessary conditions for its existence. For that, we need the following theorem given in [1, Chap 4]

Theorem 2.2 (Quasi-hyperbolic attractors): *Let f be a piecewise smooth map of $\mathbb{R}^2$ and L be the union of points of discontinuity of f and Df. Let $D = \bigcup_{i=0}^{\infty} f^{-i}L$. If $x_0 \notin D$ then $f^n(x_0) \notin L$, $n \in \mathbb{Z}_+$. Assume that R is an absorbing region, i.e. $f(x) \in IntR$ if $x \in R$ and $x \notin D$. Then an attractor A is the set of all limit points of all orbits $f^i(x_0)$, $f \in \mathbb{Z}_+$, $x_0 \notin D$, i.e. $A = \text{closure}((\bar{R}\, D) \cap f^2(\bar{R}\, D) \cap \ldots)$. We call it quasi-hyperbolic (or generalized hyperbolic, following Ya. Pesin) if for $x_0 \notin D$ every orbit $x_0, f(x_0), f^2(x_0), \ldots$ is hyperbolic.*

We will apply this result to the map (2) to get

Theorem 2.3: *There exists a quasi-hyperbolic regime in the 2-D piecewise linear smooth map F given in Equation (2).*

Proof: Let Ω be the domain defined by

$$\Omega(A, B, C, D) = \{x, y \in \mathbb{R},\ x_3 \le x \le x_1,\ y_3 \le y \le y_1\}.$$

Let us consider the points $A(x_1, y_1)$, $B(x_2, y_2)$, $C(x_3, y_3)$, and $D(x_4, y_4)$. We will determine under what conditions those points, their images by the formula (2), and the images of the connecting lines between the points are in our domain Ω, by taking $x_1 > 0, x_2 > 0, x_3 < 0$ and $x_4 < 0$.

As a first step, the points $A(x_1, y_1)$, $B(x_2, y_2)$, $C(x_3, y_3)$ and $D(x_4, y_4)$ are in Ω if

$$x_3 \leq x_2 \leq x_1, \quad y_3 \leq y_2 \leq y_1,$$
$$x_3 \leq x_4 \leq x_1, \quad y_3 \leq y_4 \leq y_1.$$

Next, we will determine under what conditions the images of the points A, B, C, and D by the formula (2) are in Ω. Firstly,

$$F(A) = \begin{cases} \tau_R y_1 + \alpha x_1 + \mu, \\ -\delta_R y_1, \end{cases}$$

the image $F(A)$ is in Ω if

$$x_3 \leq \tau_R y_1 + \alpha x_1 + \mu \leq x_1,$$

and

$$y_3 \leq -\delta_R y_1 \leq y_1,$$

which means that

$$\frac{\tau_R y_1 + \mu}{1 - \alpha} \leq x_1, \tag{9}$$

and

$$y_1 > 0. \tag{10}$$

Secondly,

$$F(B) = \begin{cases} \tau_R y_2 + \alpha x_2 + \mu, \\ -\delta_R y_2, \end{cases}$$

the image $F(B)$ is in Ω if

$$x_3 \leq \tau_R y_2 + \alpha x_2 + \mu \leq x_1,$$

and

$$y_3 \leq -\delta_R y_2 \leq y_1,$$

which means that

$$\frac{x_3 - \tau_R y_2 - \mu}{\alpha} \leq x_2 \leq \frac{x_1}{\alpha}, \tag{11}$$

and

$$\frac{-y_1}{\delta_R} < y_2 < \frac{-y_3}{\delta_R}. \tag{12}$$

Thirdly,

$$F(C) = \begin{cases} \tau_L y_3 + \alpha x_3 + \mu, \\ -\delta_L y_3, \end{cases}$$

the image $F(C)$ is in Ω if

$$x_3 \leq \tau_L y_3 + \alpha x_3 + \mu \leq x_1,$$

and

$$y_3 \leq -\delta_L y_3 \leq y_1,$$

which means that

$$x_3 < \max\left\{\frac{\mu}{1-\alpha}, \frac{x_1 - \tau_L y_3 - \mu}{\alpha}\right\}, \tag{13}$$

and

$$y_3 < 0. \tag{14}$$

Finally,

$$F(D) = \begin{cases} \tau_L y_4 + \alpha x_4 + \mu, \\ -\delta_L y_4, \end{cases}$$

the image $F(D)$ is in Ω if

$$x_3 \leq \tau_L y_4 + \alpha x_4 + \mu \leq x_1,$$

and

$$y_3 \leq -\delta_L y_4 \leq y_1,$$

which means that

$$\frac{x_3 - \tau_L y_4 - \mu}{\alpha} \leq x_4 < \frac{x_1 - \tau_L y_4 - \mu}{\alpha}, \tag{15}$$

and

$$\frac{-y_1}{\delta_R} < y_4 < \frac{-y_3}{\delta_R}. \tag{16}$$

Now, we will find the necessary conditions in which the images of the lines (AB), (BC), (AD), and (DC) are in Ω. Firstly, the line segment [AB] has an equation in the form $y = ax + b$ which is,

$$y = \left(\frac{y_2 - y_1}{x_2 - x_1}\right)x + \frac{x_2 y_1 - x_1 y_2}{x_2 - x_1}.$$

F(AB) the image of (AB) by (2) is defined by

$$F(AB) = \begin{cases} \tau_k\left(\left(\frac{y_2 - y_1}{x_2 - x_1}\right)x + \frac{x_2 y_1 - x_1 y_2}{x_2 - x_1}\right) + \alpha x + \mu, \\ -\delta_k\left(\left(\frac{y_2 - y_1}{x_2 - x_1}\right)x + \frac{x_2 y_1 - x_1 y_2}{x_2 - x_1}\right), \end{cases}$$

so, F(AB) is in Ω if

$$x_3 \leq \tau_k\left(\left(\frac{y_2 - y_1}{x_2 - x_1}\right)x + \frac{x_2 y_1 - x_1 y_2}{x_2 - x_1}\right) + \alpha x + \mu \leq x_1,$$

and

$$y_3 \leq -\delta_k\left(\left(\frac{y_2 - y_1}{x_2 - x_1}\right)x + \frac{x_2 y_1 - x_1 y_2}{x_2 - x_1}\right) \leq y_1.$$

Starting with the first inequality

$$x_3 - \mu - \tau_k \left(\frac{x_2 y_1 - x_1 y_2}{x_2 - x_1} \right) \leq \left(\tau_k \left(\frac{y_2 - y_1}{x_2 - x_1} \right) + \alpha \right) x \leq x_1 - \mu - \tau_k \left(\frac{x_2 y_1 - x_1 y_2}{x_2 - x_1} \right),$$

which means

$$\frac{(x_3 - \mu)(x_2 - x_1) - \tau_k (x_2 y_1 - x_1 y_2)}{\tau_k (y_2 - y_1) + \alpha (x_2 - x_1)} \leq x \leq \frac{(x_1 - \mu)(x_2 - x_1) - \tau_k (x_2 y_1 - x_1 y_2)}{\tau_k (y_2 - y_1) + \alpha (x_2 - x_1)}. \tag{17}$$

Passing to the second inequality, we have

$$\frac{-y_1}{\delta_k} - \left(\frac{x_2 y_1 - x_1 y_2}{x_2 - x_1} \right) \leq \left(\frac{y_2 - y_1}{x_2 - x_1} \right) x \leq \frac{-y_3}{\delta_k} - \left(\frac{x_2 y_1 - x_1 y_2}{x_2 - x_1} \right),$$

which means

$$\frac{-y_1}{\delta_k} \left(\frac{x_2 - x_1}{y_2 - y_1} \right) - \left(\frac{x_2 y_1 - x_1 y_2}{y_2 - y_1} \right) \leq x \leq \frac{-y_3}{\delta_k} \left(\frac{x_2 - x_1}{y_2 - y_1} \right) - \left(\frac{x_2 y_1 - x_1 y_2}{y_2 - y_1} \right). \tag{18}$$

Secondly, the line segment [BC] has an equation in the form $y = ax + b$ which is

$$y = \left(\frac{y_3 - y_2}{x_3 - x_2} \right) x + \frac{x_3 y_2 - x_2 y_3}{x_3 - x_2}.$$

F(BC) the image of (BC) by (2) is defined by

$$F(BC) = \begin{cases} \tau_k \left(\left(\frac{y_3 - y_2}{x_3 - x_2} \right) x + \frac{x_3 y_2 - x_2 y_3}{x_3 - x_2} \right) + \alpha x + \mu, \\ -\delta_k \left(\left(\frac{y_3 - y_2}{x_3 - x_2} \right) x + \frac{x_3 y_2 - x_2 y_3}{x_3 - x_2} \right). \end{cases}$$

$F(BC)$ is in Ω if

$$x_3 \leq \tau_k \left(\left(\frac{y_3 - y_2}{x_3 - x_2} \right) x + \frac{x_3 y_2 - x_2 y_3}{x_3 - x_2} \right) + \alpha x + \mu \leq x_1,$$

and

$$y_3 \leq -\delta_k \left(\left(\frac{y_3 - y_2}{x_3 - x_2} \right) x + \frac{x_3 y_2 - x_2 y_3}{x_3 - x_2} \right) \leq y_1.$$

which means

$$\frac{(x_3 - \mu)(x_3 - x_2) - \tau_k (x_3 y_2 - x_2 y_3)}{\tau_k (y_3 - y_2) + \alpha (x_3 - x_2)} \leq x \leq \frac{(x_1 - \mu)(x_3 - x_2) - \tau_k (x_3 y_2 - x_2 y_3)}{\tau_k (y_3 - y_2) + \alpha (x_3 - x_2)}. \tag{19}$$

and

$$\frac{-y_1}{\delta_k} \left(\frac{x_3 - x_2}{y_3 - y_2} \right) - \left(\frac{x_3 y_2 - x_2 y_3}{y_3 - y_2} \right) \leq x \leq \frac{-y_3}{\delta_k} \left(\frac{x_3 - x_2}{y_3 - y_2} \right) - \left(\frac{x_3 y_2 - x_2 y_3}{y_3 - y_2} \right). \tag{20}$$

Thirdly, the line segment [DC] has an equation in the form $y = ax + b$ which is

$$y = \left(\left(\frac{y_3 - y_4}{x_3 - x_4}\right)x + \frac{x_3y_4 - x_4y_3}{x_3 - x_4}\right).$$

F(DC) the image of (DC) by (2) is defined by

$$F(DC) = \begin{cases} \tau_k\left(\left(\frac{y_3 - y_4}{x_3 - x_4}\right)x + \frac{x_3y_4 - x_4y_3}{x_3 - x_4}\right) + \alpha x + \mu, \\ -\delta_k\left(\left(\frac{y_3 - y_4}{x_3 - x_4}\right)x + \frac{x_3y_4 - x_4y_3}{x_3 - x_4}\right), \end{cases}$$

so, $F(DC)$ is in Ω if

$$\frac{(x_3 - \mu)(x_3 - x_4) - \tau_k(x_3y_4 - x_4y_3)}{\tau_k(y_3 - y_4) + \alpha(x_3 - x_4)} \leq x \leq \frac{(x_1 - \mu)(x_3 - x_4) - \tau_k(x_3y_4 - x_4y_3)}{\tau_k(y_3 - y_4) + \alpha(x_3 - x_4)}, \tag{21}$$

and

$$\frac{-y_1}{\delta_k}\left(\frac{x_3 - x_4}{y_3 - y_4}\right) - \left(\frac{x_3y_4 - x_4y_3}{y_3 - y_4}\right) \leq x \leq \frac{-y_3}{\delta_k}\left(\frac{x_3 - x_2}{y_3 - y_2}\right) - \left(\frac{x_3y_2 - x_2y_3}{y_3 - y_2}\right). \tag{22}$$

Finally, the line segment [AD] has an equation in the form $y = ax + b$ which is

$$y = \left(\left(\frac{y_4 - y_1}{x_4 - x_1}\right)x + \frac{x_4y_1 - x_1y_4}{x_4 - x_1}\right).$$

F(AD) the image of (AD) by (2) is defined by :

$$F(AD) = \begin{cases} \tau_k\left(\left(\frac{y_4 - y_1}{x_4 - x_1}\right)x + \frac{x_4y_1 - x_1y_4}{x_4 - x_1}\right) + \alpha x + \mu, \\ -\delta_k\left(\left(\frac{y_4 - y_1}{x_4 - x_1}\right)x + \frac{x_4y_1 - x_1y_4}{x_4 - x_1}\right), \end{cases}$$

so, F(AD) is in Ω if

$$\frac{(x_3 - \mu)(x_4 - x_1) - \tau_k(x_4y_1 - x_1y_4)}{\tau_k(y_4 - y_1) + \alpha(x_4 - x_1)} \leq x \leq \frac{(x_1 - \mu)(x_4 - x_1) - \tau_k(x_4y_1 - x_1y_4)}{\tau_k(y_4 - y_1) + \alpha(x_4 - x_1)}, \tag{23}$$

and

$$\frac{-y_1}{\delta_k}\left(\frac{x_4 - x_1}{y_4 - y_1}\right) - \left(\frac{x_4y_1 - x_1y_4}{y_4 - y_1}\right) \leq x \leq \frac{-y_3}{\delta_k}\left(\frac{x_4 - x_1}{y_4 - y_1}\right) - \left(\frac{x_4y_1 - x_1y_4}{y_4 - y_1}\right). \tag{24}$$

As a last step, for every point $(x, y) \in \Omega$, the image $F(x, y)$ of this point by the formula (2) will be in Ω if

$$x_3 \leq \tau_k y + \alpha x + \mu \leq x_1,$$

and

$$y_3 \leq -\delta_k y \leq y_1,$$

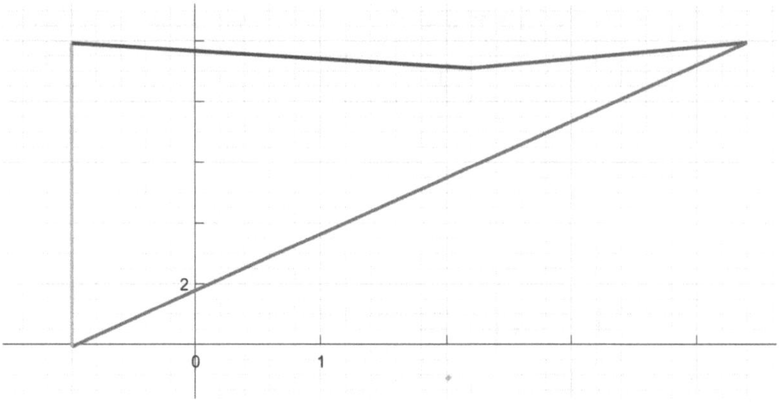

Figure 1. An approximate shape for the absorbing region in the following conditions $\alpha = 0.5$, $\delta_k = 1.2$, $\tau_L = -1$ and $\tau_R = 1$.

which means

$$\frac{x_3 - \tau_k y - \mu}{\alpha} \leq x \leq \frac{x_1 - \tau_k y - \mu}{\alpha}, \tag{25}$$

and by using the condition (8) on the second inequality

$$-y_1 < y < -\alpha y_3. \tag{26}$$

By considering all of the previous results with the hyperbolicity condition obtained in (8), Ω presents an absorbing region for the map (2). To show that the set of the parameters of the map (2) satisfying the above conditions is not empty, a numerical approximation of an absorbing region is shown in Figure 1. In this case, all the conditions of Theorem 2.1 are satisfied. Indeed, $\alpha = 0.5$ and $\delta = 1.2$ are verifying the condition (8).

■

3. Conclusion

In this work, we have proved the existence of a quasi-hyperbolic attractor for a particular family of 2-D piecewise linear maps. The main analysis is based on finding an absorbing region and then proving that it contains a quasi-hyperbolic attractor for the form (2) under some specific conditions.

Acknowledgments

The authors wish to acknowledge the editor and anonymous reviewers for their insightful comments, which have improved the quality of this publication.

Disclosure statement

No potential conflict of interest was reported by the author(s).

Funding

This work is supported by the Directorate-General for Scientific Research and Technological Development of Algeria (DGRSDT).

References

[1] V. Afraimovich and S.-B. Hsu, *Lectures on Chaotic Dynamical Systems*, AMS/IP Studies in Advanced Mathematics, Vol. 28. AMS and International Press of Boston, 2003.

[2] V. Afraimovich and L.P. Shilnikov, *Strange attractors and quasi-attractors*, in *Nonlinear Dynamics and Turbulence*, G.I. Barenblatt, G. Iooss, and D.D. Joseph, eds., Pitman, NY, 1983, pp. 1–28.

[3] V.S. Anishchenko and G.I. Strelkova, *Attractors of dynamical systems*, in *1997 1st International Conference, Control of Oscillations and Chaos Proceedings (Cat. No. 97TH8329)*, Vol. 3, IEEE, 1997, August, pp. 498–503.

[4] D.V. Anosov, *Geodesic flows on closed Riemannian manifolds of negative curvature*, Proc. Steklov Math. Inst. 90 (1967), pp. 1–235.

[5] S. Banerjee and C. Grebogi, *Border collision bifurcations in tow-dimensional piecewise smooth maps*, Phys. Rev. E 59 (1999), p. 4052.

[6] Z. Elhadj, *Lozi Mapping: Theory and Applications.* Taylor & Francis Group, Boca Raton, CRC Press, 2013.

[7] H. Fareh and Z. Elhadj, *Necessary and sufficient conditions for the occurrence of bisecting bifurcations in the general 2D piecewise-Linear mapping*, Int. J. Bifurcat. Chaos 27(5) (2017), Article ID 1750079.

[8] P. Glendinning, *Robust chaos*, Eur. Phys. J. Spec. Top. 226 (2017), pp. 1721–1738.

[9] P. Glendinning and D.J. Simpson, *Constructing robust chaos: Invariant manifolds and expanding cones*, preprint (2019). Available at arXiv:1906.11969.

[10] A. Lichtenberg and M. Lieberman, *Regular and Stochastic Motion*, Springer Verlag, Berlin, Heidelberg, New York, 1983.

[11] E.N. Lorenz, *Deterministic non-periodic flow*, J. Atmos. Sci. 20 (1963), pp. 130–141.

[12] Y. Neimark and P. Landa, *Stochastic and Chaotic Oscillations*, Nauka, Moscow, 1989.

[13] S.E. Newhouse, *Lectures on dynamical systems*, in *Dynamical Systems*, Birkhäuser, Boston, MA, 1980, pp. 209–312.

[14] R.V. Plykin, *On the problem of topological classification of strange attractors of dynamical systems*, Russ. Math. Surv. 57(6) (2002), p. 1163.

[15] M. Rabinovich and D. Trubetskov, *The Introduction to the Theory of Oscillations and Waves*, Nauka, Moscow, 1984.

[16] H. Schuster, *Deterministic Chaos*, Physik-Verlag GmbH, Weinheim (F.R.G.), 1984.

[17] L.P. Shilnikov, *Some cases of generation of period motions from singular trajectories*, Mat. Sb. 103(4) (1963), pp. 443–466.

Right fractional calculus to inverse-time chaotic maps and asymptotic stability analysis

Guo-Cheng Wu, Jia-Li Wei and Maokang Luo

ABSTRACT
Inverse or terminal value problems of fractional differential equations become popular recently. But memory effects or non-locality of fractional operators cause many difficulties for theoretical analysis. This study suggests a right fractional calculus method for inverse problem modelling and proposes a concept of inverse-time fractional chaotic maps. First, a simple right fractional linear differential equation's terminal value problem and the solutions are investigated. Then, some basics of the right discrete fractional calculus are introduced and the idea is extended to the discrete case. Right fractional sum equations are derived and numerical schemes are provided for dynamical analysis. Discrete chaos does exist in an inverse-time fractional logistic map and Hénon map, respectively. Their local stability conditions are given. It can be concluded that this right fractional calculus method is simpler but more efficient than the left one.

1. Introduction

Fractional chaotic maps [26, 29] were proposed on an isolated time scale $\{0, 1, 2, \ldots, \}$. Much attention has been paid in the nonlinear dynamics of fractional difference equations and applications, such as image encryption [18, 25], fuzzy analysis [16], synchronization [14, 27, 32], q-deformed chaotic maps [19, 28]. In viewpoint of control, if one knows the presented status, the past statuses are to be determined. This is also an important aspect of fractional difference equations or fractional discrete-time systems.

In fact, the fractional calculus [17] holds memory effects or non-locality. The inverse or terminal value problems of fractional differential equations of Caputo type were suggested in form of

$$ {}_0^C D_t^{\alpha} x = f(x, t), \quad x(T) = x_T, t \in [0, T], 0 < \alpha \leq 1 \tag{1} $$

where $t = 0$ in the Caputo derivative is the initial time and T is the current time or terminal time. Then the problem was transformed to the following integral equation

$$x(t) = x_T - \frac{1}{\Gamma(\alpha)} \int_0^T (T-s)^{\alpha-1} f(x,s)\,\mathrm{d}s + \frac{1}{\Gamma(\alpha)} \int_0^t (t-s)^{\alpha-1} f(x,s)\,\mathrm{d}s \tag{2}$$

which contains Volterra and Fredholm integral terms meanwhile. An explicit solution of fractional linear systems with non-homogenous terms was given in [30]. Much attention was paid to find the initial point, see for example, the existing conditions and numerical methods were given in [13, 20, 22].

Many terminal value problems mainly focus on a stable case and initial state can be detected accurately. Dynamical systems in chaotic statues have high sensitivity. It becomes more challenging to detect a chaotic initial state in real world applications. Generally, it is not easy to discuss the nonlinear dynamics for the past statues $[0, T]$ as that of initial value problems. So we turn to re-consider a right fractional differential equation approach as

$${}_t^C D_T^\alpha x = f(x,t), \quad x(T) = x_T, t \in [0, T], 0 < \alpha \leq 1 \tag{3}$$

whose equivalent integral equation reads

$$x(t) = x_T + \frac{1}{\Gamma(\alpha)} \int_t^T (s-t)^{\alpha-1} f(x,s)\,\mathrm{d}s. \tag{4}$$

This idea was in fact used in a time-fractional Black–Scholes equation [9] for the pricing of options in the financial market. The right Caputo derivative was adopted for the terminal value problems. The fundamental solutions were derived by use of the Fourier–Laplace transform. It is proven that the right fractional derivative is useful and efficient in terminal value problems.

Motivated by this idea [9], we propose a new concept of inverse-time fractional difference equations by the right fractional derivative in this study. More generally, it includes two cases (see Figure 1): the terminal time T is the future time and the current time, respectively. How to achieve the request target if T is the future time (see [9])? It is indeed a control problem from t. On the other hand, if T is the current time, the time domain $t < T$ means this is an inverse problem to recover the past dynamics. In both of the cases, t can be infinity. In comparison with the existing inverse problems by use of the left fractional derivatives, the inverse-time fractional order equation by the right ones has the following new features:

- Initial time t is arbitrary;
- This new inverse problem is modelled by the right fractional calculus approach which is much simpler;
- Many numerical methods or theories in the left fractional calculus can be easily extended to the right case due to the duality.

Considering these new features of the right fractional calculus, we try to address the terminal problem of fractional discrete-time systems and chaotic maps in this paper. We organize the paper as the following: Section 2 gives preliminaries of right fractional calculus and fractional differences; Section 3 defines an inverse-time fractional difference

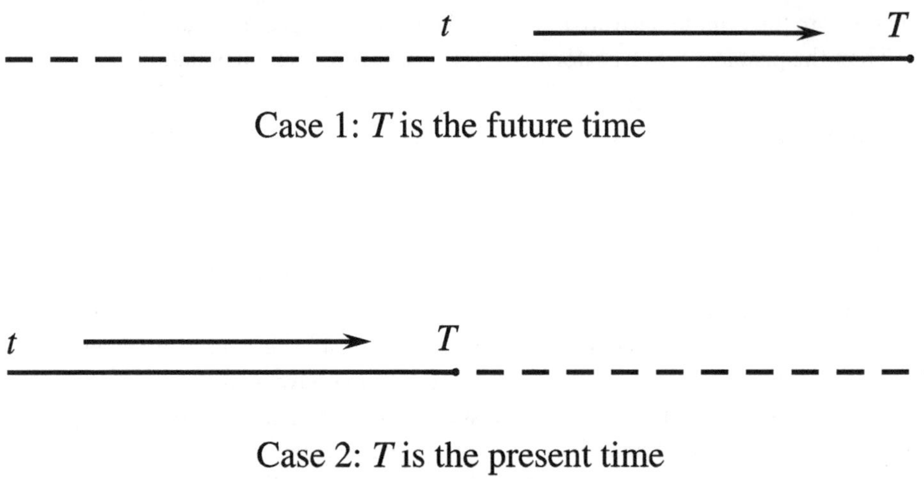

Figure 1. Physical meaning of inverse-time fractional differential equations: Case 1[9] and Case 2 (The present study).

equations. Section 4 investigates inverse-time fractional chaotic maps which is a special case of Section 3 and gives asymptotic stability analysis. Section 5 makes a conclusion.

2. Preliminaries of the right fractional calculus

We first revisit the left and right fractional calculus.

Definition 2.1 ([17]): Let $x \in L^1[a, b]$ and $0 < \alpha$. The left and right Riemann–Liouville integrals of order α are defined by

$$
\begin{aligned}
{}_aI_t^\alpha x &= \frac{1}{\Gamma(\alpha)} \int_a^t (t-s)^{\alpha-1} x(s)\,\mathrm{d}s, \\
{}_tI_b^\alpha x &= \frac{1}{\Gamma(\alpha)} \int_t^b (s-t)^{\alpha-1} x(s)\,\mathrm{d}s
\end{aligned} \tag{5}
$$

where Γ is the gamma function.

Definition 2.2 ([17]): Let $x \in AC^n[a, b]$ and $n - 1 < \alpha \leq n$. The left and right Caputo derivatives of α order are defined by

$$
\begin{aligned}
{}_a^CD_t^\alpha x &= \frac{1}{\Gamma(n-\alpha)} \int_a^t (t-s)^{n-\alpha-1} x^{(n)}(s)\,\mathrm{d}s, \\
{}_t^CD_b^\alpha x &= \frac{(-1)^n}{\Gamma(n-\alpha)} \int_t^b (s-t)^{n-\alpha-1} x^{(n)}(s)\,\mathrm{d}s.
\end{aligned} \tag{6}
$$

For $\alpha = n$, ${}_a^CD_t^\alpha x = \frac{\mathrm{d}^n x}{\mathrm{d}t^n}$ and ${}_t^CD_b^\alpha x = (-1)^n \frac{\mathrm{d}^n x}{\mathrm{d}t^n}$.

For $n-1<\alpha\leq n$, the following propositions [17] hold

$$ {}_t^C D_b^\alpha {}_t I_b^\alpha x = x $$

and

$$ x = {}_t I_b^\alpha {}_t^C D_b^\alpha x + \sum_{k=0}^{n-1} \frac{x^{(k)}(b)}{k!}(t-b)^k. $$

Then we can see that $\phi(t)$ is the solution of

$$ {}_t^C D_b^\alpha x = f(x,t), \quad x(b) = x_b,\ a \leq t \leq b, 0 < \alpha \leq 1 \tag{7} $$

if and only if $\phi(t)$ solves

$$ x = x_b + {}_t I_b^\alpha f(x,t). \tag{8} $$

If we consider a terminal value problem in the left Caputo derivative's sense

$$ {}_0^C D_t^\alpha x = \lambda x, \quad x(T) = x_T, t \in [0,T], 0 < \alpha \leq 1, \tag{9} $$

then we get

$$ x(t) = x_T - \frac{\lambda}{\Gamma(\alpha)} \int_0^T (T-s)^{\alpha-1} x \, ds + \frac{1}{\Gamma(\alpha)} \int_0^t (t-s)^{\alpha-1} \lambda x \, ds. \tag{10} $$

It becomes much complicated. In order to address this problem, we start from the following example.

Example 2.3: Consider the simple right fractional differential equation

$$ {}_t^C D_b^\alpha x = \lambda x(t), \quad x(b) = c,\ a \leq t \leq b, 0 < \alpha \leq 1. \tag{11} $$

We start from the fractional integral equation

$$ x = x(b) + \lambda {}_t I_b^\alpha x(t) \tag{12} $$

and Picard's method is used to give

$$ x_{k+1} = x_0 + \lambda {}_t I_b^\alpha x_k, \quad k = 0, 1, \ldots, x_0 = c. \tag{13} $$

With the proposition ${}_t I_b^\beta \frac{(b-t)^\alpha}{\Gamma(\alpha+1)} = \frac{(b-t)^{\alpha+\beta}}{\Gamma(\alpha+\beta+1)}$, we give the exact solution in form of Mittag–Leffler function

$$ x(t) = \lim_{n\to\infty} x_{n+1} = \lim_{n\to\infty} c \sum_{j=0}^{n} \frac{\lambda^j (b-t)^{j\alpha}}{\Gamma(j\alpha+1)} = cE_\alpha(\lambda, (b-t)^\alpha). \tag{14} $$

Recall the fractional chaotic maps [26, 29] which have exact chaos and there are no numerical errors, see for example, the fractional logistic map is presented in the recurrence relationship

$$ x_{n+1} = x_0 + \frac{\mu}{\Gamma(\nu)} \sum_{j=0}^{n} \frac{\Gamma(n-j+\nu)}{\Gamma(n-j+1)} x_j(1-x_j). \tag{15} $$

It is an explicit scheme and x_{n+1} has exact values from $x_0, \ldots, x_n$.

We can use this idea in the discrete fractional calculus. Before that, let us introduce definitions of right fractional sum and difference. We denote by ${}_{b-1}\mathbb{Z} = \{\cdots, b-2, b-1\}$ and $b \in \mathbb{R}$.

Definition 2.4 ([2, 3]): The right fractional sum of order $\alpha > 0$ is defined by

$$ {}_t\nabla_b^{-\alpha} x(t) = \frac{1}{\Gamma(\alpha)} \sum_{s=t}^{b-1} (s - \rho(t))^{\overline{\alpha-1}} x(s), \quad t \in_{b-1} \mathbb{Z}. \tag{16} $$

Definition 2.5 ([2, 3]): The right fractional difference of order $n-1 < \alpha \leq n$ is defined by

$$ {}_t^C\nabla_b^{\alpha} x(t) = {}_t\nabla_b^{-(n-\alpha)} {}_{\ominus}\Delta^n x(s), \quad t \in_{b-1} \mathbb{Z} \tag{17} $$

where ${}_{\ominus}\Delta^n x(t) = (-1)^n \Delta^n x(t)$.

For $0 < \alpha \leq 1$, we have a discrete-time version for the Caputo derivative on the isolated time scale

$$ {}_t^C\nabla_b^{\alpha} x(t) = \frac{1}{\Gamma(1-\alpha)} \sum_{s=t}^{b-1} (s - \rho(t))^{\overline{-\alpha}} {}_{\ominus}\Delta x(s), \quad t \in_{b-1} \mathbb{Z}. \tag{18} $$

For more details about the discrete fractional calculus and fractional difference equations, readers are suggested to refer to initial value problems [6], Laplace transform [15], stability analysis [5, 7, 8, 21], boundary value problems [12], discrete Hadamard fractional calculus [23, 31], discrete tempered fractional calculus [4, 11], fractional Mandelbrot sets [24] and neural networks [32].

3. Inverse-time fractional difference equations

We consider the general fractional difference equation as the following

$$ \begin{cases} {}_t^C\nabla_b^{\alpha} x(t) = f(x(t+1), t+1), t \in_{b-1} \mathbb{Z}, 0 < \alpha \leq 1, \\ x(b) = C. \end{cases} \tag{19} $$

Here the time t can be negative with respect to the terminal time b. So we use the notation of $\mathbb{Z}$ instead of $\mathbb{N}$. We need some lemmas to give an equivalent fractional sum equation.

3.1. Fractional sum equations

Lemma 3.1 (Leibniz sum law [2]): *Suppose* $x : {}_{b-1}\mathbb{Z} \to \mathbb{R}$ *and* $0 < \alpha \leq 1$. *Then*

$$ {}_t\nabla_b^{-\alpha} {}_t^C\nabla_b^{\alpha} x(t) = x(t) - x(b). \tag{20} $$

Lemma 3.2 (Semi-group property [2]): *Suppose* $x : {}_{b-1}\mathbb{Z} \to \mathbb{R}$ *and* $\alpha, \beta > 0$. *Then*

$$ {}_t\nabla_b^{-\alpha} {}_t\nabla_b^{-\beta} x(t) = {}_t\nabla_b^{-(\alpha+\beta)} x(t), \quad t \in {}_{b-1}\mathbb{Z}. \tag{21} $$

Lemma 3.3 ([2]): *Suppose $f : {}_{b-1}\mathbb{Z} \to \mathbb{R}$ and $0 < \alpha$. Then the following equality holds*

$$ {}_t\nabla_b^{-\alpha} {}_{\ominus}\Delta f(t) = {}_{\ominus}\Delta_t \nabla_b^{-\alpha} f(t) - \frac{(b-t)^{\overline{\alpha-1}}}{\Gamma(\alpha)} f(b), \quad t \in_{b-1} \mathbb{Z}. \tag{22} $$

Lemma 3.4: *Suppose $x : {}_{b-1}\mathbb{Z} \to \mathbb{R}$ and $0 < \alpha \leq 1$, then*

$$ {}_t^C\nabla_b^{\alpha} {}_t\nabla_b^{-\alpha} x(t) = x(t). \tag{23} $$

Proof: According to the definition of the right nabla fractional difference, we get

$$ {}_t^C\nabla_b^{\alpha} {}_t\nabla_b^{-\alpha} x(t) = {}_t\nabla_b^{-(1-\alpha)} {}_{\ominus}\Delta {}_t\nabla_b^{-\alpha} x(t). \tag{24} $$

Using Lemma 3.3, we obtain

$$ {}_t\nabla_b^{-(1-\alpha)} {}_{\ominus}\Delta_t \nabla_b^{-\alpha} x(t) = {}_{\ominus}\Delta_t \nabla_b^{-(1-\alpha)} {}_t\nabla_b^{-\alpha} x(t) - \frac{(b-t)^{\overline{\alpha-1}}}{\Gamma(\alpha)} {}_t\nabla_b^{-\alpha} x(b). $$

Since ${}_t\nabla_b^{-\alpha} x(b) = 0$ and Lemma 3.2, it follows that

$$ \begin{aligned} &{}_t^C\nabla_b^{\alpha} {}_t\nabla_b^{-\alpha} x(t) \\ &= {}_{\ominus}\Delta \, {}_t\nabla_b^{-(1-\alpha)} \nabla_b^{-\alpha} x(t) \\ &= {}_{\ominus}\Delta \, {}_t\nabla_b^{-1} x(t) \\ &= x(t) \end{aligned} $$

which completes the proof. ■

Theorem 3.5: *Suppose $\phi : {}_{b-1}\mathbb{Z} \to \mathbb{R}$ and $0 < \alpha \leq 1$. $\phi(t)$ is the solution of (19) if and only if $\phi(t)$ solves*

$$ x(t) = x(b) + {}_t\nabla_b^{-\alpha} f(x(t+1), t+1), \quad t \in_{b-1} \mathbb{Z}. \tag{25} $$

Proof: From (19), we give

$$ {}_t^C\nabla_b^{\alpha} \phi(t) = f(\phi(t+1), t+1), \quad \phi(b) = x(b) $$

and

$$ {}_t\nabla_b^{-\alpha} {}_t^C\nabla_b^{\alpha} \phi(t) = {}_t\nabla_b^{-\alpha} f(\phi(t+1), t+1), \quad t \in_{b-1} \mathbb{Z}. $$

From Lemma 3.1, we give

$$ \phi(t) = \phi(b) + {}_t\nabla_b^{-\alpha} f(\phi(t+1), t+1) = x(b) + {}_t\nabla_b^{-\alpha} f(\phi(t+1), t+1), \quad t \in_{b-1} \mathbb{Z}. $$

On the other hand, from Equation (25),

$$ {}^C\nabla_b^{\alpha} \phi(t) = {}^C\nabla_b^{\alpha} x(b) + {}^C\nabla_b^{\alpha} {}_t\nabla_b^{-\alpha} f(\phi(t+1), t+1), \quad t \in_{b-1} \mathbb{Z}. $$

By use of Lemma 3.4 and ${}^C\nabla_b^{\alpha} x(b) = 0$, we get

$$ {}^C\nabla_b^{\alpha} \phi(t) = f(\phi(t+1), t+1), \quad \phi(b) = x(b) $$

which completes the proof. ■

Example 3.6: Use the right fractional difference equation

$$^{C}\nabla_b^{\alpha} x(t) = \lambda x(t), x(b) = c, t \in_{b-1} \mathbb{Z}, \quad 0 < \alpha \leq 1. \tag{26}$$

By use of a discrete right Mittag–Leffler function[1]

$$\varepsilon_{\alpha}(\lambda, (\rho(b) - t)^{\overline{\alpha}}) := \sum_{k=0}^{+\infty} \lambda^k \frac{(b - t - k(1 - \alpha))^{\overline{\frac{k\alpha}{h}}}}{\Gamma(k\alpha + 1)}, \quad t \in_{b-1} \mathbb{Z}, \tag{27}$$

the exact solution also can be given as

$$x(t) = x(b)\varepsilon_{\alpha}(\lambda, (\rho(b) - t))^{\overline{\alpha}}), \quad t \in_{b-1} \mathbb{Z}. \tag{28}$$

The solution is convenient for mathematical analysis. But it is given in an infinite series and difficult to evaluate the numerical values directly. However, we note that the fractional difference equation is also a discrete equation and we can directly given the numerical formula through the fractional sum equation. Let $t = b-1-n$ and $s = b-1-j$. We give the numerical version as

$$x(b - (n+1)) = x(b) + \frac{\lambda}{\Gamma(\alpha)} \sum_{j=0}^{n} \frac{\Gamma(n-j+\alpha)}{\Gamma(n-j+1)} x(b-j), \quad t \in_{b-1} \mathbb{Z}$$

which is an explicit scheme with respect to $x(b - (n + 1))$.

We also can use right Delta Caputo difference [2, 3] to define the fractional difference Equation (19). It leads to the same fractional sum Equation (25) and numerical scheme. So there is no need to repeat the similar work.

4. Inverse-time fractional chaotic maps

In this section, we discuss dynamics of one and high dimensional inverse-time fractional chaotic maps, respectively.

4.1. Inverse-time fractional logistic map

We suggest an inverse problem of fractional logistic map

$$^{C}\nabla_b^{\alpha} u(t) = \mu u(t+1)(1 - u(t+1)), u(b) = 0.3, t \in_{b-1} \mathbb{Z}, \quad 0 < \alpha \leq 1 \tag{29}$$

and its equivalent recurrent form

$$u(b - (n+1)) = u(b) + \frac{\mu}{\Gamma(\alpha)} \sum_{j=0}^{n} \frac{\Gamma(n-j+\alpha)}{\Gamma(n-j+1)} u(b-j)(1 - u(b-j)). \tag{30}$$

The terminal value is known as $b = 20$ and $u(b) = 0.3$. We try to detect the initial time $t = 14$. Suppose other parameter values $\alpha = 0.9$ and $\mu = 2.6$. Through Equation (30), we can obtain $u(14) = 1.025$ (see Figure 2). In other words, starting from $u(14) = 1.025$, we can arrive at the target or terminal value $u(b)$. This idea is simpler than the left derivative method [20, 22].

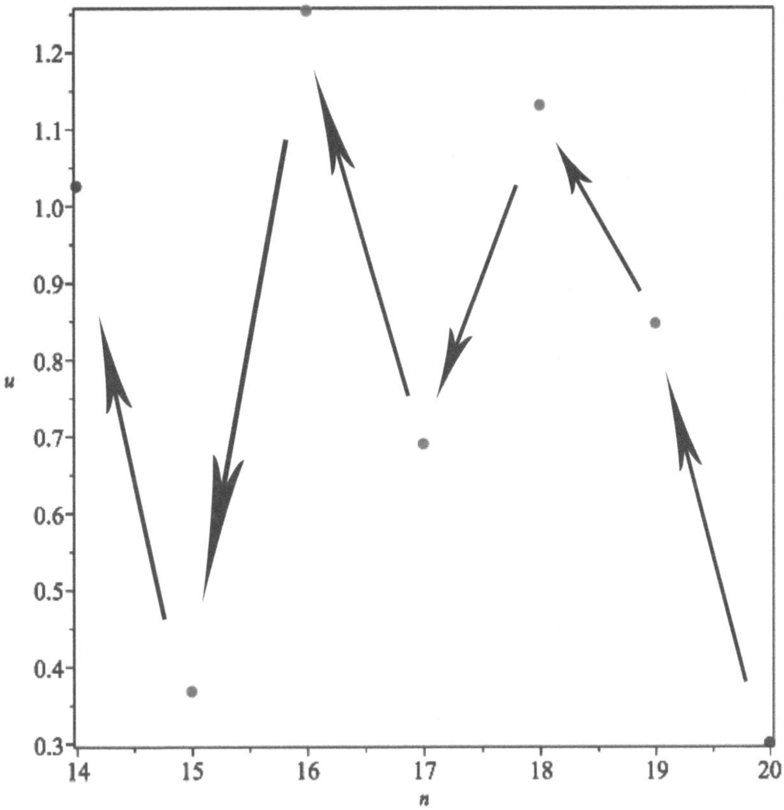

Figure 2. Inverse-time fractional logistic map: $\alpha = 0.9$ and $\mu = 2.6$.

We also can observe rich dynamics with varied μ (see Figure 3). Assume $b = 400$, $\alpha = 0.9$ and $n = 500$. The bifurcation diagram with respect to the coefficient μ is illustrated in Figure 3. Rich dynamics in the inverse-time fractional logistic map (30) are displayed such as stability, bifurcation and chaos. If the fractional order α plays a role as a bifurcation parameter, the dynamics with respect to α are demonstrated in Figure 4.

It can be concluded that both the fractional order and the coefficient can change dynamics. For $\alpha = 1$, the inverse-time fractional logistic map (30) can be reduced to

$$u(b-(n+1)) = u(b) + \mu \sum_{j=0}^{n} u(b-j)(1-u(b-j)). \tag{31}$$

It also can be rewritten as

$$u(b-(n+1)) = (\mu+1)u(b-n) - \mu u^2(b-n) \tag{32}$$

which can be considered as an inverse-time version of the classical logistic map (see (1.4) on pp. 3 of [10]).

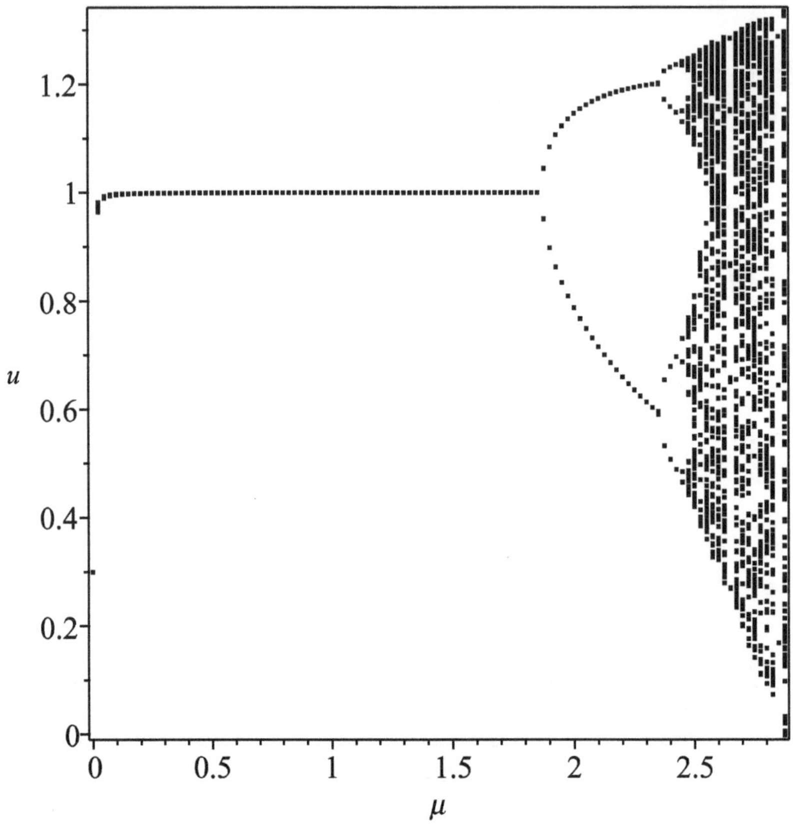

Figure 3. Bifurcation diagram of inverse-time fractional logistic map: $\alpha = 0.9$.

4.2. Inverse-time fractional Hénon map

We consider an inverse-time fractional Hénon map as

$$\begin{cases} {}^{C}\nabla_{b}^{\alpha}x(t) = 1 - \beta x^2(t+1) + y(t+1) - x(t+1), \\ {}^{C}\nabla_{b}^{\alpha}y(t) = \omega x(t+1) - y(t+1),\ t \in_{b-1} \mathbb{Z}, 0 < \alpha \leq 1. \end{cases} \tag{33}$$

Suppose $b = 20$, $x(b) = y(b) = 0.3$, $t = -380$, $\alpha = 0.9$ and $\omega = 0.3$. Take β as a bifurcation parameter. Rich dynamics are numerically illustrated in Figure 5 and chaotic attractor is shown in Figure 6, respectively.

In order to study the asymptotic stability of inverse-time fractional Hénon map, we first linearize it at the equilibrium point. Suppose (x^*, y^*) is the equilibrium point of Equation (33), then

$$\begin{cases} 1 - \beta x^{*2} + y^* - x^* = 0, \\ \omega x^* - y^* = 0. \end{cases} \tag{34}$$

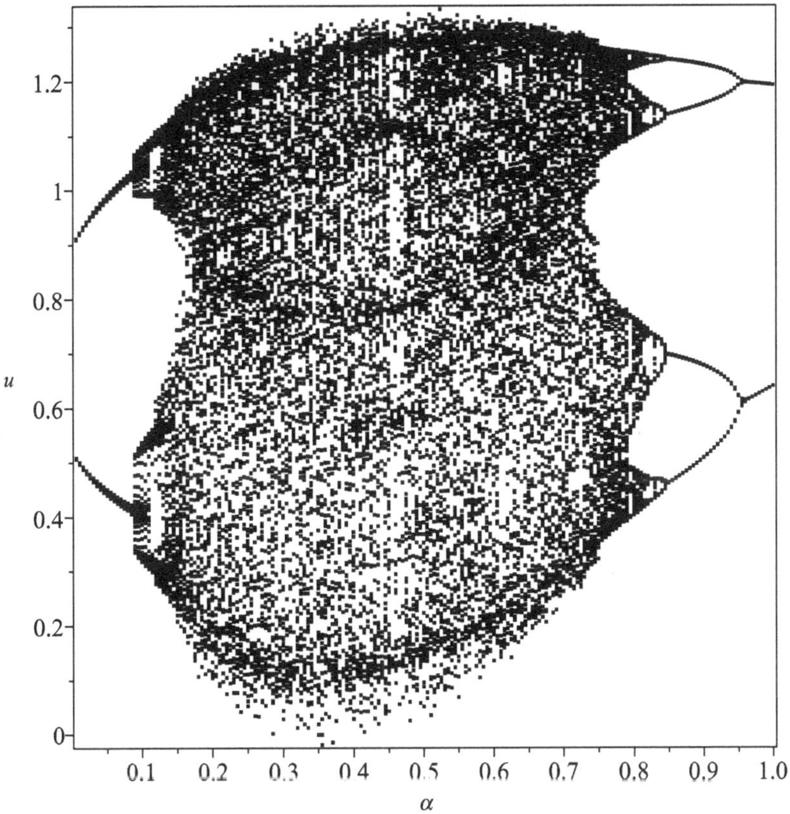

Figure 4. Bifurcation diagram of inverse-time fractional logistic map: $b = 200$, $n = 250$ and $\mu = 2.4$.

Solve Equation (34) to obtain

$$\begin{cases} (x_1^*, y_1^*) = \left(\dfrac{\omega - 1 + \sqrt{(1-\omega)^2 + 4\beta}}{2\beta}, \dfrac{\omega(\omega - 1 + \sqrt{(1-\omega)^2 + 4\beta})}{2\beta} \right), \\ (x_2^*, y_2^*) = \left(\dfrac{\omega - 1 - \sqrt{(1-\omega)^2 + 4\beta}}{2\beta}, \dfrac{\omega(\omega - 1 - \sqrt{(1-\omega)^2 + 4\beta})}{2\beta} \right), \end{cases}$$

where $(1-\omega)^2 + 4\beta \geq 0$.

Consider one of the equilibrium points and suppose $u_1(t) = x(t) - x_1^*$ and $u_2(t) = y(t) - y_1^*$, then Equation (33) is linearized as

$$^{C}\nabla_b^{\alpha} X(t) = AX(t+1), \tag{35}$$

where $X = [u_1, u_2]^T$ and $A = \begin{pmatrix} -\omega - \sqrt{(1-\omega)^2 + 4\beta} & 1 \\ \omega & -1 \end{pmatrix}$. Equation (35) can be rewritten in form of an iteration formula

$$X(b - (n+1)) = X(b) + \frac{1}{\Gamma(\alpha)} \sum_{j=0}^{n} \frac{\Gamma(n-j+\alpha)}{\Gamma(n-j+1)} AX(b-j). \tag{36}$$

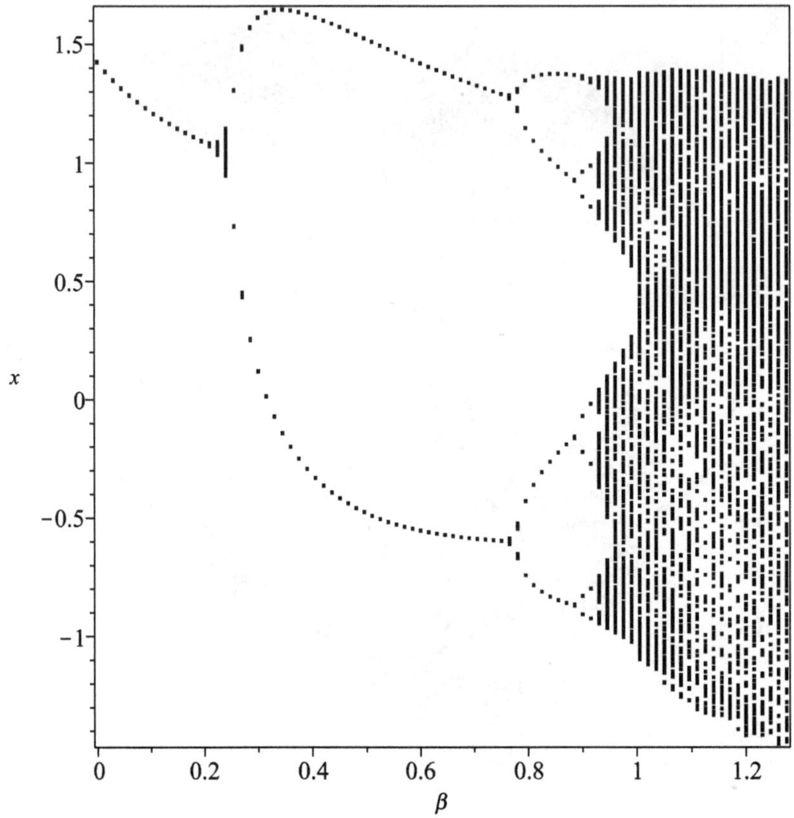

Figure 5. Bifurcation diagram of an inverse-time fractional Hénon map with respect to β.

Let $X_{n+1} = X(b - (n+1))$ and $X_0 = X(b)$. Equation (36) is simplified as

$$X_{n+1} = X_0 + \frac{1}{\Gamma(\alpha)} \sum_{j=0}^{n} \frac{\Gamma(n-j+\alpha)}{\Gamma(n-j+1)} AX_j. \tag{37}$$

We use the z-transform $\mathcal{Z}[X_n](z) = \sum_{k=0}^{\infty} X_k z^{-k}$ and give

$$\tilde{X} = \frac{z}{z-1}\left(I - \frac{1}{z}\left(\frac{z}{z-1}\right)^{\alpha} A\right)^{-1} X_0. \tag{38}$$

We need the following theorem for asymptotic stability analysis.

Theorem 4.1 ([5, 7, 21]): *Let R be the set of all roots of the equation*

$$det\left(I - \frac{1}{z}\left(\frac{z}{z-1}\right)^{\alpha} A\right) = 0. \tag{39}$$

if all elements from R are strictly inside the unit circle, then the equilibrium point (x^*, y^*) *of the inverse-time fractional Hénon map* (33) *is asymptotically stable.*

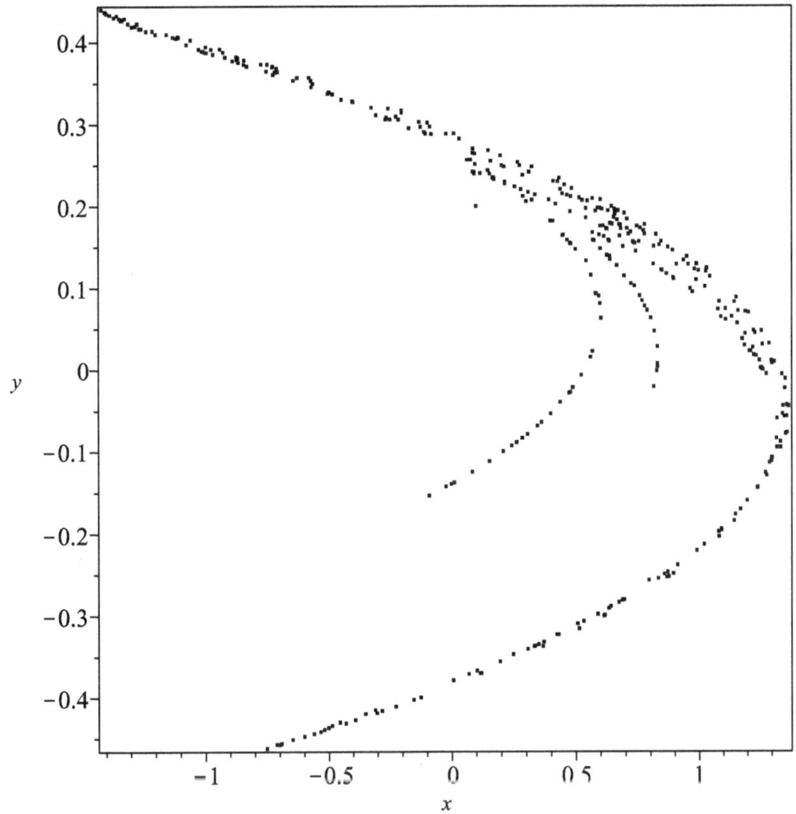

Figure 6. Chaotic attractors: $\alpha = 0.9$, $\beta = 1.275$ and $\omega = 0.3$.

Corollary 4.2 ([5, 7, 21]): *If $\lambda \in Spec(A)$ and $-2^{\alpha} < \lambda < 0$, then the system (33) is asymptotically stable.*

Case I: Assume $\Delta_1 = 2 + 2\omega^2 + 4\beta - 2(1-\omega)\sqrt{(1-\omega)^2 + 4\beta} \geq 0$ for simplicity. The characteristic roots of the coefficient matrix A are real ones

$$\lambda_1 = \frac{-1 - \omega - \sqrt{(1-\omega)^2 + 4\beta} - \sqrt{\Delta_1}}{2} \quad \text{and}$$

$$\lambda_2 = \frac{-1 - \omega - \sqrt{(1-\omega)^2 + 4\beta} + \sqrt{\Delta_1}}{2}.$$

According to Corollary 4.2, in order to guarantee the system (33) asymptotically stable, the parameters should satisfy these inequalities

$$\begin{cases} (1-\omega)^2 + 4\beta \geq 0, \\ 2 + 2\omega^2 + 4\beta - 2(1-\omega)\sqrt{(1-\omega)^2 + 4\beta} \geq 0, \\ \omega < 2^{\alpha} - 1, \\ \beta < \left(\dfrac{2^{\alpha-1}}{2^{\alpha} - 1}\right)^2 (2^{\alpha} - 1 - \omega)^2 - \dfrac{(1-\omega)^2}{4}. \end{cases} \tag{40}$$

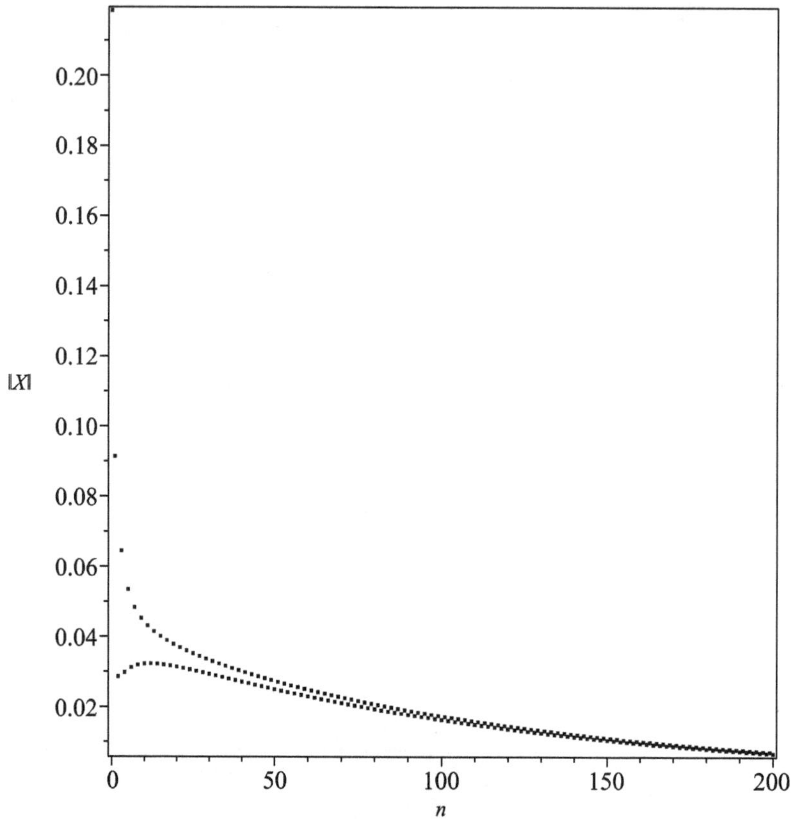

Figure 7. The system (33) is asymptotically stable from the terminal value (1.2, 0.6).

Select the parameters $\alpha = 0.8$, $\beta = 0.24$, $\omega = 0.2$, and $(x(b), y(b)) = (0.9, 0.2)$ near the equilibrium point (x^*, y^*). we can check $\lim_{n\to\infty} X(n) = 0$ or $\lim_{t\to\infty}(x(t), y(t)) = (x_1^*, y_1^*)$ (see Figure 7). Here we use the norm as $\sqrt{x^2 + y^2}$.

Case II: If the system starts from the terminal value near another equilibrium point (x_2^*, y_2^*). Suppose $v_1(t) = x(t) - x_2^*$ and $v_2(t) = y(t) - y_2^*$, then Equation (33) is linearized as

$$ {}^{C}\nabla_b^{\alpha} Y(t) = BY(t+1), \tag{41} $$

where $Y = [v_1, v_2]^T$ and $B = \begin{pmatrix} -\omega+\sqrt{(1-\omega)^2+4\beta} & 1 \\ \omega & -1 \end{pmatrix}$. The calculated characteristic roots of matrix B are $\lambda = \frac{-1-\omega+\sqrt{(1-\omega)^2+4\beta}\pm\sqrt{\Delta_2}}{2}$, where $\Delta_2 = (1+\omega-\sqrt{(1-\omega)^2+4\beta})^2 + 4\sqrt{(1-\omega)^2+4\beta}$ and $\Delta_2 \geq 0$ obviously.

Using the analysis method of **Case I**, we have the following stability condition

$$ \begin{cases} (1-\omega)^2 + 4\beta \geq 0, \\ \omega < 2^{\alpha} - 1, \\ \beta < \left(\dfrac{2^{\alpha-1}}{1-2^{\alpha}}\right)^2 (2^{\alpha} - 1 - \omega)^2 - \dfrac{(1-\omega)^2}{4}. \end{cases} \tag{42} $$

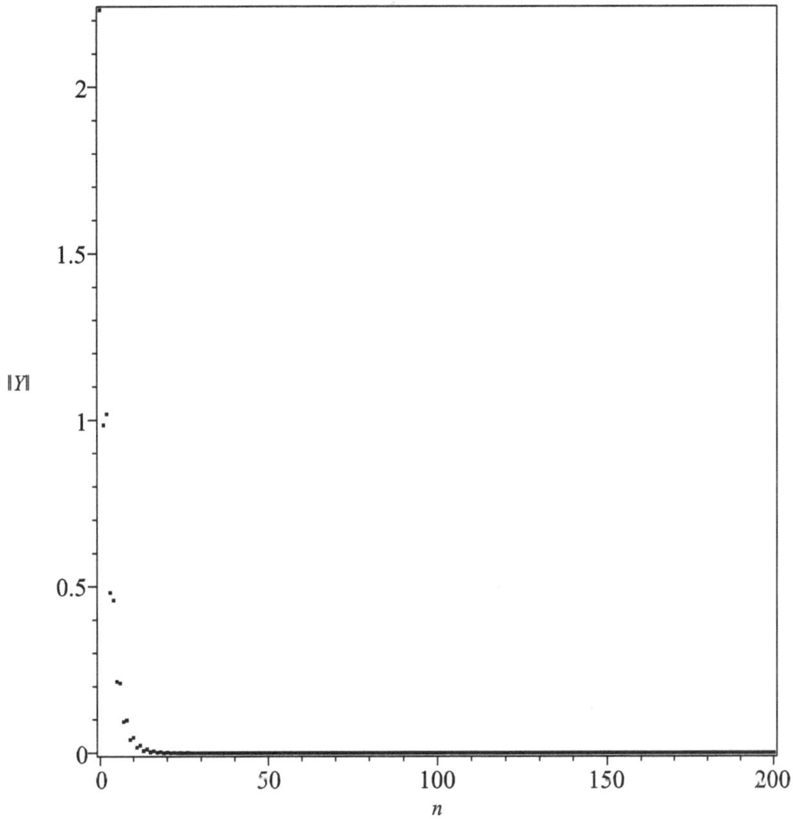

Figure 8. The system (33) is asymptotically stable from the terminal value $(-5, -1)$.

The parameters are set to $\alpha = 0.8, \beta = 0.24, \omega = 0.2$, and the terminal value $(x(b), y(b)) = (-5, -0.5)$ near the equilibrium point (x^*, y^*). The system is also asymptotically stable (see Figure 8).

5. Conclusion

Fractional calculus holds memory effects such that analysis of terminal value problems becomes complicated. A right fractional calculus approach is suggested in fractional difference equations. The inverse-time fractional chaotic maps are presented in form of a recurrent formula and this feature leads to no numerical errors. Then accurate inverse-time chaotic series are obtained. By using the duality of the left and right fractional calculus, stability conditions of an inverse-time fractional Hénon map are given. It can be concluded that the right fractional calculus is efficient in both theoretical analysis and inverse problem modelling.

Acknowledgments

The authors appreciate the Editor's valuable time and referees' sincere suggestions.

Disclosure statement

No potential conflict of interest was reported by the author(s).

Funding

This work is financially supported by the National Natural Science Foundation of China (NSFC) [grant numbers 62076141 and 12101338] and Sichuan Youth Science and Technology Foundation [grant number 2022JDJQ0046].

Author contributions

All authors completed the paper together. All authors read and approved the final manuscript.

Ethics approval

All authors approve ethics.

Consent to participate

All authors consent to participate.

Consent for publication

All authors consent for publication.

ORCID

Guo-Cheng Wu http://orcid.org/0000-0002-1946-6770

References

[1] T. Abdeljawad, *On Riemann and Caputo fractional differences*, Comput. Math. Appl. 62 (2011), pp. 1602–1611.
[2] T. Abdeljawad, *Dual identities in fractional difference calculus within Riemann*, Adv. Differ. Equ.2013 (2013), p. Article ID. 36, 16pp.
[3] T. Abdeljawad and F.M. Atici, *On the definition of Nabla fractional operators*, Abstr. Appl. Anal.2012 (2012), p. Artical ID 406757, 13pp.
[4] T. Abdeljawad, S. Banerjee, and G.C. Wu, *Discrete tempered fractional calculus for new chaotic systems with short memory and image encryption*, Optik 203 (2020), p. 163698.
[5] R. Abu-Saris and Q. Al-Mdallal, *On the asymptotic stability of linear system of fractional-order difference equations*, Fract. Calc. Appl. Anal. 16 (2013), pp. 613–629.
[6] F.M. Atici and P.W. Eloe, *Initial value problems in discrete fractional calculus*, Proc. Amer. Math. Soc. 137 (2009), pp. 981–989.
[7] J. Cermak, I. Gyori, and L. Nechvatal, *On explicit stability conditions for a linear fractional difference system*, Fract. Calc. Appl. Anal. 18 (2105), pp. 651–672.
[8] F. Du and B. Jia, *Finite-time stability of a class of nonlinear fractional delay difference systems*, Appl. Math. Lett. 98 (2019), pp. 233–239.
[9] J.S. Duan, L. Lu, L. Chen, and Y.L. An, *Fractional model and solution for the Black–Scholes equation*, Math. Method. Appl. Sci. 41 (2018), pp. 697–704.

[10] S.N. Elaydi, Discrete Chaos, Chapman and Hall/CRC, New York, 2007.

[11] H. Fu, L.L. Huang, T. Abdeljawad, and C. Luo, *Tempered fractional calculus on time scale for discrete-time systems*, Fractals 29 (2021), p. 2140033.

[12] C. Goodrich and A.C. Peterson, Discrete Fractional Calculus, Springer, 2015.

[13] C.Y. Gu, G.C. Wu, and B. Shiri, *An inverse problem approach to determine possible memory length of fractional differential equations*, Fract. Calc. Appl. Anal 24 (2021), pp. 1919–1936.

[14] Y.J. Gu, H. Wang, and Y. Yu, *Synchronization for fractional-order discrete-time neural networks with time delays*, Appl. Math. Comput. 372 (2020), p. 124995.

[15] M.T. Holm, *The Laplace transform in discrete fractional calculus*, Comput. Math. Appl. 62 (2011), pp. 1591–1601.

[16] L.L. Huang, G.C. Wu, D. Baleanu, and H.Y. Wang, *Discrete fractional calculus for interval-valued systems*, Fuzzy Set. Syst. 404 (2020), pp. 141–158.

[17] A.A. Kilbas, H.M. Srivastava, and J.J. Trujillo, Theory and Applications of Fractional Differential Equations, Elsevier Science B. V., Amsterdam, 2006.

[18] Z.Y. Liu, T. Xia, and J. Wang, *Image encryption technique based on new two-dimensional fractional-order discrete chaotic map and Menezes–Vanstone elliptic curve cryptosystem*, Chin. Phys. B 27 (2018), p. 030502.

[19] C. Luo, B.Q. Liu, and H.S. Hou, *Fractional chaotic maps with q-deformation*, Appl. Math. Comput.393 (2021), p. 125759.

[20] M.L. Morgado and M. Rebelo, *Well-posedness and numerical approximation of tempered fractional terminal value problems*, Fract. Calc. Appl. Anal. 20 (2017), pp. 1239–1262.

[21] D. Mozyrska and M. Wyrwas, *The $\mathcal{Z}$-transform method and delta type fractional difference operators*, Discr. Dyn. Nat. Soc. 2015 (2015), p. 852734.

[22] B. Shiri, G.C. Wu, and D. Baleanu, *Terminal value problems for the nonlinear systems of fractional differential equations*, Appl. Numer. Math. 170 (2021), pp. 162–178.

[23] T.T. Song, G.C. Wu, and J.L. Wei, *Hadamard fractional calculus on time scales*, Fractals 30 (2022), p. 2250145.

[24] Y.P. Wang, X. Li, D. Wang, and S. Liu, *A brief note on fractal dynamics of fractional Mandelbrot sets*, Appl. Math. Comput. 432 (2022), p. 127353.

[25] Z. Wang, B. Shiri, and D. Baleanu, *Discrete fractional watermark technique*, Front. Infor. Tech. Electr. Engr. 21 (2020), pp. 880–883.

[26] G.C. Wu and D. Baleanu, *Discrete fractional logistic map and its chaos*, Nonlinear Dyn. 75 (2014), pp. 283–287.

[27] G.C. Wu, D. Baleanu, H.P. Xie, and F.L. Chen, *Chaos synchronization of fractional chaotic maps based on stability results*, Phys. A. 460 (2016), pp. 374–383.

[28] G.C. Wu, M. Cankaya, and S. Banerjee, *Fractional q-chaotic maps: A weight function approach*, Chaos12 (2020), p. 112106.

[29] G.C. Wu, Z.G. Deng, D. Baleanu, and D.Q. Zeng, *New variable-order fractional chaotic systems for fast image encryption*, Chaos 29 (2019), p. 083103.

[30] G.C. Wu, B. Shiri, Q. Fan, and H.R. Feng, *Terminal value problems of non-homogeneous fractional linear systems with general memory kernels, J*, Nonlinear Math. Phys. (2022). DOI: 10.1007/s44198-022-00085-2

[31] G.C. Wu, T.T. Song, and S.Q. Wang, *Caputo-Hadamard fractional differential equation on time scales: Numerical scheme, stability, and chaos*, Chaos 32 (2022), p. 093143.

[32] X.L. Zhang, H.L. Li, Y. Kao, L. Zhang, and H. Jiang, *Global Mittag–Leffler synchronization of discrete-time fractional-order neural networks with time delays*, Appl. Math. Comput. 433 (2022), p. 127417.

Search for invariant sets of the generalized tent map

Kimberly Ayers, Dmitriy Dmitrishin, Ami Radunskaya, Alexander Stokolos and Kostyantyn Stokolos

ABSTRACT
This paper describes a predictive control method to search for unstable periodic orbits of the generalized tent map. The invariant set containing periodic orbits is a repelling set with a complicated Cantor-like structure. Therefore, a simple local stabilization of the orbit may not be enough to find a periodic orbit, due to the small measure of the basin of attraction. It is shown that for certain values of the control parameter, both the local behaviour and the global behaviour of solutions change in the controlled system; in particular, the invariant set enlarges to become an interval or the entire real axis. The computational particularities of using the control system are considered, and necessary conditions for the orbit to be periodic are given. The question of local asymptotic stability of subcycles of the controlled system's stable cycles is fully investigated, and some statistical properties of the subset of the classical Cantor middle thirds set that is determined by the periodic points of the generalized tent map are described.

Introduction

In his seminal paper *R. Lozi* noted that numerical computations using computers play a central role in analysing solutions of nonlinear dynamical systems, that computer-aided proofs are complex and necessarily require additional special validation of the results [22]. Nevertheless, numerous studies in fields related to chaotic dynamical systems are confident in the numerical solutions that they found using popular software, sometimes without carefully checking the reliability of these results. Computationally, computers store numbers in registers and memory cells with a limited number of digits. Thus, the set of real numbers represented in the machine is discrete and finite – irrational numbers, and rational numbers with infinite decimal expansions are rounded to decimal expansions that terminate. This can lead to problems when attempting to numerically find unstable periodic orbits of discrete chaotic dynamical systems.

Consider the family of chaotic dynamical systems given by the *generalized tent map*:

$$x_{n+1} = f(x_n), \quad n = 1, 2, \ldots, \tag{1}$$

where

$$f(x) = H\left(\frac{1}{2} - \left|x - \frac{1}{2}\right|\right) = \begin{cases} Hx, \ x \leq \frac{1}{2}, \\ H(1-x), \ x > \frac{1}{2}, \end{cases} \tag{2}$$

$x \in (-\infty, \ +\infty)$, $H \geq 2$. Note that, despite the relative simplicity of function (2), Equation (1) is of great theoretical importance and has appeared in several applications [6, 16]. Consider the classical tent map, given by $H = 2$. Note that $x_0 = \frac{2}{3}$ is a fixed point: $f(\frac{2}{3}) = \frac{2}{3}$. Since $|f'(\frac{2}{3})| = 2 > 1$, $\frac{2}{3}$ is an unstable fixed point.

Moreover, all orbits converge towards the other unstable fixed point at 0, which is also unstable, due to the finite representation of numbers in the computer, and to rounding errors. This phenomenon is well-studied: see, for example, [32]. Thus, any amount of rounding error will cause a numerically computed orbit to either eventually diverge from the non-zero fixed point, or to, surprisingly, converge to the unstable fixed point at zero.

In this article, we show how we can correct the computational procedure in the problem of finding unstable periodic orbits of a nonlinear discrete system using the tent map as an illustrative example.

The dynamics of even the simplest nonlinear discrete systems can be quite complex [3, 5, 7]. Such systems are often characterized by extremely unstable motions in phase space, which are defined as chaotic [7]. In dissipative systems, these motions define invariant sets, which can be strange attractors or repellers. Trajectories on such invariant sets have positive Lyapunov exponents; therefore, these trajectories are exponentially sensitive to the initial conditions. Unstable periodic orbits are canonical examples of repelling invariant sets, and we can get insight into the general study of repellers by considering unstable periodic orbits. Periodic orbits have a hierarchical structure determined by their length, which makes it possible to calculate various characteristics of invariant sets and their subsets, for example, topological dimension and entropy [4]. However, when applying numerical methods to search for points along a periodic orbits we encounter a number of fundamental problems. Due to sensitivity to initial conditions and rounding errors, after several calculation steps the results can vary greatly depending on the chosen calculation accuracy: the so-called 'butterfly effect' occurs. Even with a real possibility to choose a very high accuracy of calculations, we will never be able to say with certainty what we actually found: a long cycle, a pseudo-cycle or a strange attractor [22].

There are several methods of searching for periodic orbits of a nonlinear discrete system, which can be divided into two groups: methods that do not use the correction of the original discrete dynamical system, for example, the method of interval arithmetic analysis [14, 15, 33], or the method connected with the construction of special Hamiltonian systems [4]; and methods based on local stabilization of an unknown periodic orbit of a given length [2, 11, 23, 25, 28, 31]. The second group of methods is more preferable in the sense that their accuracy increases with the number of iterations, due to the correction of the original dynamical system. If we can locally stabilize the orbit with the help of the control action, then the trajectories of the system will remain in the neighbourhood of the orbit and will be attracted to it, i.e. the periodic orbit will be 'found'. By choosing different initial points, different periodic orbits can be found. To solve the problems of stabilization in the search of periodic orbits, various control schemes were proposed that use information about the states of the controlled system at previous points in time [8, 10, 24, 27] (delayed control) or about the states of the initial system at future points in time [12, 26, 29, 30] (predictive control).

The purpose of this paper is to illustrate the effectiveness of the predictive control method using the example of the generalized tent map. The invariant sets of the generalized tent map are repellers that have the structure of a Cantor set. We show that in a controlled system, along with a change in the local behaviour of solutions, the global behaviour also changes. We describe the invariant sets of the controlled system, which are either a single interval or a union of intervals. Locally asymptotically stable periodic orbits are subsets of the invariant set. The basins of attraction of these orbits are also discussed.

The paper is organized as follows. In Section 2 we present the main result from previous work [12], which substantiates the generalized predictive control scheme. Section 1 poses the problem of finding a cycle of length T, and provides necessary and sufficient conditions for the local asymptotic stability of this cycle.

Section 1 presents the main theoretical results in the paper: it provides conditions under which all solutions of the controlled system are bounded (Theorem 2.1), or solutions with initial values from a given interval are bounded (Theorem 2.2). We also show that the cycles do not lose the property of local asymptotic stability. In Section 2, some computational particularities of using the control system for finding cycles are considered, and necessary conditions for the found sequence to be a cycle are given. In Section 3, using the example of stabilization of cycles of small lengths, we consider the relationship between the graphical properties of the original map and the map determined by the controlled system. In Section 4, the question about local asymptotic stability of subcycles of the controlled system's stable cycles is fully investigated (Theorem 5.1). Section 5 studies subsets of the classical Cantor middle thirds set. Since any point in the Cantor set is a limit of periodic points, we can use the controlled system to visualize the Cantor set itself. We compare the distribution of periodic points of very long cycles to the distribution of points of the first type, i.e. endpoints of the intervals that make up the complement of the Cantor set. The images in this paper were created using Maple, and the files that generated them can be found at [1].

1. Background

We consider the discrete system

$$x_{n+1} = f(x_n),\ x_n \in \mathbb{R}^m, \quad n = 1, 2, \ldots, \tag{3}$$

where $f(x)$ is, in general, a nonlinear differentiable function from $\mathbb{R}^m$ to $\mathbb{R}^m$. It is assumed that this system has one or more unstable T-cycles $(\eta_1, \ldots, \eta_T)$, where all the vectors $\eta_1, \ldots, \eta_T$ are distinct, i.e. $\eta_{j+1} = f(\eta_j)$, $j = 1, \ldots, T-1$, $\eta_1 = f(\eta_T)$. Note that in the case of the generalized tent map, we are guaranteed the existence of cycles of any length [20].

Vectors in the cycles are called *cyclic points*, and each cycle of length T constitutes a *T-periodic orbit*. The *multipliers* of the unstable cycles are defined as the eigenvalues of the products of Jacobi matrices $\prod_{j=1}^{T} f'(\eta_{T-j+1})$ of dimension $m \times m$ at the points of the cycle. The matrix $\prod_{j=1}^{T} f'(\eta_{T-j+1})$ is called the Jacobi matrix of the cycle $(\eta_1, \ldots, \eta_T)$. The collection of all multipliers $\{\mu_1, \ldots, \mu_m\}$ is called the spectrum of the Jacobi matrix. As a rule, the cycles $(\eta_1, \ldots, \eta_T)$ of system (3) are not known a priori. Consequently, the spectrum is not known either.

Consider the control system

$$x_{n+1} = f\left(\vartheta_1 x_n + \sum_{j=2}^{N} \vartheta_j f^{((j-1)T)}(x_n)\right). \tag{4}$$

The numbers $\vartheta_1, \ldots, \vartheta_N$ are real. We can verify that when $\sum_{j=1}^{N} \vartheta_j = 1$, system (4) also has a cycle $\{\eta_1, \ldots, \eta_T\}$. The problem is to choose a parameter N and coefficients $\vartheta_1, \ldots, \vartheta_N$ so that the cycle $\{\eta_1, \ldots, \eta_T\}$ of system (4) is locally asymptotically stable.

The following result gives a criterion for stability of the cycles in terms of the multipliers. We will see how to use this result to determine a suitable range for the control parameter, ϑ.

Proposition 1.1 ([12]): *Suppose $f \in C^1$ and that system (3) has an unstable T-cycle with multipliers $\{\mu_1, \ldots, \mu_m\}$. Then this cycle will be a locally asymptotically stable cycle of system (4) if*

$$\mu_j \left[r(\mu_j)\right]^T \in D, \quad j = 1, \ldots, m,$$

where $D = \{z \in C: |z| < 1\}$ is an open central unit disc on the complex plane, $r(\mu) = \sum_{j=1}^{N} \vartheta_j \mu^{j-1}$.

A set U is called *invariant* for Equation (3) if for any $x_0 \in U$ it follows that $f^{(k)}(x_0) \in U$, $k = 1, 2, \ldots$. It is shown in [19] that for $H = 3$ the invariant set of Equation (1) is the classical Cantor set. Analogously, it can be shown that when $H > 2$, the invariant set for Equation (1) is a set of the Cantor type, that is an uncountable, closed set with zero Lebesgue measure.

In fact, we can say a bit more. For $H > 2$, the open interval $(\frac{1}{H}, 1 - \frac{1}{H})$ is mapped outside of $[0, 1]$, therefore it can contain no periodic points of any period, including fixed points. The remaining points are in the union: $[0, \frac{1}{H}] \cup [1 - \frac{1}{H}, 1]$, thus the invariant set, U is contained in the union:

$$I_0 \cup I_1 =\subset \left[0, \frac{1}{H}\right] \cup \left[1 - \frac{1}{H}, 1\right].$$

The points that are mapped outside of the unit interval under the second iterate of the tent map are open intervals taken from the centre of each of I_0 and I_1:

$$\left(\frac{1}{H^2}, \frac{1}{H} - \frac{1}{H^2}\right) \cup \left(1 - \frac{1}{H} + \frac{1}{H^2}, 1 - \frac{1}{H^2}\right)$$

Therefore, the invariant set, U is contained in the union of four disjoint closed intervals:

$$U \subset \left[0, \frac{1}{H^2}\right] \cup \left[\frac{1}{H} - \frac{1}{H^2}, \frac{1}{H}\right] \cup \left[1 - \frac{1}{H}, 1 - \frac{1}{H} + \frac{1}{H^2}\right] \cup \left[1 - \frac{1}{H^2}, 1\right]$$
$$= I_{00} \cup I_{01} \cup I_{10} \cup I_{11}.$$

All periodic points, including fixed points, must be in the union of these four intervals. We can continue in this manner to show that, for all N the invariant sent (and hence any periodic point) is contained in a disjoint union of 2^N closed intervals of length $\frac{1}{H^N}$.

This construction of U allows us to locate the periodic points for any given n. At each stage, we remove an open interval from the centre of the closed interval $I_{\alpha_1 \dots \alpha k}$ The closed interval that remains to the left of the removed interval is labelled $I_{\alpha_1 \dots \alpha k,0}$, while the closed interval that remains to the right of the removed interval is labelled $I_{\alpha_1 \dots \alpha k,1}$. Looking at the intervals that remain at each stage in the construction, we note that a point in the interval $I_{\alpha_1 \alpha_2 \dots \alpha_k}$ is in the interval $[\sum_{j=1}^{k} \frac{\alpha_j \cdot (H-1)}{H^j}, \sum_{j=1}^{k} \frac{\alpha_j \cdot (H-1)}{H^j} + \frac{1}{H^k}]$. Hence, each point of the invariant set can be represented in the form $\sum_{j=1}^{\infty} \frac{\alpha_j}{H^j}$, where $\alpha_j \in \{0, H-1\}$ by identifying each point with a closed interval $I_{\alpha_1 \alpha_2 \dots}$. The periodic points of period n can be identified as points where the sequence of α_j's are periodic with the same period, n. For example, the period-2 points of the tent map with $H = 3$ are precisely:

$$x_1 = \frac{0}{3} + \frac{2}{3^2} + \frac{0}{3^3} + \frac{2}{3^4} + \dots \quad \text{and} \quad x_2 = \frac{2}{3} + \frac{0}{3^2} + \frac{2}{3^3} + \frac{0}{3^4} + \dots$$

in addition to the two fixed points whose corresponding sequences of α_j's are all zeros or all 2's. Thus, if we are searching for period-2 points, we should focus on the intervals: $[\frac{2}{9}, \frac{3}{9}]$ and $[\frac{6}{9}, \frac{7}{9}]$, using the first two digits in the expansions, or we could get a more refined estimate by using four digits, and focus on the intervals $[\frac{8}{27}, \frac{9}{27}]$ and $[\frac{24}{27}, \frac{25}{27}]$

If x_0 does not belong to the invariant set, then the corresponding sequence $\{f^{(k)}(x_0)\}_{k=1}^{\infty}$ tends to $-\infty$. Such invariant sets are called repellers of map (2).

The problem we address in this paper is the following: for given T, can we numerically find T-periodic points of the tent map (2) as limit points of some iterative scheme.

2. Main results: the control system and global behaviour of the control system trajectories

To solve the problem of finding unstable periodic orbits of the tent map, we will use the predictive control method, which we now describe explicitly. Along with Equation (1), consider the equation

$$x_{n+1} = F(x_n), \quad n = 1, 2, \dots, \tag{5}$$

where $F(x) = f(\vartheta x + (1 - \vartheta) f^{(T)}(x))$, ϑ is some real number, called the *control parameter*, whose value will be determined later. We will call Equation (5) the *control system* for Equation (1). The idea is to find the values of ϑ that stabilize our target periodic cycles.

Note that Equation (5) might have pre-periodic points outside $[0, 1]$. For instance, let

$$f(x) = 3\left(\frac{1}{2} - \left|x - \frac{1}{2}\right|\right), \quad \text{and} \quad F(x) = f\left(\frac{17}{24}x + \frac{7}{24}f(x)\right).$$

After simplification one gets

$$F(x) = 3\left(\frac{1}{2} - \left|\frac{17}{24}x - \frac{1}{16} - \frac{7}{8}\left|x - \frac{1}{2}\right|\right|\right).$$

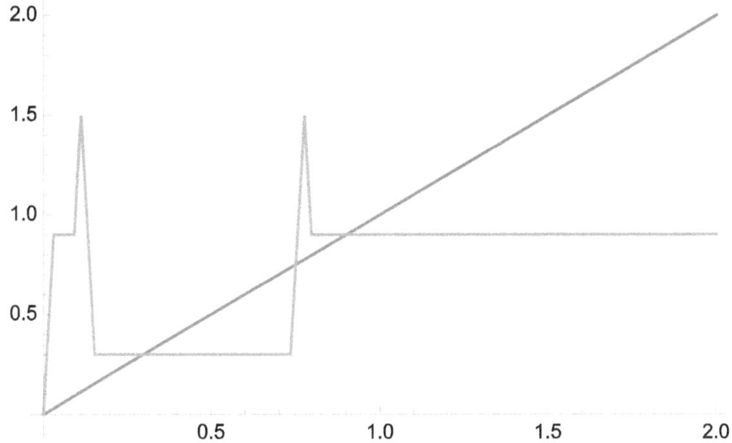

Figure 1. The graph of $F^2(x)$, where $F(x) = f(\frac{9}{10}x + \frac{1}{10}f(f(x)))$. Notice that $F^2(x) = \frac{9}{10}$ for all $x > \frac{9}{10}$, so all these points are pre-periodic for F, as $\frac{9}{10}$ is a period 2 point.

Note that the map $f(x)$ has only two fixed points, namely 0 and 3/4. However, the map $F(x)$ has many eventually fixed points, e.g.

$$\frac{375}{76} \to \frac{3}{19} \to \frac{3}{4},$$
$$\frac{142887}{27436} \to \frac{144}{6859} \to \frac{36}{361} \to \frac{9}{19} \to \frac{3}{4},$$

and others.

Similarly, for any period, T, let x^* be a period-T point of f, hence a period-T point of F. Choose ϑ in Equation (5) so that $\frac{d}{dx}(F^T(x^*)) = 0$. Since f is piecewise linear, $\frac{d}{dx}F^T(x) = 0$ on some interval containing x^*, i.e. F^T is constant on this interval, which will extend beyond $[0, 1]$ for the upper-most periodic point. Thus, x^* has infinitely many pre-images under F^T, and it is possible that these pre-images lie outside the unit interval. We illustrate this in Figure 1, where $T = 2$, $x^* = \frac{9}{10}$ and $\vartheta = \frac{9}{10}$.

Thus, even though the tent map, $f(x)$ has no periodic points outside of the unit interval, we will consider orbits of (5) with initial values outside the interval [0,1].

Let $\{\eta_1, \ldots, \eta_T\}$ be a cycle of Equation (1) of length T. Since $\vartheta\eta_k + (1-\vartheta)f^{(T)}(\eta_k) = \eta_k$, then $F(\eta_k) = f(\eta_k)$, which means that the cycle of Equation (1) will also be a cycle of Equation (5). Note that the converse statement is generally not true.

The multiplier of Equation (1) cycle is defined by the formula

$$\mu = f'(\eta_T) \cdot \ldots \cdot f'(\eta_1).$$

Since $|f'(\eta_j)| = H$, then $|\mu| = H^T > 1$, that is, any cycle of Equation (1) is unstable. Let us find the value of the multiplier λ of the same cycle $\{\eta_1, \ldots, \eta_T\}$, but for Equation (5). From Proposition 1.1 we get that

$$\lambda = \mu\,(\vartheta + (1-\vartheta)\mu)^T.$$

In what follows, we will consider two cases separately: $\mu > 0$ and $\mu < 0$. Let $\mu = H^T$. Then the condition for local asymptotic stability of the T-cycle of Equation (5) is: $|\lambda| = |H^T(\vartheta + (1-\vartheta)H^T)^T| < 1$, from which it follows that

$$1 < \frac{H^T - \frac{1}{H}}{H^T - 1} < \vartheta < \frac{H^T + \frac{1}{H}}{H^T - 1} < \frac{5}{2}. \tag{6}$$

If $\mu = -H^T$, then the condition for local asymptotic stability of Equation (5) cycle is: $|\lambda| = |H^T(\vartheta - (1-\vartheta)H^T)^T| < 1$, from which

$$\frac{1}{2} < \frac{H^T - \frac{1}{H}}{H^T + 1} < \vartheta < \frac{H^T + \frac{1}{H}}{H^T + 1} < 1. \tag{7}$$

Thus, Proposition 1.1 yields the following conditions for asymptotic stability of cycles:

Claim 2.1: *Given a T-cycle of Equation* (1) *with the multiplier μ. This cycle will be a locally asymptotically stable cycle of Equation* (5) *if inequalities* (6), *in the case $\mu > 0$, or inequalities* (7), *when $\mu < 0$, are satisfied.*

If there are locally asymptotically stable cycles in Equation (5), then the invariant set of this equation also includes the basins of attraction of these cycles. Since the basin of attraction of a cycle is an open set, its measure is positive. However, there remains the question about how large this measure is. This question is important when doing numerical experiments, because we must judiciously choose the initial point to be in the basin of attraction the desired cycle. Thus, as we pass from Equations (1) to (5), we are ensured that the nature of the periodic orbit (the set of points of the cycle) changes from a repeller into an attractor. For numericaly tractability, it is also important to know that we have a basin of attraction of sufficiently large measure.

Next, we will establish the properties of invariant sets of Equation (5) and the global behaviour of its solutions. In particular, we will show that under the conditions on the control parameter, θ, given in Claim 2.1, we have globally attracting invariant sets made up of intervals.

We state these properties as two theorems, one for the case when the multiplier μ is positive, and one when μ is negative. The proofs of the two theorems are broken down into several lemmas, each dealing with a specific range of ϑ values. The two main theorems follow from these lemmas, allowing us to find all cycles of arbitrary lengths with any given accuracy.

Theorem 2.1: *(Case $\mu > 0$)*

If inequalities $\frac{H^T - \frac{1}{H}}{H^T - 1} < \vartheta \leq \frac{H^T + \frac{1}{H}}{H^T - 1}$ are satisfied, then any solution of Equation (5) *is bounded, and its limit set is contained in* $[0, \frac{H}{2}]$. *Moreover, any T-cycle $\{\eta_1, \ldots, \eta_T\}$ of this equation, for which the quantity $\mu = f'(\eta_T) \cdot \ldots \cdot f'(\eta_1)$ is positive, is locally asymptotically stable.*

Theorem 2.2: *(Case $\mu < 0$) If the inequalities $\frac{H^T - \frac{1}{H}}{H^T + 1} < \vartheta \leq \frac{H^T}{H^T + 1}$ are satisfied, the set* $[0, \frac{H}{2}]$ *is an invariant set of Equation* (5). *Moreover, any T-cycle $\{\eta_1, \ldots, \eta_T\}$ of this* *equation, for which the quantity $\mu = f'(\eta_T) \cdot \ldots \cdot f'(\eta_1)$ is negative, is locally asymptotically stable.*

Theorems 2.1 and 2.2 will be proved through a sequence of five lemmas, each one dealing with a different sub-case. Lemmas 2.1–2.3 divide the positive multiplier case into three sub-cases: one where ϑ is exactly in the middle of the allowed interval (Lemma 2.1), one where ϑ is in the lower half of the allowed interval (Lemma 2.2) and one where ϑ is in the upper half of the allowed interval (Lemma 2.3). The remaining two lemmas (Lemmas 2.4 and 2.5) deal with the case of a negative multiplier, $\mu < 0$.

2.1. Proof of Theorem 2.1

Throughout the proof, to break down the calculations, we let the function

$$\zeta(x) = \vartheta x + (1 - \vartheta) f^{(T)}(x)$$

be the intermediary linear combination that appears in the definition of the control function, $F(x)$ given in Equation (5). Thus, $F(x) = f(\zeta(x))$, and $\zeta_0 = \zeta(x_0)$.

Suppose $\mu > 0$, and let inequalities (6) be satisfied. Divide the conditions (6) into three cases: (i) $\vartheta = \frac{H^T}{H^T-1}$, (*ii*) $\frac{H^T-\frac{1}{H}}{H^T-1} < \vartheta < \frac{H^T}{H^T-1}$, and (iii) $1 < \frac{H^T}{H^T-1} < \vartheta < \frac{H^T+\frac{1}{H}}{H^T-1}$.. Each case is treated in a separate lemma.

Lemma 2.1: *Let $\vartheta = \frac{H^T}{H^T-1}$. Then, if $x_0 \leq 0$, it follows that $F(x_0) = 0$, hence, $F^{(k)}(x_0) = 0$ when $k = 1, 2, \ldots$; if $x_0 \geq 1$, then $F(x_0) < 0$, hence, $F^{(k)}(x_0) = 0$ when $k = 2, 3, \ldots$; if $x_0 \in (0, 1)$, then $\{F^{(k)}(x_0)\}_{k=2}^{\infty} \in [0, 1]$.*

Proof: Let $x_0 \leq 0$, then $f(x_0) = Hx_0 \leq 0$, $f^{(2)}(x_0) = H^2 x_0 \leq 0$, …, $f^{(T)}(x_0) = H^T x_0$. Find

$$\zeta_0 = \zeta(x_0) = \vartheta x_0 + (1 - \vartheta) f^{(T)}(x_0) = \frac{1}{H^T - 1}\left(H^T x_0 - H^T x_0\right) = 0.$$

Hence, $F(x_0) = f(\zeta_0) = 0$.

Let $x_0 \geq 1$, then

$$f(x_0) = H(1 - x_0) \leq 0, f^{(2)}(x_0) = H^2(1 - x_0) \leq 0, \ldots, f^{(T)}(x_0) = H^T(1 - x_0),$$

and

$$\begin{aligned}\zeta_0 = \vartheta x_0 + (1 - \vartheta) f^{(T)}(x_0) &= \frac{1}{H^T - 1}\left(H^T x_0 - H^T(1 - x_0)\right) \\ &= \frac{H^T}{H^T - 1}(2x_0 - 1) > 1,\end{aligned}$$

so that $F(x_0) = f(\zeta_0) = H(1 - \zeta_0) < 0$. Hence, $F^{(2)}(x_0) = 0$.

The last case remains: if $x_0 \in (0, 1)$, then there is a k_0 such that $F^{(k)}(x_0) = 0$, for all $k > k_0$; or $\{F^{(k)}(x_0)\}_{k=2}^{\infty} \in (0, 1)$ for all k. ∎

The case $\frac{H^T-\frac{1}{H}}{H^T-1} < \vartheta < \frac{H^T}{H^T-1}$ is considered in a similar way.

Lemma 2.2: *Let* $\frac{H^T-\frac{1}{H}}{H^T-1} < \vartheta < \frac{H^T}{H^T-1}$. *Then,*
if $x_0 \leq 0$ *or* $x_0 \geq 1$, $F^{(k)}(x_0) \xrightarrow[k\to\infty]{} 0$;
when $x_0 \in (0, 1)$, *either* $F^{(k)}(x_0) \xrightarrow[k\to\infty]{} 0$ *or* $\{F^{(k)}(x_0)\}_{k=1}^{\infty} \in (0, 1)$.

Proof: Let $x_0 \leq 0$, then $f(x_0) = Hx_0 \leq 0, \ldots, f^{(T)}(x_0) = H^T x_0$. Find

$$\zeta_0 = \vartheta x_0 + (1-\vartheta)f^{(T)}(x_0) = \left(H^T - \vartheta(H^T-1)\right)x_0.$$

It follows from the inequality $\frac{H^T-\frac{1}{H}}{H^T-1} < \vartheta < \frac{H^T}{H^T-1}$ that, first, $\zeta_0 \leq 0$, and

$$F(x_0) = f(\zeta_0) = H\zeta = H\left(H^T - \vartheta(H^T-1)\right)x_0 \leq 0,$$

and, second:

$$0 < \alpha = H\left(H^T - \vartheta(H^T-1)\right) < 1,$$

from which $F(x_0) = -\alpha|x_0| > -|x_0|$. Then $|F^{(k)}(x_0)| = \alpha^k|x_0| \xrightarrow[k\to\infty]{} 0$.

Let $x_0 \geq 1$, then $f(x_0) = H(1-x_0) \leq 0, \ldots, f^{(T)}(x_0) = H^T(1-x_0)$,

$$\begin{aligned}\zeta_0 &= \vartheta x_0 + (1-\vartheta)f^{(T)}(x_0)\\ &= x_0 + (\vartheta-1)\left(x_0 + H^T(x_0-1)\right)\\ &> x_0 > 1.\end{aligned}$$

Then $F(x_0) = f(\zeta_0) = H(1-\zeta_0) < 0$. Hence,

$$F^{(2)}(x_0) = -\alpha\,|F(x_0)| > -\,|F(x_0)|\,, \left|F^{(k)}(x_0)\right| = \alpha^{k-1}\,|F(x_0)| \xrightarrow[k\to\infty]{} 0.$$

Let $x_0 \in (0, 1)$. If for some k_0, $F^{(k_0)}(x_0) \leq 0$ or $F^{(k_0)}(x_0) \geq 1$, then $F^{(k)}(x_0) \xrightarrow[k\to\infty]{} 0$. Otherwise, $\{F^{(k)}(x_0)\}_{k=1}^{\infty} \in (0, 1)$. The lemma is proved. ■

The case $\frac{H^T}{H^T-1} < \vartheta < \frac{H^T+\frac{1}{H}}{H^T-1}$ remains.

Lemma 2.3: *Let* $\frac{H^T}{H^T-1} < \vartheta < \frac{H^T+\frac{1}{H}}{H^T-1}$. *In this case, if* $-H^2 + \frac{H}{2} \leq x_0 \leq \frac{H}{2}$, *then*

$$-H^2 + \frac{H}{2} \leq F^{(k)}(x_0) \leq \frac{H}{2}, \quad k = 1, 2, \ldots. \tag{8}$$

If $x_0 > \frac{H}{2}$ *or* $x_0 < -H^2 + \frac{H}{2}$, *there exists a number* $k_0 \geq 0$ *such that inequalities (8) are satisfied for all k greater than* k_0.

Proof: Let $x_0 \le 0$, then $f(x_0) = Hx_0 \le 0, \ldots, f^{(T)}(x_0) = H^T x_0$,

$$\zeta_0 = \left(H^T - \vartheta(H^T - 1)\right) x_0.$$

The inequality $\frac{H^T}{H^T-1} < \vartheta < \frac{H^T+\frac{1}{H}}{H^T-1}$ is equivalent to

$$-\frac{1}{H} < H^T - \vartheta(H^T - 1) < 0,$$

from which it follows that $0 \le \zeta_0 \le \frac{|x_0|}{H}$. If $-H \le x_0 \le 0$, then $0 \le \zeta_0 \le 1$ and $0 \le F(x_0) \le \frac{H}{2}$.

If $x_0 < -H$ and $\zeta_0 \ge 1$, then $0 \ge F(x_0) = H(1 - \zeta_0) > H - |x_0|$. Moreover, if $F^{(k)}(x_0) < -H$ and $(H^T - \vartheta(H^T - 1))F^{(k)}(x_0) \ge 1$, then $0 \ge F^{(k+1)}(x_0) > H - |F^{(k)}(x_0)| > (k+1)H - |x_0|$. It means that there exists such k_0 that $-H \le F^{(k_0)}(x_0) \le 0$. And hence, $0 \le F^{(k_0+1)}(x_0) \le \frac{H}{2}$.

Let $0 \le x_0 \le \frac{H}{2}$, then $H^T(1 - \frac{H}{2}) \le f^{(T)}(x_0) \le \frac{H}{2}$. Since $\vartheta > 1$, then

$$\begin{aligned}
-\frac{3}{4}H &< \vartheta x_0 - (\vartheta - 1)\frac{H}{2} \\
&\le \zeta_0 \\
&\le \vartheta x_0 + (\vartheta - 1)H^T\left(\frac{H}{2} - 1\right) \\
&< \vartheta\frac{H}{2} + (\vartheta - 1)H^T\left(\frac{H}{2} - 1\right) \\
&< \frac{1}{H^T - 1}\left(\left(H^T + \frac{1}{H}\right)\frac{H}{2} + \left(1 + \frac{1}{H}\right)H^T\left(\frac{H}{2} - 1\right)\right) \\
&= \frac{1}{H^T - 1}\left(H^{T+1} - \frac{H^T}{2} - H^{T-1} + \frac{1}{2}\right).
\end{aligned}$$

Note that $1 < H - 1 < \frac{1}{H^T-1}(H^{T+1} - \frac{H^T}{2} - H^{T-1} + \frac{1}{2}) < H + \frac{1}{2}$. Then

$$F(x_0) > H\left(1 - \frac{1}{H^T - 1}\left(H^{T+1} - \frac{H^T}{2} - H^{T-1} + \frac{1}{2}\right)\right) > -H\left(H - \frac{1}{2}\right).$$

Therefore, if $0 \le x_0 \le \frac{H}{2}$, then $-H^2 + \frac{H}{2} \le F(x_0) \le \frac{H}{2}$. Moreover, if $F(x_0) < 0$ then for some k_0 the inequality $0 \le F^{(k_0)}(x_0) \le \frac{H}{2}$ is satisfied.

Let $x_0 > \dfrac{H}{2}$, then

$$f^{(T)}(x_0) = H^T(1 - x_0), \quad \zeta_0 = \vartheta x_0 + (\vartheta - 1)H^T(x_0 - 1) > 1, \quad \text{and} \quad F(x_0) < 0.$$

This means that for some k_0 the inequality $0 \le F^{(k_0)}(x_0) \le \frac{H}{2}$ is satisfied.

Summing up the four cases above, we get that when $-H^2 + \frac{H}{2} \le x_0 \le \frac{H}{2}$, inequalities (8) hold. If $x_0 > \frac{H}{2}$ or $x_0 < -H^2 + \frac{H}{2}$, then inequalities (8) will be satisfied starting from some iterate, x_k. The lemma is proved. ■

The assertion of Theorem 2.1 follows from Claim 2.1, and Lemmas 2.1–2.3,.

2.2. Proof of Theorem 2.2

Let us now pass to the study of the global behaviour of the T-cycles $\{\eta_1, \dots, \eta_T\}$ of Equation (5) for which the quantity $\mu = f'(\eta_T) \cdot \dots \cdot f'(\eta_1)$ is negative, assuming that conditions (7) are satisfied.

Lemma 2.4: *Let conditions* (7) *be satisfied and* $x_0 < 0$. *Then* $F^{(k)}(x_0) \xrightarrow[k\to\infty]{} -\infty$.

Proof: Since $x_0 < 0$, then $f(x_0) = Hx_0 \le 0, \dots, f^{(T)}(x_0) = H^T x_0$, and

$$\zeta_0 = \vartheta x_0 + (1-\vartheta) f^{(T)}(x_0) = \left(\vartheta + (1-\vartheta)H^T\right) x_0.$$

Since $\frac{1}{2} < \vartheta < 1$, then $\zeta_0 < x_0$, $F(x_0) = H\zeta_0 < Hx_0$, $F^{(k)}(x_0) < H^k x_0$ when $k = 1, 2, \dots$, from which the conclusion of the lemma follows. ■

Lemma 2.5: *Let the inequalities* $\frac{H^T - \frac{1}{H}}{H^T+1} < \vartheta \le \frac{H^T}{H^T+1}$ *be satisfied, and* $0 \le x_0 \le \frac{H}{2}$. *Then* $0 \le F(x_0) \le \frac{H}{2}$.

Proof: Let $1 \le x_0 \le \frac{H}{2}$, then $f(x_0) = H(1-x_0) \le 0, \dots, f^{(T)}(x_0) = H^T(1-x_0)$,

$$\zeta_0 = \vartheta x_0 + (1-\vartheta) f^{(T)}(x_0) = \frac{1}{H^T+1}\left((H^T+\alpha)x_0 + (1-\alpha)H^T(1-x_0)\right),$$

where $\vartheta = \frac{H^T+\alpha}{H^T+1}$, $-\frac{1}{H} < \alpha \le 0$. Since $0 \le x_0 \le \frac{H}{2}$,

$$\begin{aligned}
\zeta_0 &= \frac{1}{H^T+1}\left(H^T + \alpha(x_0 + x_0 H^T - H^T)\right) \\
&> \frac{1}{H^T+1}\left(H^T - \frac{1}{H}\left(\frac{H}{2} + \frac{H}{2}H^T - H^T\right)\right) \\
&= \frac{1}{H^T+1}\left(\frac{1}{2}H^T - \frac{1}{2} + H^{T-1}\right) \\
&> \frac{1}{2}.
\end{aligned}$$

On the other side, $\zeta_0 \le \frac{H^T}{H^T+1} < 1$. Hence, $0 < F(x_0) \le \frac{H}{2}$.

Let $0 \le x_0 < 1$, $p \in \{1, \dots, T\}$ is the smallest number at which $f^{(p-1)}(x_0) < 1$, but $f^{(p)}(x_0) > 1$ (here we mean that $f^{(0)}(x) := x$). It is clear that $f^{(p)}(x_0) \le \frac{H}{2}$.

Note that

$$
\begin{aligned}
&\frac{1}{H}f^{(p)}(x_0) \le f^{(p-1)}(x_0) \le 1 - \frac{1}{H}f^{(p)},\\
&\frac{1}{H^2}f^{(p)}(x_0) \le \frac{1}{H}f^{(p-1)}(x_0) \le f^{(p-2)}(x_0)\\
&\qquad\qquad \le 1 - \frac{1}{H}f^{(p-1)}(x_0) \le 1 - \frac{1}{H^2}f^{(p)}(x_0),\\
&\qquad\qquad \vdots \qquad ,\\
&\frac{1}{H^p}f^{(p)}(x_0) \le x_0 \le 1 - \frac{1}{H^p}f^{(p)}(x_0).
\end{aligned}
\tag{9}
$$

If $p = T$, then $\zeta_0 > 0$ and, since $x_0 - f^{(T)}(x_0) < 0$,

$$
\zeta_0 = \vartheta x_0 + (1-\vartheta)f^{(T)}(x_0) < \frac{1}{H^T+1}\left(\left(H^T - \frac{1}{H}\right)x_0 + \left(1+\frac{1}{H}\right)f^{(T)}(x_0)\right),
$$

and because of (9),

$$
\begin{aligned}
&\left(H^T - \frac{1}{H}\right)x_0 + \left(1+\frac{1}{II}\right)f^{(T)}(x_0)\\
&\quad \le \left(H^T - \frac{1}{H}\right)\left(1 - H^{-T}f^{(T)}(x_0)\right) + \left(1+\frac{1}{H}\right)f^{(T)}(x_0)\\
&\quad = H^T - H^{-1} \quad f^{(T)}(x_0) + H^{-T-1}f^{(T)}(x_0) + f^{(T)}(x_0) + H^{-1}f^{(T)}(x_0)\\
&\quad < H^T + \frac{1}{2}H \quad ^{T} + \frac{1}{2}\\
&\quad < H^T + 1.
\end{aligned}
$$

This implies $\zeta_0 < 1$.

Let $p < T$. Then $\zeta_0 < 1$. Check the inequality $\zeta_0 > 0$. Since $f^{(T)}(x_0) = H^{T-p}(1 - f^{(p)}(x_0)) \le 0$, then $x_0 - f^{(T)}(x_0) > 0$ and

$$
\zeta_0 = \vartheta x_0 + (1-\vartheta)f^{(T)}(x_0) > \frac{1}{H^T+1}\left(\left(H^T - \frac{1}{H}\right)x_0 + \left(1+\frac{1}{H}\right)f^{(T)}(x_0)\right).
$$

Because of (9),

$$
\begin{aligned}
&\left(H^T - \frac{1}{H}\right)x_0 + \left(1+\frac{1}{H}\right)f^{(T)}(x_0)\\
&\quad \ge \left(H^T - \frac{1}{H}\right)H^{-p}f^{(p)}(x_0) + \left(1+\frac{1}{H}\right)H^{T-p}\left(1 - f^{(p)}(x_0)\right)
\end{aligned}
$$

$$= H^{T-p}f^{(p)}(x_0) - H^{T-p-1}f^{(p)}(x_0) + \left(1 + \frac{1}{H}\right) H^{T-p}$$
$$- \left(H^{T-p} + H^{T-p-1}\right) f^{(p)}(x_0)$$
$$= \left(1 + \frac{1}{H}\right) H^{T-p} - 2H^{T-p-1}f^{(p)}(x_0)$$
$$\geq \left(1 + \frac{1}{H}\right) H^{T-p} - H^{T-p}$$
$$> 0.$$

Thus, when $0 \leq x_0 < 1$, the inequalities $0 < \zeta_0 < 1$ hold and, therefore, $0 < F(x_0) \leq \frac{H}{2}$. The lemma is proved. ∎

Theorem 2.2 follows from Lemmas 2.4 and 2.5.

Note that when the inequalities $\frac{H^T}{H^T+1} < \vartheta < \frac{H^T+\frac{1}{H}}{H^T+1}$ are satisfied, the invariant set of Equation (5) will no longer be a segment, but will be the union of a finite or countable number of intervals, and the measure of this set may be small.

Theorems 2.1 and 2.2 are illustrated in Section 3 for the case $H = 3$ and $T = 2$.

3. Computational particularities of using the control system

In this section we introduce the residuals:

$$U_n = ||f(\vartheta x_n + (1 - \vartheta)f^{(T)}(x_n)) - f(x_n)||$$

and

$$\hat{U}_n = ||x_{n+T} - x_n||,$$

whose rate of decay allows us to compare solutions of system (5) to solutions of system (1), as well as the T-periodicity of the solutions.

Let us consider the computational particularities of the iterative scheme (5) for finding the cycles of Equation (1). For cycles with positive multipliers, it is theoretically possible to take any number from the interval $(\frac{H^T-\frac{1}{H}}{H^T-1}, \frac{H^T+\frac{1}{H}}{H^T-1})$ as the control parameter; however, if this parameter belongs to the half-interval $(\frac{H^T-\frac{1}{H}}{H^T-1}, \frac{H^T}{H^T-1}]$, then the fixed point will have a fairly large basin of attraction. This will be illustrated geometrically below. Therefore, if we want to find true period-T cycles, it is reasonable to choose the control parameter closer to $\dfrac{H^T+\frac{1}{H}}{H^T-1}$. For cycles with negative multipliers, the control parameter can be taken from the interval $(\dfrac{H^T-\frac{1}{H}}{H^T+1}, \dfrac{H^T}{H^T+1})$. Since the basin of attraction of a given cycle can be quite small, the initial value x_0 should belong to the nodes of a sufficiently dense grid of the interval (0, 1) in order to find the largest possible number of cycles.

Note that for large values of T, the control parameter ϑ is close to one, and the value $1 - \vartheta$ is close to zero. In this case, the lengths of the intervals of possible changes in the control parameter are equal to $\frac{1}{H(H^T-1)}$ or $\frac{1}{H(H^T+1)}$, i.e. as T grows, they tend to zero exponentially.

Therefore, although the method suggested above for determining cycles theoretically allows us to solve the stated problem numerically, practical questions remain: when can we rely on numerical solutions? How can we control numerical results? What calculation accuracy should be chosen?

In practice, intermediate calculations should be introduced to control the results. Let the sequence $\{x_n\}_{n=1}^{\infty}$ be an orbit of the system given by Equation (5), and let the quantity $1-\vartheta$ have the order of magnitude 10^{-p}, where p is large enough. Then the first checkpoint will be the estimate of the residual $U_n = \|f(\vartheta x_n + (1-\vartheta)f^{(T)}(x_n)) - f(x_n)\|$. If the sequence $\{x_n\}$ tends to an orbit of system (1), then the sequence $\{U_n\}$ tends to zero. However, if the sequence $\{x_n\}$ does not tend to an orbit of system (1), then the residual can have the order of magnitude $1-\vartheta \sim 10^{-p}$, i.e. be close to zero. To be sure that the residual tends to zero, we have to choose the calculation accuracy $\delta = 10^{-p_1}$, where p_1 should be significantly greater than p. Then the first point of control will be the condition $U_n \sim 10^{-p_1}, n \geq n_1$.

The second checkpoint is the check of periodicity of the numerical solution: $\hat{U}_n = \|x_{n+T} - x_n\| \sim 10^{-p_1}, n \geq n_1$. Of course, it is also necessary to check that T is a proper cycle of system (1), and not a subcycle, i.e. a cycle of shorter length.

The checkpoints give necessary conditions for the sequence $\{x_n\}$ to be a T-cycle of Equation (1). The effectiveness of these necessary conditions is that they are quite simple to check, which we illustrate in the following example.

Example: Consider the problem of finding 5-cycles of Equation (1) when $H = 4$. Choose two initial conditions and for each of them find one 5-cycle with a positive and negative multiplier respectively. Set $x_0 \subset \{0.25, 0.85\}$, $\vartheta \in \{\frac{H^T - \frac{0.4}{H}}{H^T+1}, \frac{H^T + \frac{0.4}{H}}{H^T-1}\}$. Thus, we got four iterative schemes. Choose one colour for each scheme in order to visualize the results:

x_0/ϑ	$\frac{H^T - \frac{0.4}{H}}{H^T+1} \approx 0.9989$	$\frac{H^T + \frac{0.4}{H}}{H^T-1} \approx 1.0010$
0.25	red	green
0.85	blue	black

Since $(1-\vartheta)$ is on the order of 10^{-3}, we choose a calculation accuracy of 10^{-15}. The corresponding cyclic points are shown in Figure 2. Figure 3 shows the graphs of the residuals U_n and $\hat{U}_n$ as n increases. The periodicity condition, $\hat{U} = \|x_{n+5} - x_n\| < 10^{-15}$ holds for all four schemes, starting from $n = 43$.

We note that there are 6 distinct cycles of period 5, but we only show four of these here. We list the exact values of all six 5-cycles in Table 1, along with successive errors in the numerical estimates of the first point of the first four cycles. To find the remaining two cycles, we would judiciously select two more values of x_0 and ϑ.

All cycles of any length can be found in a similar fashion. The limitation is the calculation accuracy, which should be chosen to be approximately $H^{1.05T}$, since $H = |f'(x)|$. Figure 4 (Left Panel) shows the cyclic points of four cycles of length 100. The accuracy of calculations was taken 10^{65}. For negative multipliers, the control parameter is chosen to be $\vartheta = \dfrac{H^T - \frac{0.4}{H}}{H^T+1} \approx 1 - 1.78 \cdot 10^{-61}$; for positive multipliers, $\vartheta = \dfrac{H^T + \frac{0.4}{H}}{H^T - 1} \approx$

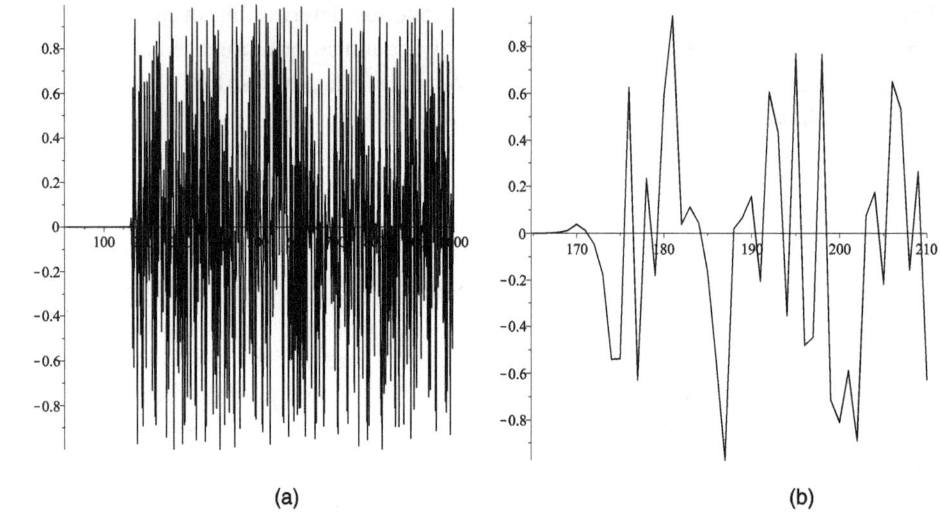

Figure 2. Four 5-cycles of the tent map. The upper left panel shows two cycles with positive multipliers (in red and blue), and the upper right panel shows two cycles with negative multipliers (in green and black). The lower graph shows all four cycles superimposed on each other, showing how the basins of attraction are intertwined. The exact values and some errors are shown in Table 1.

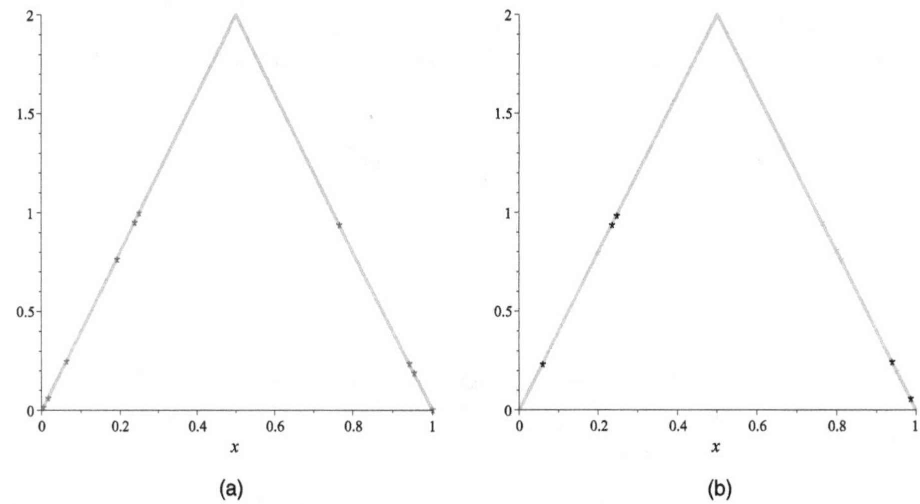

Figure 3. Residuals for the numerically estimated period-5 points. Left Panel: residuals U_n where n increases from 10 to 50, with $\vartheta \approx 0.99893$; Right Panel: the residuals U_n with $\vartheta \approx 1.0010$ for n in the range $38, \ldots 50$. The control point condition is satisfied for $n \geq 43$.

$1 + 0.68 \cdot 10^{-62}$. Note that the lengths of the intervals of possible changes in the control parameter have the order of magnitude $H^{-(T+1)} \approx 1.5 \cdot 10^{-61}$. The necessary bounds on the residuals are attained at the 250th step (Figure 4, Right Panel).

Table 1. All six cycles of length 5 for the tent map with $H = 4$.

Cycle	Values					colour
L_1	$x^* = \frac{16}{1025}$	$\frac{64}{1025}$	$\frac{256}{1025}$	$\frac{1024}{1025}$	$\frac{4}{1025}$	red
L_2	$x^* = \frac{784}{1025}$	$\frac{964}{1025}$	$\frac{244}{1025}$	$\frac{976}{1025}$	$\frac{196}{1025}$	blue
L_3	$x^* = \frac{68}{341}$	$\frac{272}{341}$	$\frac{276}{341}$	$\frac{260}{341}$	$\frac{324}{341}$	green
L_4	$x^* = \frac{20}{341}$	$\frac{80}{341}$	$\frac{320}{341}$	$\frac{84}{341}$	$\frac{336}{341}$	black
L_5	$\frac{4}{341}$	$\frac{16}{341}$	$\frac{64}{341}$	$\frac{256}{341}$	$\frac{340}{341}$	
L_6	$\frac{208}{1025}$	$\frac{832}{1025}$	$\frac{772}{1025}$	$\frac{1012}{1025}$	$\frac{52}{1025}$	

Cycle	errors: $e_k = x_k - x^*$		
L_1	$e_{18} = 1.6 \cdot 10^{-11}$	$e_{23} = 1.6 \cdot 10^{-13}$	$e_{28} = 1.0 \cdot 10^{-18}$
L_2	$e_{21} = 3.9 \cdot 10^{-11}$	$e_{26} = 4.0 \cdot 10^{-13}$	$e_{31} = 2.0 \cdot 10^{-15}$
L_3	$e_{18} = -3.7 \cdot 10^{-8}$	$e_{23} = 3.7 \cdot 10^{-10}$	$e_{28} = -3.8 \cdot 10^{-12}$
	$e_{33} = 4.0 \cdot 10^{-14}$	$e_{38} = 0.$	
L_4	$e_{20} = -1.8 \cdot 10^{-7}$	$e_{25} = 1.8 \cdot 10^{-9}$	$e_{30} = -1.9 \cdot 10^{-11}$
	$e_{35} = 1.9 \cdot 10^{-13}$	$e_{40} = -1.9 \cdot 10^{-15}$	

Notes: For the first four cycles, we show the error in the numerical estimates in terms of the distance from the first point on the cycle, x^*, for several iterates. These errors decrease along the orbit. For the third cycle, L_3, the difference between the numerical estimate of x_{38} and the true value of the period-5 point is zero, up to machine precision. These first four cycles are shown in Figure 2 with L_1 shown in red, L_2 in blue, L_3 in green and L_4 in black.

4. Graphical study of the control system map

For some choices of the control parameter ϑ the invariant set is visible on the graph of $y = f(\vartheta x + (1-\vartheta)f(x))$. Several choice of T and ϑ are illustrated below.

Equation (1) has two fixed points $\eta = 0$ and $\eta = \frac{H}{H+1}$, both of which are unstable since their multipliers are H and $-H$ respectively. Consider Equation (5)

$$x_{n+1} = f\left(\vartheta x_n + (1-\vartheta)f(x_n)\right)$$

for different values of $\vartheta \in [0, \frac{H^T + \frac{1}{H}}{H^T - 1}]$. For definiteness, set $H = 3$. Then $[0, \frac{H^T + \frac{1}{H}}{H^T - 1}] = [0, \frac{5}{3}]$. According to Theorem 2.1, to stabilize the fixed point at $\eta = 0$, we need to choose the control parameter from the interval $(\frac{4}{3}, \frac{5}{3})$. By Proposition 1.1 the multiplier of this fixed point is $\lambda = 9 - 6\vartheta$, which decreases from 1 to -1 as the control parameter ϑ increases from $\frac{9}{3}$ to $\frac{5}{3}$, and $\lambda = 0$ when $\vartheta = \frac{3}{2}$. The graphs of the function $F(x) = f(\vartheta x + (1-\vartheta)f(x))$ for different values of ϑ are shown in Figure 5.

According to Theorem 2.2, the fixed point at $\eta = \frac{3}{4}$ will be a locally asymptotically stable fixed point of Equation (5) if $\vartheta \in (\frac{2}{3}, \frac{5}{6})$. As the control parameter increases, the multiplier of this equilibrium decreases from 1 to -1. When $\vartheta = \frac{3}{4}$, the multiplier equals zero. The graphs of the function $F(x) = f(\vartheta x + (1-\vartheta)f(x))$ for different values of ϑ are depicted in Figure 6.

If we represent the Lamerey diagram on the graphs of Figure 5, then we can observe that for any initial value x_0 and $\vartheta \in (\frac{4}{3}, \frac{5}{3})$, the corresponding solution of Equation (5) will tend to zero. Similarly, Figure 6 shows that the segment $[0, \frac{3}{2}]$, is mapped to itself under $F(x) = f(\vartheta x + (1-\vartheta)f(x))$, and for $\vartheta \in (\frac{2}{3}, \frac{5}{6}))$ it is mapped strictly *into* itself. Moreover,

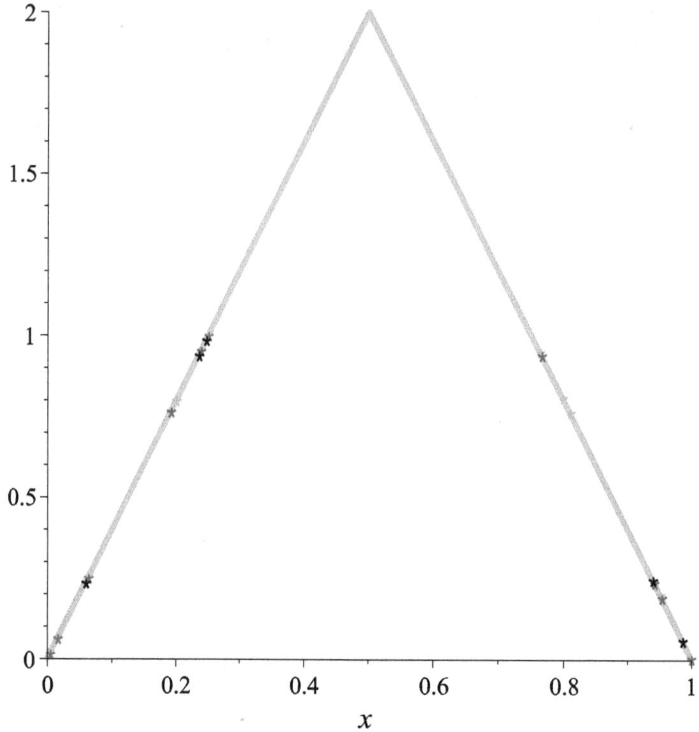

Figure 4. Four sets of cyclic points of the tent map 100-cycles. Left Panel: the four cycles are coloured red, blue, green and black. Notice that many points in the four cycles are tightly clustered. Right Panel: the residuals, U_n, for each of the four cycles as a function of n, where n increases from 250 to 300. The control point conditions are satisfied for $n \geq 250$ in all four cases.

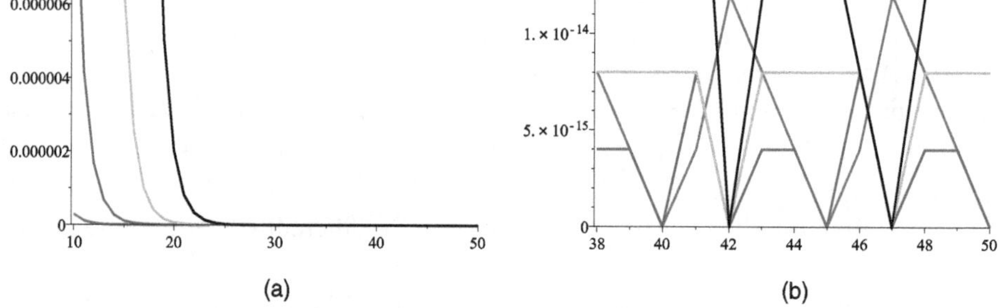

Figure 5. Graphs of the function $y = f(\vartheta x + (1 - \vartheta)f(x))$ at $\vartheta = \frac{4}{3}$ (red); at $\vartheta = \frac{3}{2}$ (black); at $\vartheta = \frac{5}{3}$ (blue); graphs of the functions $y = x$ and $y = f(x)$ are marked in grey.

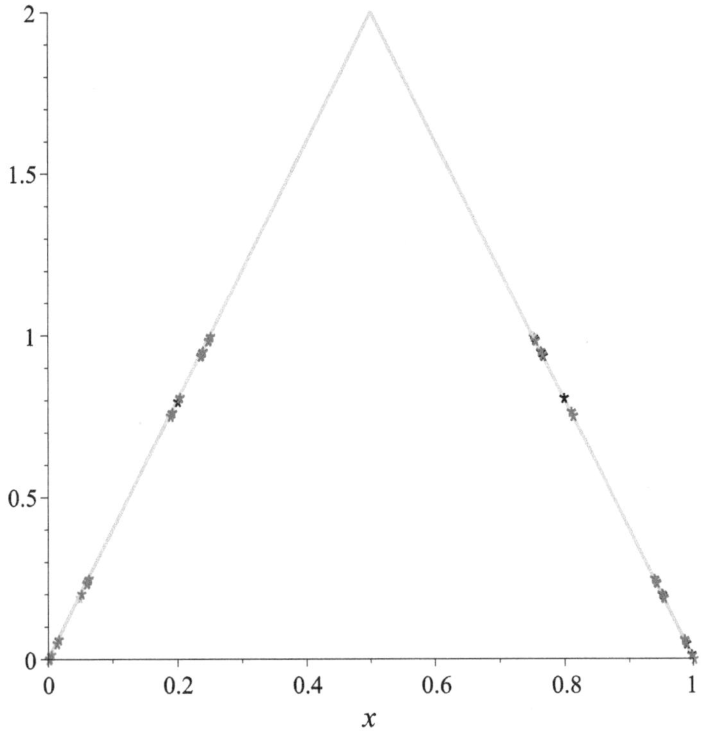

Figure 6. Graphs of the function $y = f(\vartheta x + (1-\vartheta)f(x))$ at $\vartheta = \frac{2}{3}$ (red); at $\vartheta = \frac{3}{4}$ (black); at $\vartheta = \frac{5}{6}$ (blue); graphs of the functions $y = x$ and $y = f(x)$ are marked in grey.

for any $x_0 \in (0, \frac{3}{2})$, the solution tends to the fixed point $\eta = \frac{3}{4}$. These facts follow from the proofs of Theorems 2.1 and 2.2.

Let $T = 2$. Equation (1) has only one 2-cycle, and its multiplier is negative. In this case, Equation (5) takes the form

$$x_{n+1} = F(x) = f\left(\vartheta x_n + (1-\vartheta)f^{(2)}(x_n)\right).$$

Consider the graphs of the functions $y = F(x)$, and $y = F^{(2)}(x)$, when $\vartheta = \dfrac{H^T}{H^T - 1} = \dfrac{9}{8}$ and $\vartheta = \dfrac{H^T}{H^T + 1} = \dfrac{9}{10}$, shown in Figures 7 and 8, respectively.

From Figure 7, we can see that the set $[0, \frac{3}{2}]$ is invariant under the mapping $y = F(x)$. In Figure 8(a), we see that the 2-cycle of Equation (5) becomes locally asymptotically stable with the multiplier equal to zero. In Figure 8(b), we see that the 2-cycle of Equation (5) is unstable, but both fixed points are locally asymptotically stable, with zero multipliers. We give an explanation for this fact in the next few paragraphs.

Consider the global behaviour of solutions of Equation (5) at $\vartheta = \frac{H^T}{H^T+1}$. The function

$$\zeta(x) = \frac{H^T}{H^T+1}x + \frac{1}{H^T+1}f^{(T)}(x)$$

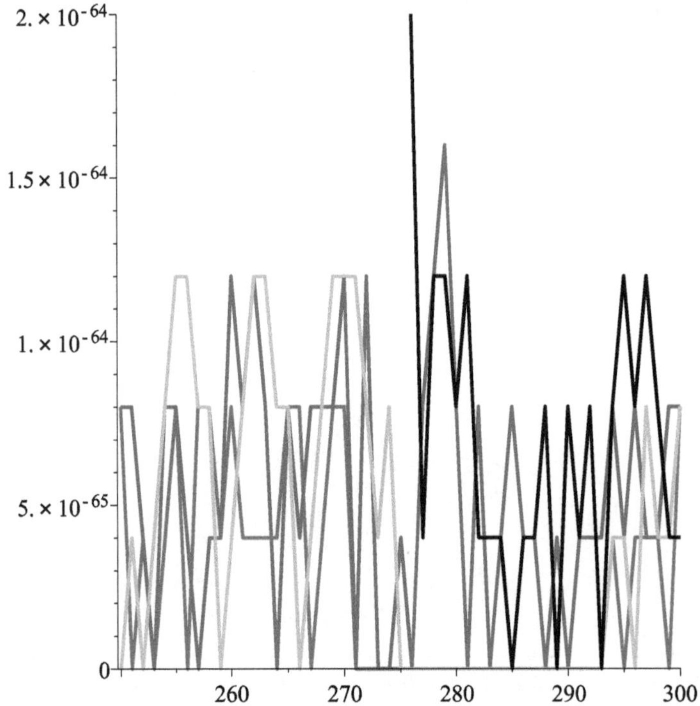

Figure 7. Graphs of the function $y = F(x)$ at $\vartheta = \frac{9}{10}$ (black dashed line); $y = x$ and $y = f(x)$ (grey). Both fixed points are unstable, while $F'(x) = 0$ at both points on the period-2 cycle: $\{\frac{3}{10}, \frac{9}{10}\}$, so the period-2 cycle is super-stable.

does not decrease on $(-\infty, \infty)$. Indeed,

$$\zeta'(x) = \frac{H^T \pm H^T}{H^T + 1} = \begin{cases} 0, \, x \in \Sigma, \\ \dfrac{2H^T}{H^T + 1}, \, x \notin \Sigma, \end{cases}$$

where Σ is the set on which the function $f^{(T)}(x)$ decreases. Note that the sets on which $f^{(T)}(x)$ and $\zeta(x)$ are increasing coincide. In particular, the function $\zeta(x)$ increases when $x \in (\frac{1}{2}, 1 - \frac{1}{H} + \frac{1}{2H^{T-1}})$ from $-\frac{1}{2}H^{T-1}(H-1)$ to $\frac{H}{2}$. The point $(\frac{1}{2}, -\frac{1}{2}H^{T-1}(H-1))$ is the minimum of the function $f^{(T)}(x)$ on $[0, 1]$. Figure 9 shows the graphs of the functions $f^{(T)}(x)$ and $\zeta(x)$ in the case $H = 3$, $T = 3$.

Let $x \in (\frac{1}{2}, 1 - \frac{1}{H} + \frac{1}{2H^{T-1}})$, $T > 1$. Denote $x = 1 - \frac{1}{H} + \frac{\alpha}{H^T}$. Then $\alpha \in (-\frac{1}{2}H^{T-1}(H-2), \frac{H}{2})$. Compute $f^{(T)}(x)$: the first iterate remains greater than 1/2, the second iterate is less than 1/2, and all subsequent iterates increase to α:

$$f(x) = 1 - \frac{\alpha}{H^{T-1}} > \frac{1}{2}, \quad f^{(2)}(x) = \frac{\alpha}{H^{T-2}} < \frac{1}{2}, \ldots, f^{(T)}(x) = \alpha.$$

This implies that the equation $\frac{H^T}{H^T+1}(1 - \frac{1}{H} + \frac{\alpha}{H^T}) + \frac{1}{H^T+1}\alpha = \frac{1}{2}$ has only one root on $(-\frac{1}{2}H^{T-1}(H-2), \frac{H}{2})$, which we denote by $\widehat{\alpha} = \frac{1}{4} - \frac{1}{4}H^T + \frac{1}{2}H^{T-1}$. Since $\zeta(\frac{1}{2}) = \frac{H^{T-1}}{H^T+1} < \frac{1}{2}$, and $\zeta(1 - \frac{1}{H} + \frac{1}{2H^{T-1}}) = \frac{H^T - H^{T-1} + H}{H^T+1} > \frac{1}{2}$, then in the interval $x \in (\frac{1}{2}, 1 -$

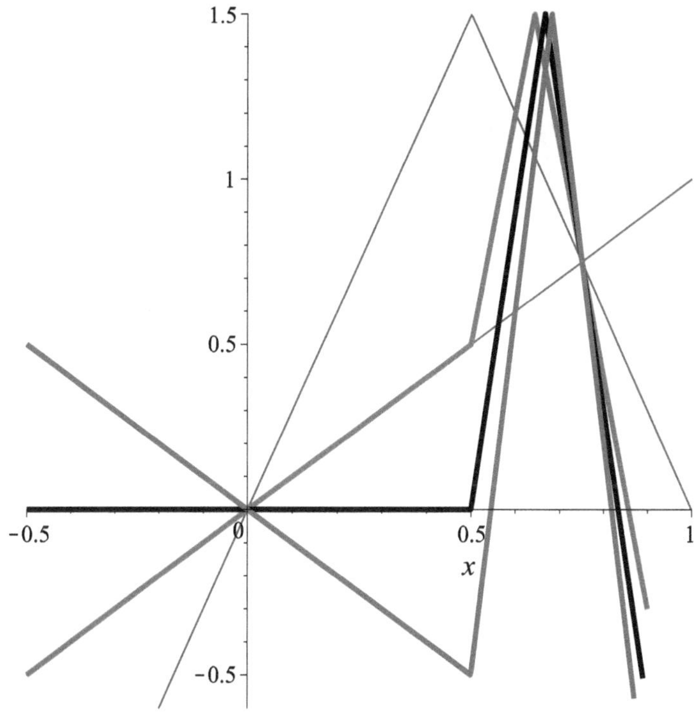

Figure 8. Graphs of the function $y = F^{(2)}(x)$ are drawn in black. In the Left Panel a): $\vartheta = \frac{9}{10}$; in the Right Panel b): $\vartheta = \frac{9}{8}$. Graphs of the functions $y = x$ and $y = f^{(2)}(x)$ are shown in grey. In a) we see that the derivative of $F^{(2)}(x)$ is 0 at the two fixed points of $F^{(2)}$ that are in the proper period-2 cycle of f, showing that these points are *super-stable* fixed points of $F^{(2)}$. By contrast, in b) the period-2 points of f are now unstable, while the two fixed points of f are super-stable.

$\frac{1}{H} + \frac{1}{2H^{T-1}}$) there is exactly one root of the equation $\zeta(x) = \frac{1}{2}$, namely $\hat{x} = 1 - \frac{1}{H} + \frac{\widehat{\alpha}}{H^T} = \frac{3}{4} - \frac{1}{2H} + \frac{1}{4H^T}$. This means that the function $F(x) = f(\zeta(x))$ does not decrease on the interval $[0, \hat{x}]$, does not increase on $[\hat{x}, 1]$, $F(\hat{x}) = \frac{H}{2}$ and $F(x) = F(1 - \frac{1}{2H^{T-1}}) = \frac{H^T}{H^T+1}$ when $x \geq 1 - \frac{1}{2H^{T-1}}$. The graph of the function $F(x)$ when $H = 3$ and $T = 4$ is shown in Figure 10. On the same figure we plot the corresponding function, $\zeta(x)$.

On the interval $[0, 1]$, there are 2^{T-1} disjoint intervals on which the function $F(x)$ is constant. This means that there are 2^{T-1} periodic points, where $F'(x) = 0$. These points are T-periodic points of the map $f(x)$ with negative multipliers. Generally speaking, the proper period of these points may be less than T if T is not a prime number. The question of stability of these so-called subcycles is considered in the next section. The fact that the derivative is zero means that the sequence defined by iterating Equation (5) converges very quickly to these cyclic points, i.e. they are *super-stable*. In particular, if $x_0 \geq \hat{x}$, then $x_1 = F(x_0) = \frac{H^T}{H^T+1} > \frac{1}{2}, x_2 = F(x_1) = f(x_1) = \frac{H}{H^T+1} < \frac{1}{2}, \ldots, x_T = F(x_{T-1}) = f(x_{T-1}) = \frac{H^{T-1}}{H^T+1} < \frac{1}{2}$, $x_{T+1} = F(x_T) = f(x_T) = \frac{H^T}{H^T+1} = x_1$, i.e. $\{x_1, \ldots, x_T\}$ is a T-cycle of both Equations (1) and (5).

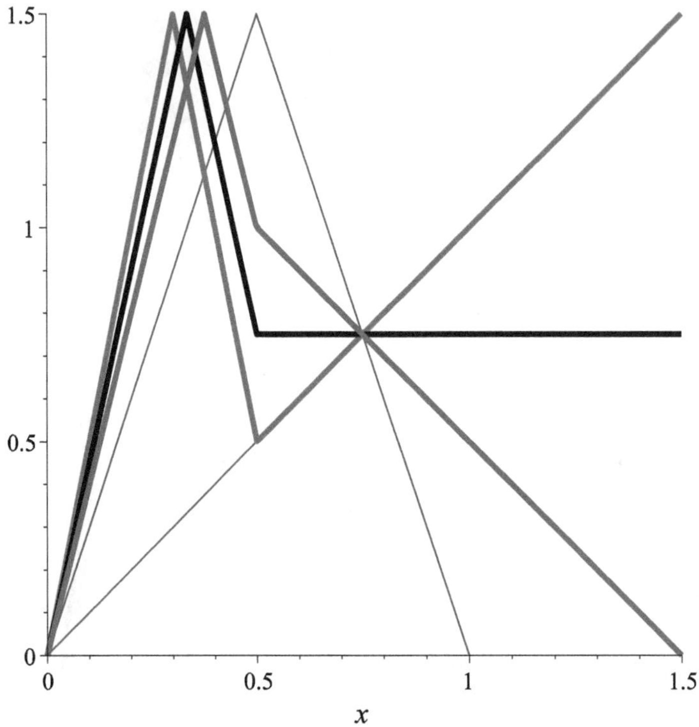

Figure 9. Graphs of the functions $f^{(T)}(x)$ (sea green) and $\zeta(x)$ (black) when $H = 3$ and $T = 3$. Note that $\zeta(x)$ and $f^{(T)}$ are increasing on the same intervals, and $\zeta(x)$ is constant where $f^{(T)}(x)$ is decreasing.

5. Stabilization of subcycles

The system given by Equation (1) has T-cycles for any $T \geq 1$. Moreover, it is not difficult to calculate the number of T-cycles of any given period. Since T-periodic points are fixed points of $f^T(x)$, they will be points of intersection of the graph of $y = f^T(x)$ and the graph $y = x$. These two graphs intersect exactly 2^T times for $x \in [0, 1]$. Two points of intersection correspond to the fixed points of f at $x = 0$, and $x = 1 - \frac{1}{H+1}$. If T is a prime number, then there are exactly $\frac{2^T-2}{T}$ cycles of length T. (By Fermat's Little Theorem, if T is prime, $2^T - 2$ is divisible by T, and thus, $\frac{2^T-2}{T}$ is an integer.) If $T = \prod_{j=1}^{s} \tau_j^{\rho_j}$, where $\tau_1, \ldots, \tau_s$ are distinct prime numbers, then there are exactly

$$\frac{1}{T}\left[2^T - \sum_{j=1}^{s} 2^{\frac{T}{\tau_j}} + \sum_{\substack{i,j=1 \\ i<j}}^{s} 2^{\frac{T}{\tau_i\tau_j}} + \ldots + (-1)^s 2^{\frac{T}{\tau_1 \cdot \ldots \cdot \tau_s}}\right]$$

cycles of the length T. The fact that the given fraction is an integer is a special case of Gauss's theorem, see for example [9, p.84].

Let τ be a factor of the number T, and let the T-cycle of the system given by Equation (5) be locally asymptotically stable. *The task* is: to find out which τ-cycles of the same system

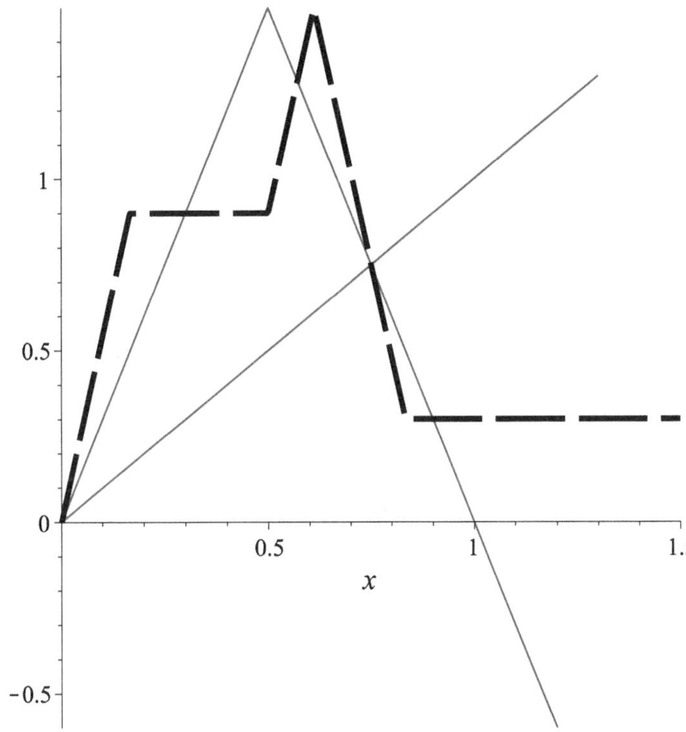

Figure 10. Graphs of the functions $F(x)$ (black) and $\varsigma(x)$ (blue) when $H = 3$ and $T = 4$. Since $F(x) = f(\varsigma(x))$, F and ς are constant on the same intervals, and F is increasing when ς is less than $\frac{1}{2}$, decreasing when ς is greater than $\frac{1}{2}$.

will be locally asymptotically stable? It turns out that the answer depends on the parity of T/τ.

Theorem 5.1: *Let condition* (7) *be satisfied, i.e. T-cycles of Equation* (1), *for which the multipliers are positive, are locally asymptotically stable cycles of Equation* (5). *Then all the* τ*-cycles of Equation* (1), *for which the multipliers are positive, and all the* τ*-cycles of Equation* (1), *for which the multipliers are negative and the number* $\frac{T}{\tau}$ *is even, are locally asymptotically stable cycles of Equation* (5).

Let condition (7) *be satisfied, i.e. T-cycles of Equation* (1), *for which the multipliers are negative, are locally asymptotically stable cycles of Equation* (5). *Then all the* τ*-cycles of Equation* (1), *for which the multipliers are negative and the number* $\frac{T}{\tau}$ *is odd, are locally asymptotically stable cycles of Equation* (5).

Proof: Let $\{\eta_1, \ldots, \eta_T\}$ be a T-cycle and $\{\widehat{\eta}_1, \ldots, \widehat{\eta}_\tau\}$ a τ-cycle of the system given by Equation (1). Let $\mu_T = f'(\eta_T) \cdot \ldots \cdot f'(\eta_1)$, $\mu_\tau = f'(\widehat{\eta}_\tau) \cdot \ldots \cdot f'(\widehat{\eta}_1)$ be the corresponding multipliers of these cycles. These cycles will also be cycles of Equation (5). We assume that the condition for local asymptotic stability of the T-cycle of Equation (5)

$$\left|\mu_T\left(\vartheta + (1-\vartheta)\mu_T\right)^T\right| < 1 \tag{10}$$

is satisfied. Let us find the multiplier of the τ-cycle of Equation (5). Let, as before, $F(x) = f(\vartheta x + (1-\vartheta)f^{(T)}(x))$, and $p = \frac{T}{\tau}$. Calculate

$$F'(\hat{\eta}_j) = f'(\hat{\eta}_j)\left(\vartheta + (1-\vartheta)\left(f'(\hat{\eta}_\tau)\cdot\ldots\cdot f'(\hat{\eta}_1)\right)^p\right) =$$
$$= f'(\hat{\eta}_j)\left(\vartheta + (1-\vartheta)(\mu_\tau)^p\right), \quad j = 1,\ldots,\tau.$$

Then the multiplier of the τ-cycle of Equation (5) equals $F'(\widehat{\eta_\tau})\cdot\ldots\cdot F'(\widehat{\eta_1}) = \mu_\tau(\vartheta + (1-\vartheta)(\mu_\tau)^p)^\tau$, and, accordingly, the condition for local asymptotic stability of the τ-cycle of Equation (5) is:

$$\left|\mu_\tau\left(\vartheta + (1-\vartheta)(\mu_\tau)^p\right)^\tau\right| < 1 \tag{11}$$

Since $\mu_T = \pm H^T$, $\mu_\tau = \pm H^\tau$, the following cases are possible:

(a) $\mu_T < 0, \mu_\tau < 0$, then $\begin{cases} \mu_T = (\mu_\tau)^p, \ p \text{ is odd,} \\ \mu_T = -(\mu_\tau)^p, \ p \text{ is even,} \end{cases}$
(b) $\mu_T < 0, \mu_\tau > 0$, then $\mu_T = -(\mu_\tau)^p$ for any p,
(c) $\mu_T > 0, \mu_\tau < 0$, then $\begin{cases} \mu_T = -(\mu_\tau)^p, \ p \text{ is odd,} \\ \mu_T = (\mu_\tau)^p, \ p \text{ is even,} \end{cases}$
(d) $\mu_T > 0, \mu_\tau > 0$, then $\mu_T = (\mu_\tau)^p$ for any p.

Inequality (10) implies inequality (11) if and only if one of the conditions holds: $\mu_T < 0$, $\mu_\tau < 0, p$ is odd; $\mu_T > 0, \mu_\tau < 0, p$ is even; $\mu_T > 0, \mu_\tau > 0$ for any p. The first condition is possible for the parameters ϑ that satisfy inequalities (6), the second and third conditions are possible for the parameters ϑ that satisfy inequalities (7), whence the conclusion of the Theorem follows. ■

Consider again the example of stabilization of 5-cycles from Section 3. Then, besides locally asymptotically stable 5-cycles, the fixed points at $\eta = 0$ when $\vartheta = \frac{H^T + \frac{0.4}{H}}{H^T - 1}$ and $\eta = 0.8$ when $\vartheta = \frac{H^T - \frac{0.4}{H}}{H^T + 1}$ will also be stable. The initial points: $x_0 = 0.001$ and $x_0 = 0.801$ are in the basins of attraction of the corresponding fixed points, and their orbits will therefore converge to the fixed points rather than to one of the true period-5 cycles. Periodic and fixed points are shown in Figure 11.

6. Distribution of cyclic points and visualization of the cantor set

As noted above, the invariant set of Equation (1) at $H = 3$ is the classical Cantor middle thirds set. However, due to its strong instability, it is impossible to visualize the points of this set using Equation (1). Recall that the Cantor set can be defined as those real numbers in $[0, 1]$ whose ternary expansions consist only of 0's and 2's. The Cantor set is characterized by two types of points: the points that are end-points of the open intervals that are adjacent to the Cantor set are called points of the first type (these are points which terminate in all 0's or all 2's when written in ternary expansion – they take the form $\frac{p}{3^k}$ for some natural numbers p and k), while all the other points of the set are called points of the second type. Points of the first type are solutions of the equations $f^{(k)}(x) = 1$, for $k = 1, 2, \ldots$. The set of points of the first type is countable. When $x < 1/2$ and in the Cantor set (so that is, it's

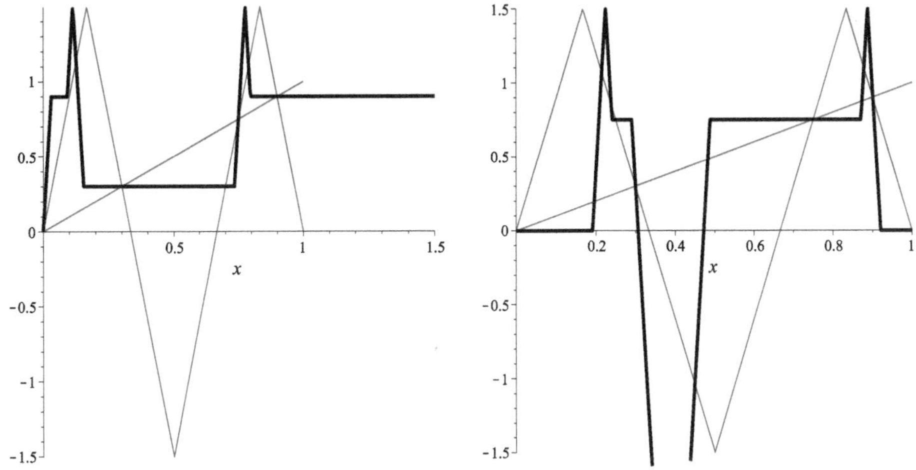

(a) (b)

Figure 11. Points of 5-cycles and fixed points of the tent map where $H = 4$. In the Left Panel, a 5-cycle with positive multiplier (blue) and the non-zero fixed point (red) are stabilized using $\vartheta = \frac{H^T + \frac{0.4}{H}}{H^T - 1}$. In the Right Panel, a 5-cycle with negative multiplier (green) and the fixed point at 0 are stabilized using $\vartheta = \frac{H^T - \frac{0.4}{H}}{H^T + 1}$. To put this example in the context of Theorem 5.1, we note that in this case $T = 5$ and $\tau = 1$, so that $p = T/\tau = 5$ is odd.

first digit in ternary expansion is 0), the tent map acts as a right shift map on the ternary expansion. If $x > 1/2$, the tent map acts on the ternary expansion by swapping the 2's and 0's, and then acting as a shift map. Among the points of the second type, we can select a subset consisting of all periodic points of the system given by Equation (1). They are obtained as the union of all the roots of the equations $f^{(k)}(x) = x$, $k = 1, 2, \ldots$. The set of periodic points of the map (2) is also countable. Points whose ternary expansion is periodic will themselves be periodic because of the way the tent map acts as a shift on the ternary expansions. Thus, there are periodic points arbitrarily close to points of the first type of the Cantor set.

The following question arises: how are the periodic points of one orbit of a given period T distributed through the Cantor set? More precisely, how uniformly do the periodic points of the orbit of a given period T fill the Cantor set? We do not consider the analytical solution of this problem in this paper, but we provide some examples simulating the density functions for the distribution of periodic point, and we compare them graphically with the analogous function for a random sample of elements from the set of points of the first type in the Cantor set.

Let us take for example the period $T = 1009$, with an accuracy 10^{-525}, initial value $x_0 = 0.555$, and values for the control parameter $\vartheta = \frac{H^T \pm \frac{0.6}{H}}{H^T + 1}$. We will get $2T = 2018$ cyclic points. We then simulate the density function for the distribution of the cyclic points set (Figure 12(a)). The graph shows that the estimated periodic points are not quite evenly distributed in the Cantor set. Let us now find *two hundred* orbits with the period $T = 1009$, i.e. we get 20180 cyclic points. Surprisingly, the graph of the simulated density function for

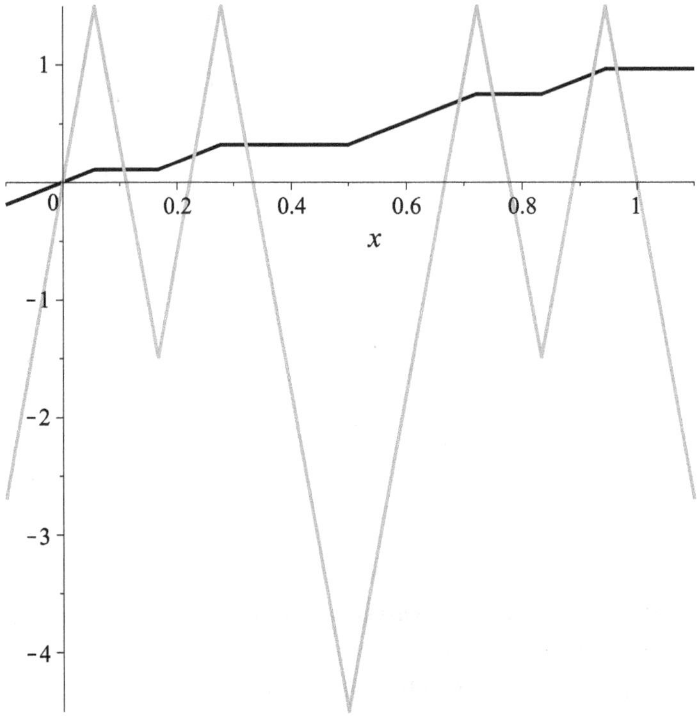

Figure 12. The graph of the distribution density function of the set of cyclic points of two and two hundred 1009-periodic orbits of map (2) at $H = 3$.

the new set of points (Figure 12(b)) does not differ much from the plot in the previous case. Finally, let us plot the density function for the distribution of randomly selected 200000 points of the first type of the Cantor set. For this purpose, represent a subset of the points of the first type of the Cantor set in their ternary expansion $s = \sum_{j=1}^{N} \alpha_j \frac{2}{3^j}$, where $\alpha_j \in \{0, 1\}$. If $N = 25$, the maximum number of possible points is 2^{25}. Let us randomly choose values for α_j: either zero or one, and thus construct 200000 points. Next, we graph the distribution of the resulting set (Figure 13). Comparison of the graphs in Figures 12, 13 shows that the points of the first type, constructed by the method outlined above, are distributed more evenly on the Cantor set.

7. Conclusions

A general predictive control framework was developed in [12] for finding a given set of periodic orbits by making them locally stable. This suggests a solution to the problem of numerically describing invariant sets [1] of nonlinear dynamical systems, since periodic orbits are often dense in these invariant sets.

Since the periodic orbits themselves are repelling sets the problem is separated in two parts: a local stabilization of periodic orbits of a given period, which could be fairly large,

[1] We consider the *largest* bounded set Γ, such that $f(\Gamma) \subseteq \Gamma$

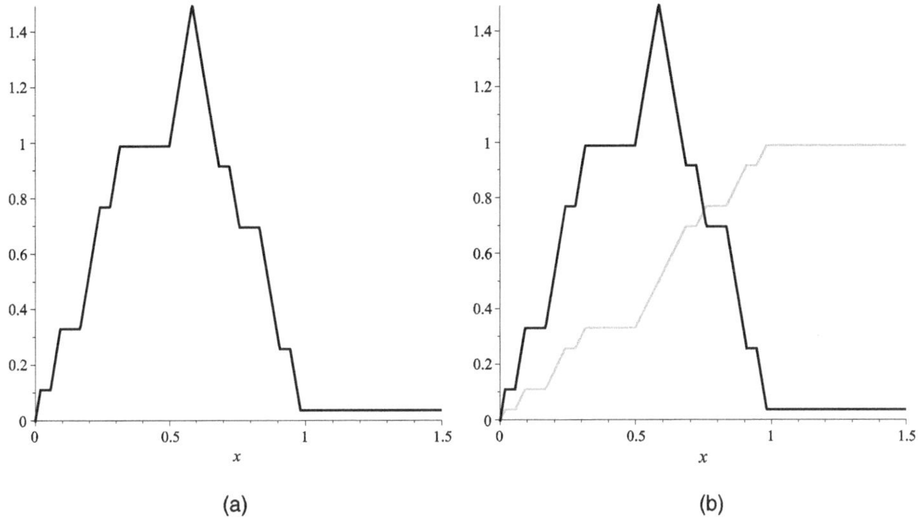

Figure 13. Graph of the distribution density function of the set of 200000 random points of the first type of the Cantor set. Note that the distribution is more uniform across the bins of the histogram.

and then selection of an initial point in the basin of attraction of the stabilized periodic orbit.

If the invariant set is a global attractor, then all trajectories are bounded, and therefore most initial points are in the basin of attraction of one of the stabilized periodic orbits.

If the invariant set is repelling, then the analogous problem turns out to be much more complicated: simple local stability of the controlled orbit is not sufficient, due to the fact that its basin of attraction can have a small measure, and a very complicated structure. For example, it might not be simply connected, or it may have a fractal boundary. The orbits of most initial points will go to infinity. Therefore, the problem of choosing an initial point in the basin of attraction of a stabilized orbit becomes significant.

However, the method of generalized predictive control, used to stabilize the orbit, has an important characteristic: in addition to the local asymptotic stability of the orbits of controlled system, the rest of the orbits remain bounded for a sufficiently large set of initial values. In this paper we showed that, for the generalized tent map, it is possible to ensure that all solutions are bounded.

The global behaviour of solutions for the generalized logistic map, the generalized Lozi [21], Hénon [17], Ikeda [18], Elhadj-Sprott [13] maps, etc. is somewhat more complicated. For such systems, we conjecture that it is necessary to choose the control parameter as a function of the current state of the system. Investigation of the global behaviour of the controlled systems solutions for these maps is the task for future research.

Acknowledgements

The authors wish to thank Andrew Sills for his help in finding the reference to Gauss' theorem.

Disclosure statement

No potential conflict of interest was reported by the author(s).

ORCID

Kimberly Ayers http://orcid.org/0000-0003-1489-5872

References

[1] Available at https://github.com/kdayers/tentmap_invariant, 2021.

[2] A.Y. Aleksandrov and A.P. Zhabko, *On stability of solutions to one class of nonlinear difference systems*, Sib. Math. J. 44(6) (2003), pp. 951–958.

[3] B.R. Andrievsky and A.L. Fradkov, *Control of Chaos: Methods and applications*, I. Methods, Avtomat. I Telemekh. 64(5) (2003), pp. 3–45.

[4] O. Biham and W. Wenzel, *Characterization of unstable periodic orbits in chaotic aitractors and repellers*, Phys. Rev. Lett. 63(8) (1989), pp. 819–822.

[5] G. Chen and X. Dong, *From chaos to Order: Methodologies, Perspectives and Application*, World Scientific, Singapore, 1999.

[6] B. Derrida, A. Gervois, and Y. Pomeau, *Iteration of endomorphisms on the real axis and representation of numbers*, Ann. Inst. H. Poincare Sect. A (N.S.) 29 (1978), pp. 305–356.

[7] R.L. Devaney, *An Introduction to Chaotic Dynamical Systems*, 2nd ed., Addison-Wesley Publ. Co., New York, 1993.

[8] S.M. de Vieira and A.J. Lichtenberg, *Controlling chaos using nonlinear feedback with delay*, Phys. Rev. E. 54(2) (1996), pp. 1200–1207.

[9] L.E Dickson, *The History of the Theory of Numbers*, Vol. I, Carnegie Institute of Washington, 1919.

[10] D. Dmitrishin and A. Khamitova, *Methods of harmonic analysis in nonlinear dynamics*, Comptes. Rendus. Math. 351(9–10) (2013), pp. 367–370.

[11] D. Dmitrishin, G. Skrinnik, I.and Lesaja, and A. Stokolos, *A new method for finding cycles by semilinear control*, Phys. Lett. A. 383(16) (2019), pp. 1871–1878.

[12] D.V. Dmitrishin, A.M. Stokolos, and I.E. Iacob, *Average predictive control for nonlinear discrete dynamical systems*, Adv. Syst. Sci. Appl. 20(1) (2020), pp. 27–49.

[13] Z. Elhadj and J.C. Sprott, *A two-dimensional discrete mapping with C^∞ multifold chaotic attractors*, Electron. J. Theor. Phys. 17(5) (2008), pp. 1–14.

[14] Z. Galias, *Interval methods for rigorous investigations of periodic orbits*, Int. J. Bifurc. Chaos. 11(9) (2001), pp. 2427–2450.

[15] Z. Galias, *Rigorous investigations of Ikeda map by means of interval arithmetic*, Nonlinearity 15(6) (2002), pp. 1759–1779.

[16] W.M.Y. Goh, *Dynamical representation of real numbers and its universality*, J. Number Theory. 33(3) (1989), pp. 334–355.

[17] M. Hénon, *A two-dimensional mapping with a strange attractor*, Commun. Math. Phys. 50(1) (1976), pp. 69–77.

[18] K. Ikeda, *Multiple-valued stationary state and its instability of the transmitted light by a ring cavity system*, Opt. Commun. 30(2) (1979), pp. 257–261.

[19] G.A. Leonov, *Strange Attractors and Classical Stability Theory*, St. Petersburg University Press, 2008.160 (ISBN 978-5-288-04500-4).

[20] T.Y. Li and J Yorke, *Period three implies chaos*, Am. Math. Mon. 82(10) (1975), pp. 985–992.

[21] R. Lozi, *Un attracteur étrange du type attracteur de Hénon*, J. Phys. Colloques. 39(C5) (1978), pp. C5-9–C5-10.

[22] R. Lozi, *Can we trust in numerical computations of chaotic solutions of dynamical systems?* Topol. Dyn. Chaos. 84(2013), pp. 63–98. Letellier, Ch. & Gilmore, R., eds, World Scientific Series in Non-linear Science Series A.

[23] J.R. Miller and J.A. Yorke, *Finding all periodic orbits of maps using newton methods: Sizes of basins*, Physica. D. 135(3–4) (2000), pp. 195–211.

[24] O. Morgul, *Further stability results for a generalization of delayed feedback control*, Nonlinear Dyn. 70 (2012), pp. 1–8.

[25] E. Ott, C. Grebodgi, and J.A. Yorke, *Controlling chaos*, Phys. Rev. Lett. 64 (1900), pp. 1196–1199.

[26] B.T. Polyak, *Stabilizing chaos with predictive control*, Autom. Remote. Control. 66(11) (2005), pp. 1791–1804.
[27] K. Pyragas, *Continuous control of chaos by self controlling feedback*, Phys. Rev. Lett. A. 170(6) (1992), pp. 421–428.
[28] Y. Qian and W. Meng, *Mixed-Mode oscillation in a class of delayed feedback system and multistability dynamic response*, Complexity 2019 (2020), pp. 4871068.
[29] L. Shalby, *Predictive feedback control method for stabilization of continuous time systems*, Adv. Syst. Sci. Appl. 17 (2017), pp. 1–13.
[30] T. Ushio and S. Yamamoto, *Prediction-based control of chaos*, Phys. Lett. A. 264(1) (1999), pp. 30–35.
[31] D. Yang and J. Zhou, *Connections among several chaos feedback control approaches and chaotic vibration control of mechanical systems*, Commun. Nonlinear Sci. Numer. Simulat. 19(11) (2014), pp. 3954–3968.
[32] G. Yuan and A. Yorke, *Collapsing of chaos in one dimensional maps*, Physica. D. 136(1–2) (2000), pp. 18–30.
[33] P. Zygliczynski, *Computer assisted proof of chaos in the Roessler equations and the Henon map*, Nonlinearity 10(1) (1997), pp. 243–252.

On Shilnikov attractors of three-dimensional flows and maps

Yulia Bakhanova, Sergei Gonchenko, Alexander Gonchenko, Alexei Kazakov and Elena Makhrova Samylina

ABSTRACT
We describe scenarios for the emergence of Shilnikov attractors, i.e. strange attractors containing a saddle-focus with two-dimensional unstable manifold, in the case of three-dimensional flows and maps. The presented results are illustrated with various specific examples.

1. Introduction

The discovery of spiral chaos, i.e. the proof of the existence of a complex structure of orbits in a neighbourhood of a homoclinic loop of a saddle-focus equilibrium state, is rightfully considered one of the most significant achievements of the theory of dynamical systems. This discovery was made by L.P. Shilnikov in his work [45], published in 1965. At that time, in both mathematics and physics, there was practically no more or less suitable concept for explaining such phenomena in models described by finite-dimensional deterministic systems. Therefore, the existence of complex structure (chaos) in a neighbourhood of a homoclinic loop of a saddle-focus equilibrium was absolutely unexpected and completely contradicted the conceptions based on the theory of two-dimensional systems. Moreover, such a significant difference in the orbit structure for a saddle and a saddle-focus was seemed very strange, since, from the topological point of view, the saddle and the saddle focus are locally equivalent. However, they differ globally. Thus, from the Shilnikov conditions of nondegeneracy for a homoclinic loop of a saddle equilibrium [43,44,47], one can deduce that in its neighbourhood there is a smooth two-dimensional global central invariant manifold, and thus the problem is effectively two-dimensional. Whereas the saddle-focus does not have such a manifold even locally, and therefore the dynamics here are principally multidimensional.

A saddle-focus differs from a saddle in that, among its leading eigenvalues (those nearest to the imaginary axis), a pair of complex-conjugate ones exists. Then, in the case of three-dimensional systems, saddle-focus equilibria are divided into two different types:

- a *saddle-focus (2,1)* with two-dimensional stable and one-dimensional unstable manifolds that has eigenvalues $\lambda \pm i\omega, \gamma$, where $\lambda < 0, \gamma > 0, \omega \neq 0$;
- a saddle-focus (1,2) with one-dimensional stable and two-dimensional unstable manifolds that has eigenvalues $\lambda, \gamma \pm i\omega$, where $\lambda < 0, \gamma > 0, \omega \neq 0$.

Back in the papers [43,44] Shilnikov showed that bifurcations of systems with a homoclinic loop of a saddle-focus (2,1) with the saddle value

$$\sigma = \lambda + \gamma$$

to be negative do not differ from the case of a saddle – here a unique stable limit cycle is born when the loop spits inward (this is due to the fact that the first return map turns out to be contracting in this case). The case of a three-dimensional saddle-focus (1,2) is obviously reduced to the case of a saddle-focus (2,1) by time reversal. Therefore, here in the case $\sigma > 0$ a unique completely unstable limit cycle is born from a homoclinic loop of a saddle-focus of type (1,2).

However, if the closest to the imaginary axis eigenvalues are complex conjugate ones, i.e. if $\sigma > 0$ in the case of a saddle-focus (2,1) or $\sigma < 0$ in the case of a saddle-focus (1,2), the situation becomes completely different – already the system itself with a homoclinic loop has a complex structure, since, in any neighbourhood of the loop, there are infinitely many saddle periodic orbits [45]. In fact, Shilnikov discovered in Shilnikov [45] that the existence of a homoclinic loop of such saddle-focus implies chaos. The notion itself did not exist then (in 1965); the Chaos Theory emerged and became popular only 10–20 years later. Moreover, chaos was found in many nonlinear models and it also occurred that strange attractors in models of various origins often have a spiral structure, i.e. the chaotic orbits seem to move near a saddle-focus homoclinic loop.

According to the classification of saddle-foci, we will also distinguish two types of spiral homoclinic attractors of three-dimensional flows:

(1) a *Figure 8 spiral attractor*, when the attractor contains a saddle-focus (2,1) (with $\sigma > 0$) and entirely both its one-dimensional unstable separatrices (composing a homoclinic–8);
(2) a *Shilnikov attractor*, when the attractor contains a saddle-focus (1,2) (with $\sigma < 0$) and entirely its two-dimensional unstable manifold.

The fact that the homoclinic loop to a saddle-focus equilibrium implies chaos is the Shilnikov theorem [45,48], see also [51], but why is the converse also often true, i.e. why is the observed chaos often spiral? This question was of great interest to Shilnikov. He discovered [49,50] that if a system depends on a parameter and evolves as it changes, from a stable (stationary) regime to a chaotic one, then in this way, a saddle-focus equilibrium arises naturally and, moreover, its stable and unstable manifolds can come close enough to each other, so that the creation of a homoclinic loop (and chaos) becomes quite expected.

Shilnikov described the corresponding scenarios in the paper [50] for the case of a one-parameter family $X_\mu : \dot{x} = X(x, \mu)$ of multidimensional systems. In Section 2, we discuss the Shilnikov scenario for the three-dimensional case. As one can see, this scenario is extremely simple (if to ignore certain fine intermediate details that themselves can be very complicated), and therefore, it should come as no surprise that it is often seen in many models.[1] We consider some of these models in Section 3 as examples of systems in which Shilnikov's scenario is implemented.

It is interesting to note that spiral chaos was not initially detected in a number of such models, the studies of chaotic dynamics ended either at the stage of Feigenbaum-type attractors (which are formed inside the Shilnikov whirlpool (see Section 2) as a result of an infinite cascade of period-doubling bifurcations with limit cycles) or a little further, when the attractor in the Poincaré section becomes ' one-component', like the Rössler attractor (an example of such attractor is shown in Figure 2(e)), or the interpretation of the process of its occurrence was inadequate etc. Of course, this situation was a consequence of the lack of a 'guiding thread' – the Shilnikov scenario in this case.

In the work [50], Shilnikov expressed one more important idea that the scenario he proposed for flows can be generalized to the case of maps. In this case, the generalization is direct: the equilibrium state is replaced by a fixed (or periodic) point, and the local and global bifurcations involved in the scenario are replaced by their analogues for maps. In Section 4, we give a phenomenological description of such scenarios for the case of orientable and nonorientable three-dimensional maps, and in Section 5, we give some examples of how such scenarios can be implemented in the case of three-dimensional generalized Hénon maps.

2. On Shilnikov scenario for three-dimensional flows

Consider a one-parameter family $X_\mu : \dot{x} = X(x, \mu)$ of three-dimensional flows such that the system X_μ has at $\mu_0 < \mu < \mu_1$ a stable equilibrium O_μ that is the only attractor in some absorbing domain D, see Figure 1(a). Let at $\mu \in (\mu_0, \mu_1)$ and close to μ_1 the equilibrium O_μ have eigenvalues $\nu_{1,2} = \lambda(\mu) \pm i\omega(\mu)$, $\nu_3 = \lambda_{ss}(\mu)$, where $\lambda_{ss} < \lambda < 0$, that is O_μ is a stable focus.

Assume that at $\mu = \mu_1$ a *supercritical* (soft) Andronov–Hopf bifurcation occurs with O_μ. Then the eigenvalues $\nu_{1,2}$ evolve as μ changes in such a way that $\lambda(\mu) < 0$ if $\mu < \mu_1$,

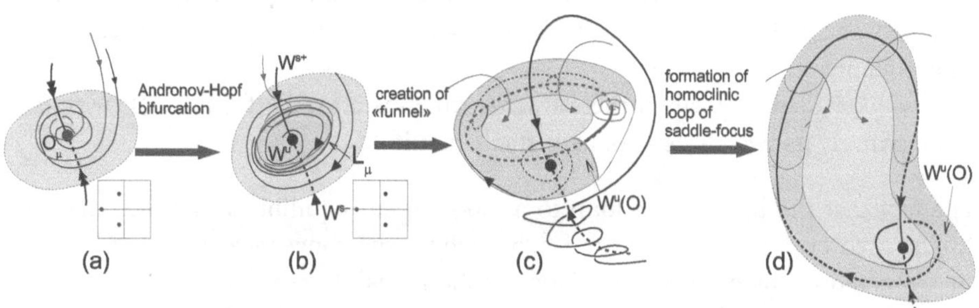

Figure 1. The main stages of the phenomenological scenario for the emergence of the Shilnikov attractor for three-dimensional flows. One of the criteria for the strangeness of such an attractor is the existence of a homoclinic (double-asymptotic) orbit to the saddle-focus O_μ.

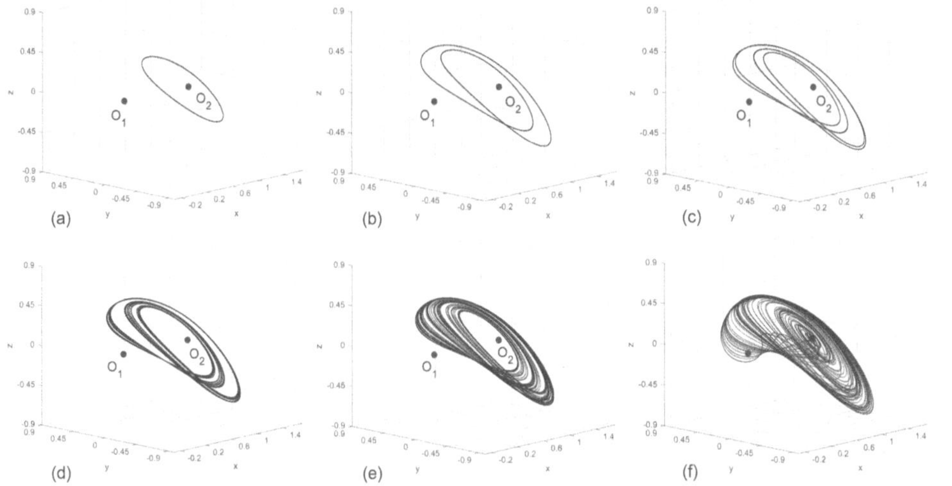

Figure 2. Scenario of the emergence of Shilnikov attractor in system (1): (a) $\mu = 0.5$, (b) $\mu = 0.75$, (c) $\mu = 0.776$, (d) $\mu = 0.782$, (e) $\mu = 0.8$, (f) $\mu = 0.863$.

$\lambda(\mu_1) = 0$, and $\lambda(\mu) > 0$ if $\mu > \mu_1$. Thus, at $\mu > \mu_1$, the equilibrium O_μ becomes a saddle-focus (1,2) and a stable limit cycle L_μ is born in its neighbourhood. Accordingly, for all sufficiently small $\mu > \mu_1$, the global attractor in D of the system X_μ is this asymptotically stable limit cycle L_μ. The type of stability of the cycle L_μ is determined by its multipliers $\rho_1(\mu)$ and $\rho_2(\mu)$, which are initially both positive and less than one. In this case, the unstable manifold $W^u(O_\mu)$ of the equilibrium O_μ is a two-dimensional disk with the boundary L_μ, and orbits starting from points on $W^u(O_\mu)$ are wound to L_μ along spirals 'from the inside', see Figure 1(b).

We assume that with a further increase in μ the multipliers $\rho_1(\mu)$ and $\rho_2(\mu)$ of L_μ become equal at some $\mu = \mu^*$ and complex conjugate for $\mu > \mu^*$. This 'smooth bifurcation' corresponds to change of type of L_μ, from nodal type for $\mu < \mu^*$ to focal one for $\mu > \mu^*$. Then, the two-dimensional unstable manifold $W^u(O_\mu)$ of the equilibrium O_μ begins to twist onto the cycle L_μ. In this case, $W^u(O_\mu)$ takes a form of roll which is a boundary of the so-called 'Shilnikov whirlpool', inside of which all orbits of system X_μ starting in D are tightened (generally speaking, except for one orbit – the stable separatrix $W^{s-}(O_\mu)$ of the equilibrium O_μ), see Figure 1(c).

When μ changes, the sizes of the whirlpool increase, the orbits absorption is preserved, but the limit cycle L_μ may lose its stability. In particular, a strange attractor can appear in its place as a result of a series of various bifurcations. The sequence of these bifurcations can be very diverse: this can be a cascade of period-doubling bifurcations followed by the appearance of a Feigenbaum attractor or, few further, a Hénon-like attractor in the Poincaré map; or a two-dimensional stable invariant torus T_μ can be born from the cycle, which can then break down, for example, according to the Afraimovich–Shilnikov scenario [1], giving rise to a strange attractor of the ' torus-chaos' type, etc. It is worth noting that yet in the paper [50], Shilnikov emphasized that chaotic dynamics inside the whirlpool can appear instantly due to a rigid (subcritical) bifurcation: this can be, for example, subcritical Andronov–Hopf or period-doubling bifurcations with stable limit cycles and so forth. In particular, an example of such subcritical period-doubling bifurcation we found in Gonchenko et al. [27] in the case of the Gaspard–Nikolis model of chemical oscillator. We illustrate this result in Section 3.2. These intermediate attractors appearing inside the Shilnikov whirlpool, in turn, can transform with varying in μ, into the Shilnikov attractor, i.e. a strange attractor that contains the saddle-focus O_μ of type (1,2) and entirely its two-dimensional unstable manifold.

The moment $\mu = \mu_\ell$ of the formation of a homoclinic loop of the saddle-focus O_μ can be traced by observing the stable separatrix $W^{s+}(O_\mu)$: at $\mu < \mu_\ell$, it leaves D (as in Figure 1(c)), at $\mu > \mu_\ell$, it enters the whirlpool, and, at $\mu = \mu_\ell$, it falls on the unstable manifold $W^u(O_\mu)$ (thus, $W^{s+}(O_\mu)$ entirely belongs to $W^u(O_\mu)$).

With a further change in μ, the separatrix $W^{s+}(O_\mu)$ enters the whirlpool, makes several (maybe quite many) turns there, then can again lie on $W^u(O_\mu)$ forming a multi-round homoclinic loop and etc. The moments of formation of such loops are discrete in the parameter μ, but they are not isolated: for each such value of the parameter μ, those values of μ are accumulated again that correspond to the formation of double- [14,20], triple- [22], and longer-round homoclinic loops.

Discrete moments with respect to the parameter μ, when the separatrix $W^{s+}(O_\mu)$ forms homoclinic loops, correspond to the existence of Shilnikov homoclinic attractors.[2] At these moments, the equilibrium O_μ enters the attractor together with its unstable manifold $W^u(O_\mu)$, and typical orbits of the attractor visit any arbitrarily small neighborhood of the point O_μ. The latter property makes it relatively easy to find the moments of formation of Shilnikov homoclinic attractors using, for example, automated plotting of the μ-dependence of the distance of attractor points from the saddle-focus [12,37].

3. Examples of Shilnikov attractors in three-dimensional flows

For a long time, the Shilnikov's works [45,46,48] on spiral chaos were practically unknown to the mathematical community. They gained world recognition due to a series of quite affordable papers by Arnéodo et al. [7], Arnéodo et al. [8], Arnéodo et al. [9], Arnéodo et al. [10], in which the importance of the Shilnikov homoclinic loop for the chaos theory was emphasized. In particular, in Arneodo et al. [8], geometric illustrations of the Shilnikov's theorem were given. Also, in this paper, it was presented a simple example of a piecewise linear oscillatory system, in which the existence of a homoclinic loop of saddle-focus (2,1) was analytically established, and strange spiral attractors were found numerically.

In this section, we consider two examples of three-dimensional systems in which the Shilnikov attractors are observed: the first example is the model proposed by Arnéodo, Coullet, and Tresser in Arnéodo et al. [9], and the second example is the Gaspard–Nicolis model of the chemical oscillator from [21].

3.1. Shilnikov attractor in the Arnéodo–Coullet–Tresser system

In the paper, [9], a smooth three-dimensional system

$$\begin{cases} \dot{x} = y \\ \dot{y} = z \\ \dot{z} = -y - \beta z + \mu x(1 - x), \end{cases} \tag{1}$$

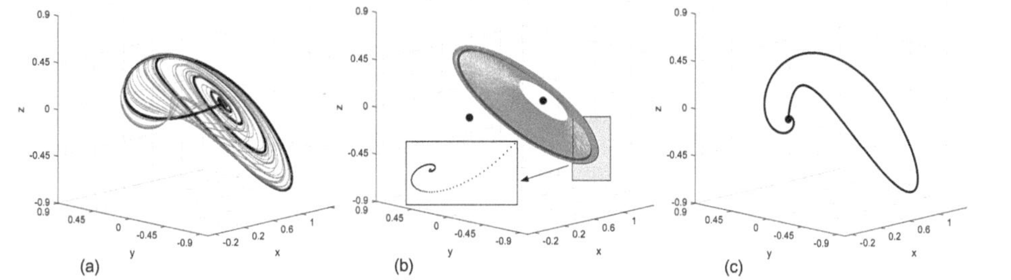

Figure 3. (a) $\beta = 0.4, \mu = 0.86311445$, Shilnikov attractor with a homoclinic loop of the saddle-focus O_2 of type (1,2); (b) $\beta = 0.01$, $\mu = 0.02$, illustration of the formation of the Shilnikov whirlpool; (c) $\beta = 0.4, \mu = \mu_5 = 1.6062$, a homoclinic loop of the saddle-focus O_1 of type (2,1), here the attractor no longer exists.

was proposed which demonstrates spiral chaos at certain regions of values of the parameters β and μ. Let us consider this system in more detail.

Since system (1) has the constant divergence equal to $-\beta$, attractors can exist only for $\beta > 0$. Note that for all values of parameters $\mu > 0$ and $\beta > 0$, system (1) has two equilibria $O_1(0,0,0)$ and $O_2(1,0,0)$. Moreover, the equilibrium O_1 for $\beta < \sqrt{3}$ is always a saddle-focus (2,1), while the equilibrium O_2 can be both stable and saddle-focus (1,2).

Following [9] we put $\beta = 0.4$. Now our goal is to illustrate the scenario of the emergence of the Shilnikov attractor containing the saddle-focus O_2 when the parameter μ changes. In Figure 2 main stages of the scenario are shown.

For $0 < \mu < \mu_1 = 0.4$, the stable equilibrium O_2 is the only attractor of the system. Note that the boundary of its absorbing domain is formed by the two-dimensional stable invariant manifold of the saddle-focus O_1. At $\mu = \mu_1$, the equilibrium O_2 undergoes a supercritical Andronov–Hopf bifurcation. As a result, for $\mu > \mu_1$, a stable limit cycle L is born in a neighborhood of the equilibrium O_2 that becomes a saddle-focus (1,2), see Figure 2(a). The stable limit cycle exists for $\mu_1 < \mu < \mu_2 \approx 0.72$. Starting with $\mu = \mu_2$, the cycle goes through a cascade of period-doubling bifurcations (see Figure 2(b,c) after the first and second period-doubling bifurcations), and, as a result, a strange attractor of Feigenbaum type appears, see Figure 2(d). Then, this attractor is transformed into a Rössler-like attractor (on a two-dimensional Poincaré section, it can be also interpreted as a Hénon-like attractor), see Figure 2(e). With a further increase in the parameter μ up to $\mu = \mu_\ell \approx 0.86311445$, the attractor changes quite smoothly and is not homoclinic, since it is separated from the saddle-focus O_2.

However, as $\mu \to \mu_\ell$, a 'hole' around O_2 decreases and, finally, disappears at $\mu = \mu_\ell$. In this case, orbits of the attractor can come arbitrarily close to the saddle-focus O_2, see Figure 2(f). As we know, this is due to the appearance of a homoclinic orbit to the saddle-focus O_2 and, hence, to the creation of the Shilnikov attractor, see Figure 3(a), where the numerically found homoclinic orbit is also shown.

Note that, using system (1), we can trace a formation of the Shilnikov whirlpool – when the cycle L becomes of focal type. An illustration of the moment of occurrence of the Shilnikov whirlpool is shown in Figure 3(b), where, for greater clarity, we take $\beta = 0.01$ (then the divergence of the system $Div = -\beta$ will be small) in order to be able to show how the unstable two-dimensional manifold of the saddle-focus O_2 winds around the stable limit cycle L.

With a further increase in the parameter μ, from $\mu = \mu_3$ to $\mu = \mu_4 \approx 0.873$, a phenomenon of complication of the structure of the Shilnikov whirlpool can be observed, see e.g. [36,37]. At $\mu = \mu_4$, a crisis of the attractor occurs: it collides with the two-dimensional stable manifold of the saddle-focus O_1, which is a natural boundary of the absorbing domain for the attractor under consideration. Note that the saddle-focus O_1 can also have homoclinic loops, see Figure 3(a) where such a loop is shown for $\mu = \mu_5 \approx 1.6062$ (it was also found numerically in Arneodo et al. [9]). However, any attractors do not exist for close values of μ.

3.2. Example of Shilnikov attractor in the Gaspard–Nicolis model of chemical oscillator.

In this section, we consider an example of an interesting system in which the Shilnikov attractor arises almost instantly as a result of a single rigid (subcritical) bifurcation. This scenario begins as usual: a stable equilibrium O undergoes a supercritical Andronov–Hopf bifurcation, as a result, it becomes a saddle-focus (1,2) and a stable limit cycle L is born. This cycle first becomes the focal one, and the Shilnikov whirlpool (with the boundary $W^u(O)$) is formed. However, this limit cycle itself remains stable for a long time (by the parameter) and, independently of it, non-attractive chaotic dynamics (metastable chaos) have time to develop inside the whirlpool. Next, at a certain value of the parameter, the cycle L undergoes a period-doubling bifurcation that is subcritical: a saddle cycle of doubled period merges with L. After this, a strange attractor is immediately observed, which at very close values of a parameter becomes homoclinic – the Shilnikov attractor appears.

Below we will show how this happens in the system

$$\begin{cases} \dot{x} = x(\beta x - fy - z + g), \\ \dot{y} = y(x + sz - \alpha), \\ \dot{z} = (x - az^3 + bz^2 - cz)/\varepsilon, \end{cases} \tag{2}$$

proposed by P. Gaspard and G. Nicolis in the paper [21], in which it was shown that strange attractors containing a saddle-focus (1,2) can exist in the system (2). The corresponding parameter regions where such attractors can be observed have been specified in Gallas [19]. Following [19] we take some parameters to be fixed

$$b = 3, \quad \varepsilon = 0.01, \quad f = 0.5, \quad g = 0.6, \quad s = 0.3, \quad c = 4.8, \quad \alpha = 0.7825, \tag{3}$$

and the parameter β choose as the governing one.

For $\beta < \beta_{AH} \approx 0.261$, the nonzero equilibrium O of system (2) is stable, see Figure 4(a) (O is the only equilibrium with positive coordinates having a physical sense). For $\beta = \beta_{AH}$ this equilibrium undergoes a supercritical Andronov–Hopf bifurcation. After this, on the interval $\beta \in (\beta_{AH}, \beta_{PD} \approx 0.3817)$, the attractor of the system is the stable limit cycle l, see Figure 4(b), and the equilibrium O becomes a saddle-focus (1,2). At $\beta \approx 0.365$ the cycle l becomes focal and a Shilnikov whirlpool is formed, further both multipliers of l become negative and one of them tends to -1 as $\beta \to \beta_{PD}$. Simultaneously, a saddle limit cycle l_s

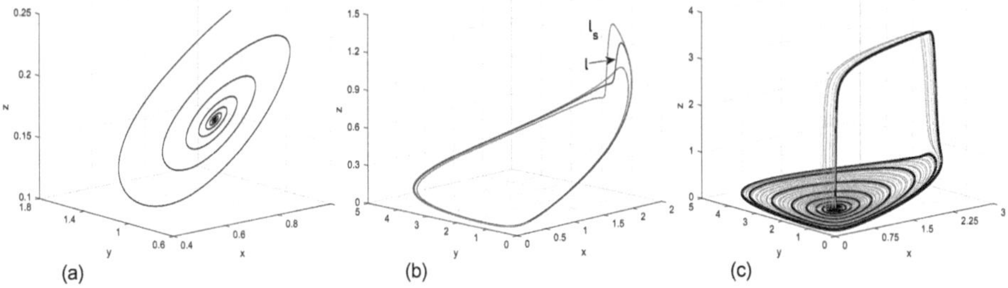

Figure 4. Scenario of the Shilnikov attractor emergence in the model (2) of the chemical oscillator: (a) $\beta = 0.1$, the attractor is the stable equilibrium state O; (b) $\beta = 0.3812$, saddle cycle l_s of period–2 approaches the stable limit cycle l; (c) $\beta = 0.39213$, Shilnikov homoclinic attractor.

of double period approaches the cycle l, see Figure 4(b) and merges with l. As a result of this *subcritical* period-doubling bifurcation, at $\beta > \beta_{PD}$ the limit cycle l becomes saddle and, instantly, a strange attractor is observed. With a further increase in the parameter β, the orbits of this attractor begin to approach closer and closer to the saddle-focus O. At $\beta = \beta_h \approx 0.3921$, a homoclinic loop of the saddle-focus O is formed, i.e. a homoclinic Shilnikov attractor arises, Figure 4(c).

4. On Shilnikov scenarios for three-dimensional orientable and nonorientable maps

We discuss now scenarios of the emergence of *discrete Shilnikov attractors*, i.e. homoclinic attractors containing a fixed (periodic) point that is a saddle-focus with two-dimensional unstable manifold, thus, it is a saddle-focus (1,2) in the case of three-dimensional maps.

We consider one-parameter families T_μ of three-dimensional diffeomorphisms in two cases, when map T_μ is orientable (orientation preserving) and when T_μ is nonorientable (orientation reversing) map. In order not to get involved with the problems of orientability of the ambient manifold, we will assume that in both cases, T_μ is given in R^3. Then the Jacobian $J(T_\mu)$ of the map T_μ will be everywhere positive in the orientable case and everywhere negative in the nonorientable case.

4.1. The orientable case

A sketch of the scenario of a typical discrete Shilnikov attractor appearance for one-parameter families T_μ of three-dimensional orientable maps is illustrated in Figure 5. This scenario starts with a stable fixed point O_μ that loses the stability at $\mu = \mu_1$ under a supercritical (soft) Andronov–Hopf bifurcation: for $\mu > \mu_1$ the point O_μ becomes a saddle-focus (1,2) and a stable closed invariant curve L_μ is born in a neighborhood of O_μ, Figure 5(a,b). Thus, the curve L_μ becomes the attractor of map T_μ. During this transition, the point O_μ changes its type from a stable point to a saddle-focus (1,2) point: for $\mu < \mu_1$ it has three multipliers inside the unit circle, for $\mu = \mu_1$, two complex conjugate multipliers of O_μ fall on the unit circle, and for $\mu > \mu_1$ they go outside it.

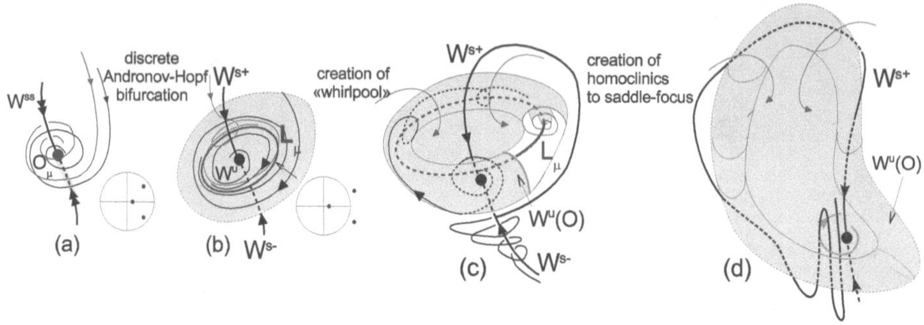

Figure 5. A sketch of scenario for the emergence of a discrete Shilnikov attractor in the case of three-dimensional orientable maps.

The next stage of the scenario is connected with changes in L_μ. Typically, this happens in the following way. Just after the Andronov–Hopf bifurcation, at small $|\mu - \mu_1|$, the unstable manifold of O_μ is a two-dimensional disk D_μ with the boundary L_μ that has a type of nodal invariant curve. We assume that, at further changing μ, the curve L_μ undergoes a 'smooth bifurcation', when it changes its type from nodal to focal and, thus, the two-dimensional manifold $W^u(O_\mu)$ begins to wind up on L_μ, like a roll. As a result, $W^u(O_\mu)$ takes the form of a kind of a roll, the Shilnikov whirlpool, inside which all orbits from the absorbing region are drawn, except for orbits of the stable separatrix $W^{s-}(O_\mu)$, see Figure 5(c).

Then, chaotic dynamics begin to develop in this whirlpool as μ changes. At first, the attractor is simple, it is the stable invariant curve L_μ, and then it loses its stability. This can happen in a variety of ways (for example, in a soft way through a cascade of doubling of invariant curves with their subsequent destruction and the formation of attractors of torus-chaos type, or in a rigid way through a subcritical bifurcation with some of the stable invariant curves, after which chaos can be observed immediately, 'by explosion', etc.). In principle, for the essence of the phenomenological scenario, it does not matter how this happens. The main thing here is that the invariant manifolds $W^u(O_\mu)$ and $W^{s+}(O_\mu)$ begin to intersect and a strange homoclinic attractor can arise containing the saddle-focus O_μ and entirely its two-dimensional unstable manifold, see Figure 5(d). We call this attractor a *discrete Shilnikov attractor*.

It should be noted that there is an almost complete similarity in the main features of this scenario with the corresponding scenario in the case of a flow, see Section 1. However, even here one can see a significant difference. The Shilnikov homoclinic flow attractor exists only for discrete values of the control parameter corresponding to the existence of homoclinic loops of the saddle-focus equilibrium. Whereas the Shilnikov attractor for maps exists on intervals of parameter values for which the invariant manifolds ($W^u(O_\mu)$ and $W^{s+}(O_\mu)$) have transversal intersections. Moreover, these intervals can be large enough and their values can reach even those at which the attractor is destroyed and disappears altogether.

There is also one more important feature of Shilnikov discrete attractors, which flow attractors do not have. This feature manifests itself in the case when the stable invariant curve L_μ is resonant. Then, on the curve itself, there are alternating saddle and stable periodic points of the same period, and the formation of the Shilnikov whirlpool occurs due

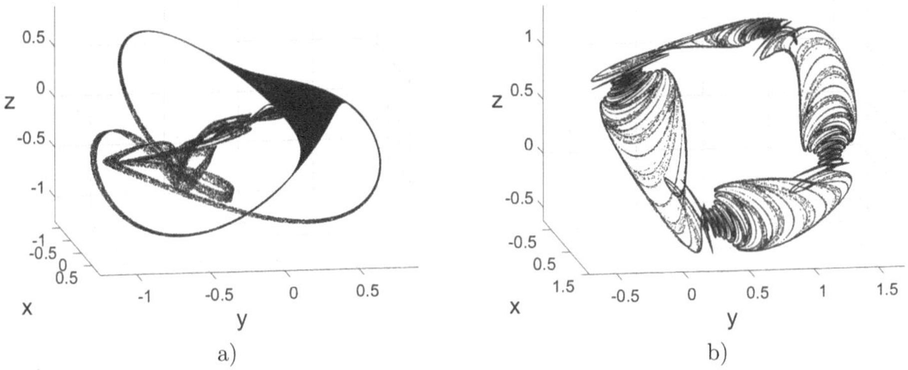

Figure 6. Discrete attractors in resonant case: (a) a 'triangular' discrete Shilnikov attractor; (b) a 'super-spiral' period–4 attractor. We found these attractors in the three-dimensional Mirá map $\bar{x} = y, \bar{y} = z, \bar{z} = M_1 + Bx + M_2 z - y^2$ for $B = 0.7$ and (a) $M_1 = 0.195, M_2 = -0.26$; (b) $M_1 = 0.35, M_2 = 0.8$. It is important to note that these attractors appear near the 1:3 and 1:4 strong resonances. Therefore they visually demonstrate something like the Z3 and Z4 symmetry. Although the map itself, of course, does not have such symmetries – they appear only locally, namely for attractors.

to the fact that the stable points become foci. Thus, the manifold $W^u(O_\mu)$ is twisted over L_μ only in some places (near stable points). This also subsequently affects the shape of the emerging Shilnikov attractor. For example, such an attractor can have a characteristic 'triangular' or 'square' shape in the case of resonances 1:3 and 1:4, when O_μ has a pair of multipliers $\lambda e^{\pm i\psi}$ with ψ close to $2\pi/3$ and $\pi/2$, respectively, see Figure 6. In addition, the resonant invariant curves of three-dimensional maps can themselves be destroyed in very interesting ways, giving rise to amusing attractors inside the whirlpool, see, for example, Figure 6(b), which shows a 'superspiral' attractor containing two orbits of period 4, which are saddle-foci of type (2,1) and (1,2). We use this term because the attractor contains saddle-foci of both types. In the case of three-dimensional flows with symmetries, such attractors can be found in the class of the so-called multispiral attractors, see e.g. [16,38].

4.2. The nonorientable case

A sketch of a typical discrete Shilnikov scenario for one-parameter families T_μ of three-dimensional nonorientable maps is illustrated in Figure 7. This scenario starts with those μ at which T_μ has a nonorientable stable fixed point O_μ with multipliers $-\lambda, \gamma^{\pm i\psi}$, where $0 < \lambda < \gamma < 1$ and $0 < \psi < \pi$, Figure 7(a). We assume that O_μ loses the stability at $\mu = \mu_1$ under a supercritical Andronov–Hopf bifurcation: for $\mu > \mu_1$ the point O_μ becomes a nonorientable saddle-focus (1,2), and a stable closed invariant curve L_μ is born in a neighborhood of O_μ, Figure 7(b). Thus, the curve L_μ becomes the attractor of map T_μ. During this transition, the point O_μ has three multipliers less than one in the absolute value for $\mu < \mu_1$, then, for $\mu = \mu_1$, two of its complex conjugate multipliers fall on the unit circle, and for $\mu > \mu_1$ they go out and the point O_μ becomes a saddle-focus (1,2) that is nonorientable since its stable multiplier is negative.

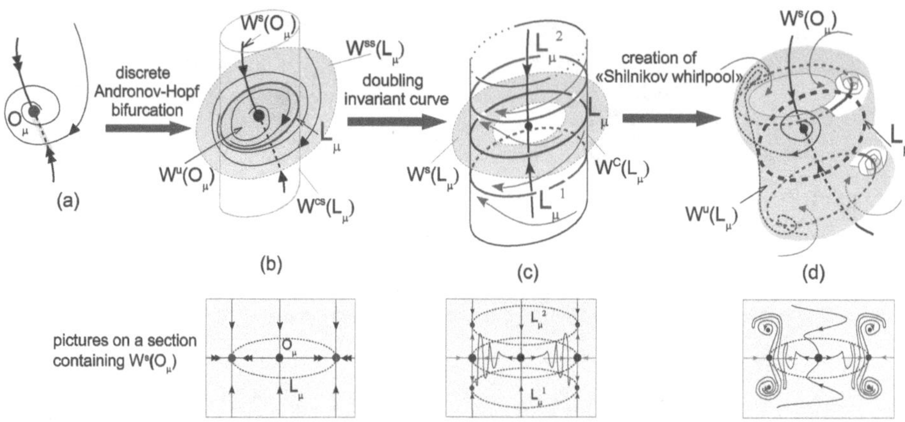

Figure 7. A sketch of scenario of the emergence of a nonorientable discrete Shilnikov attractor.

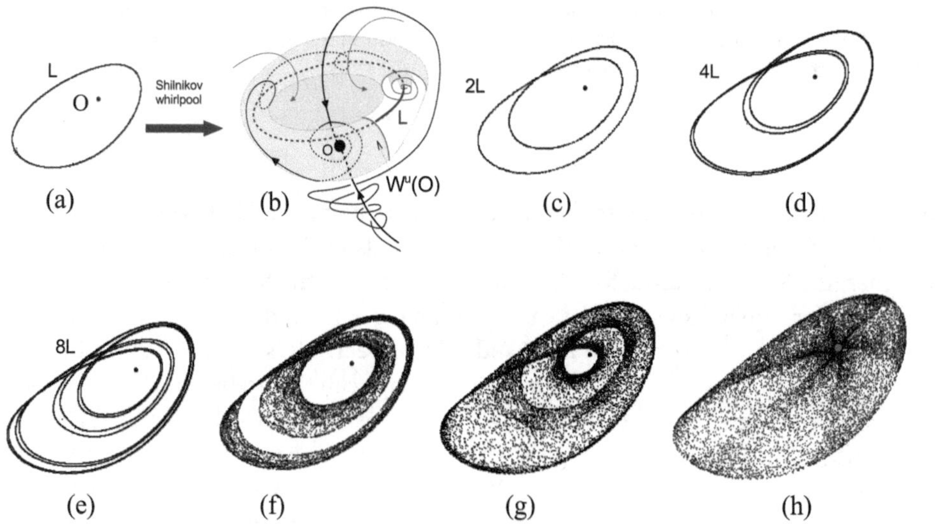

Figure 8. Stages of the emergence of the Shilnikov attractor in map (4) for $B = 0.5$, $A = 1.49$, as C changes (Figure (b) is only schematic): (a) $C = -1.7$, a stable invariant curve L ; (b) a schematic picture of a Shilnikov whirlpool, which is formed when $W^u(O)$ begins to wound onto L (the following figures (c)–(h) show what happens inside this whirlpool); (c) $C = -1.73$, the curve L is doubled, $L \to 2L$; (d) $C = -1.76$ – after the second doubling, $2L \to 4L$; (e) $C = -1.77$ – after the third doubling, $4L \to 8L$; (f) $C = -1.775$, a chaotic attractor $A2$ containing the saddle curves $4L$ and $2L$; (g) $C = -1.8$, a chaotic attractor $A1$ containing also the saddle curve L; (h) $C = -1.82$, a discrete Shilnikov attractor.

The next stage of the scenario is connected with changes in L_μ. Typically, this happens in the following way. Just after the Andronov–Hopf bifurcation, at small $|\mu - \mu_1|$, the unstable manifold of O_μ is a two-dimensional disk D_μ with the boundary L_μ where the curve L_μ has a type of nonorientable nodal invariant curve. Because the curve L_μ is nonorientable, it can not become of focal type as in the orientable case. Thus, the manifold $W^u(O_\mu)$ can not take a form of whirlpool just over L_μ. However, there exists another way to the creation a whirlpool. Namely, at further changing μ, first, the curve L_μ undergoes a doubling bifurcation: the curve L_μ itself becomes saddle and two stable period–2 invariant curves L_μ^1 and

L^2_μ originate from it here $L^2_\mu = T_\mu(L^1_\mu)$ and $L^1_\mu = T_\mu(L^2_\mu)$. Such feature of the bifurcation of doubling of an invariant curve is obtained due to the fact that, near the bifurcation, the curve L_μ possesses two invariant manifolds, strongly stable W^{ss} and central W^c, which locally are two-dimensional cylinders, see Figure 7(b), but the map T_μ in the restriction to W^c changes orientation. Accordingly, after the bifurcation, the curves L^1_μ and L^2_μ lie on W^c on opposite sides of L_μ. After this doubling bifurcation, the unstable manifold $W^u(O_\mu)$ immediately corrugates and begins to rush between the curves L^1_μ and L^2_μ but still remains inside the cylinder $W^c(L_\mu)$, see Figure 7(c).[3] Note that the curves L^1_μ and L^2_μ are both invariant and orientable for T^2_μ, therefore they can become focal ones when changing μ. In this case, the manifold $W^u(O_\mu)$ starts to wind on both L^1_μ and L^2_μ and, thus, a nonorientable Shilnikov whirlpool is created, see Figure 7(d).

Then, in this whirlpool, when μ changes, chaotic dynamics begin to develop. At first, the attractor is simple, it is the stable period–2 invariant curve (L^1_μ, L^2_μ), and then it loses its stability. Again, as for the orientable case, this can happen in a variety of ways··· The main thing here is that the invariant manifolds $W^u(O_\mu)$ and $W^s(O_\mu)$ can intersect and a strange homoclinic attractor can arise containing the saddle-focus O_μ and its two-dimensional unstable manifold. In Figure 7(d), it is shown a sketch of manifolds $W^u(O_\mu)$ and $W^s(O_\mu)$ before they crossed. One can imagine (but difficult to draw) what will happen when they intersect. However, it is clear that in this case, the manifold $W^u(O_\mu)$, since the stable multiplier of the point O_μ is negative, will accumulate towards itself from both sides, and, accordingly, the point O_μ will reside inside the attractor. In the case of a discrete orientable Shilnikov attractor, its fixed point O_μ lies on its boundary (since the global piece of $W^u(O_\mu)$ accumulates at $W^u_{loc}(O_\mu)$ only from one side (namely, from the side where $W^{s+}_{loc}(O_\mu)$ is located).

Thus, we see that orientable and nonorientable discrete Shilnikov attractors have different structures. Although they both exist for an open set of parameter values, and it is also important for their geometry whether the invariant curve L_μ is resonant or not.

5. Examples of three-dimensional maps with Shilnikov attractors

In this section, we consider two examples of discrete Shilnikov attractors in the case of three-dimensional orientable and nonorientable maps. Recall that such attractors contain a fixed point O of the saddle-focus type, with eigenvalues $\lambda, \gamma_{1,2} = \gamma e^{\pm i\psi}$, where $|\lambda| < 1, \gamma > 1, 0 < \psi < \pi$. The unstable manifold of O is two-dimensional, and it resides entirely in the attractor.

First, we consider the case of orientable (with $0 < \lambda < 1$) Shilnikov attractor in three-dimensional map of the form

$$\bar{x} = y, \quad \bar{y} = z, \quad \bar{z} = Bx + Cy + Az - y^2, \tag{4}$$

where A, B, C are parameters. Numerically found example of such attractor is shown in Figure 8(h) for map (4) with the Jacobian $B = 0.5 > 0$.

Figure 8 shows main stages of the development of a discrete Shilnikov attractor in map (4) for $B = 0.5, A = 1.49$ as C changes. The formation of the attractor proceeds in accordance with the general bifurcation scenario described in Gonchenko et al. [24], Gonchenko et al. [26], see Section 4.1:

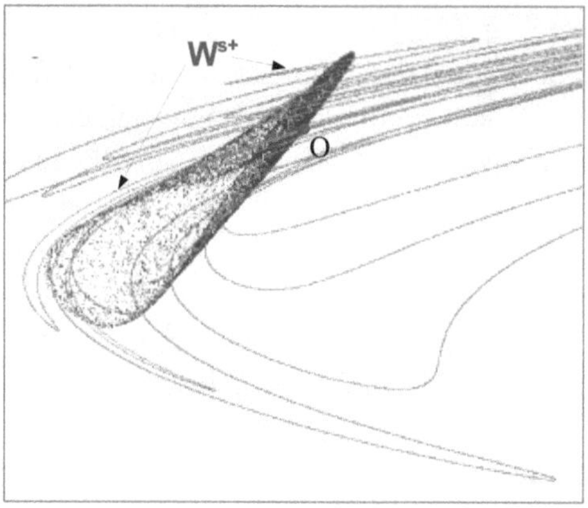

Figure 9. The Shilnikov attractor from Figure 8(h) and a part of the stable separatrix $W^{s+}(O)$ are shown in a suitable angle for viewing.

- At the beginning, the attractor of map (4) is a stable fixed point $O(0,0,0)$ (it is stable for $-0.99 > C > c_1 = -1.495$).
- Then, at $C = c_1$, the point O undergoes a supercritical Andronov–Hopf bifurcation: O becomes a saddle-focus (1,2) and a stable closed invariant curve L is born in its neighborhood (the curve L is shown in Figure 8(a) at $C = -1.7$).
- Next, the unstable manifold of O starts winding onto L and the Shilnikov whirlpool is formed (schematically, the whirlpool is shown in Figure 8(b)).
- Further, the dynamics inside the whirlpool become more and more complicated. In the case of map (4), an example of such development is shown in Figure 8(d–h). First, we observe three successive doubling bifurcations of stable invariant curves, $L \to 2L \to 4L \to 8L$, Figure 8(c–e). We did not observe the doubling of $8L$. Instead, a strange attractor appears which has at the beginning a torus-chaos type due to a destruction of the curve $8L$, then it sequentially captures the saddle curves $4L$, $2L$ (Figure 8(f)) and L Figure 8(g).
- Finally, when homoclinic intersections are created between $W^u(O)$ and $W^s(O)$, a discrete Shilnikov attractor is formed containing the fixed point O, Figure 8(h). In Figure 9, the numerically obtained stable separatrix $W^{s+}(O)$ is shown, which confirms the existence of homoclinic intersections of $W^u(O)$ and $W^{s+}(O)$ within the attractor.

Now we consider the case of the appearance of nonorientable Shilnikov attractor again in map (4), but now with the negative Jacobian B. (We note that in this map such attractors were found in Karatetskaia et al. [34]). Figure 10 shows the main stages of the development of this attractor when $B = -0.915$, $A = -2.786$ are fixed and C changes. The stages of the attractor creation follow the general bifurcation scenario described in Karatetskaia et al. [34], Gonchenko et al. [29], see also Section 4.2:

- At the beginning, the attractor is the fixed point $O(0,0,0)$ that is stable for $4.701 > C > c_1 \approx -2.712$).

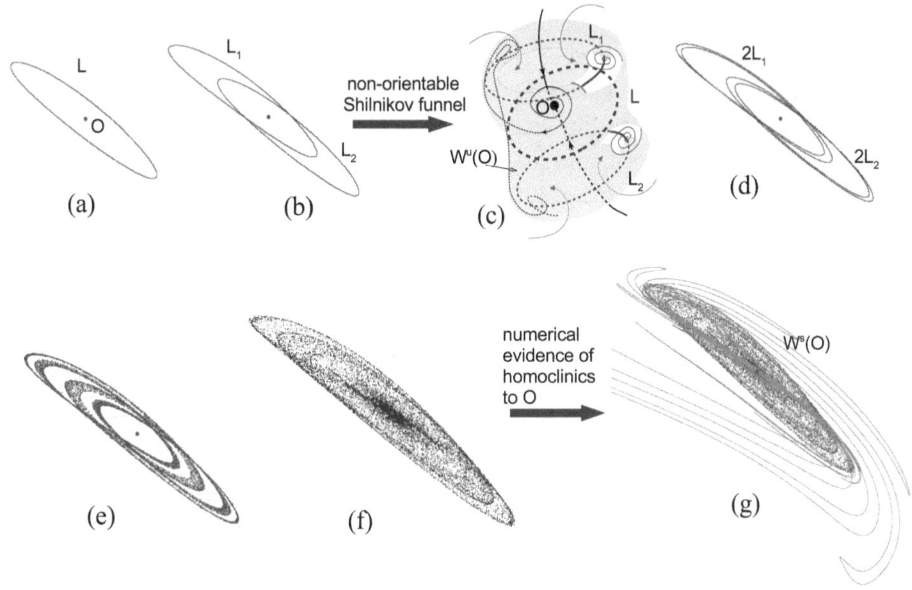

Figure 10. Stages of the emergence of a nonorientable Shilnikov attractor in map (4) for $B = -0.915, A = -2.786$ when C changes (picture (c) is only schematic).

- Then, at $C = c_1$, the point O undergoes a supercritical Andronov–Hopf bifurcation and becomes a nonorientable saddle-focus (1,2) and a stable closed invariant curve L is born (see Figure 10(a) for $C = -2.723$).
- Next, the curve L undergoes a component-doubling bifurcation [28]: a pair of stable period–2 curves L_1 and L_2 are born from L (these curves are shown in Figure 10(b) for $C = -2.733$). With the further change of C, the two-dimensional unstable manifold of O_μ begins to wind up on both the curves L_1 and L_2 and a nonorientable Shilnikov whirlpool is formed, which has a 'double roll' shape (it is shown schematically, in Figure 10(c)).
- Further, the dynamics inside the whirlpool become more complicated. In particular, several bifurcations of doubling of invariant curves occur, giving rise to strange attractors inside the whirlpool, (see Figure 10(e), where one of such attractors is shown for $C = -2.736$).
- Finally, when homoclinic intersections are created between $W^u(O)$ and $W^s(O)$, a nonorientable discrete Shilnikov attractor is formed, (see Figure 10(f) for $C = -2.743$). In Figure 10(g) a piece of $W^s(O)$ is shown, confirming that the attractor contains the point O.

6. Conclusion

We note that the Shilnikov's ideas about the importance of scenarios for the emergence of spiral attractors and, more generally, homoclinic attractors (strange attractors containing either an equilibrium or a periodic orbit and its entire unstable manifold) were implemented in a number of our works. The first one was our work with L.P. Shilnikov

[24], in which we gave a phenomenological description of scenarios for the emergence of discrete attractors of various types (discrete Lorenz attractors, Figure 8 attractors and Shilnikov attractors), and also gave examples of the implementation of these scenarios in one-parameter families of three-dimensional maps. Now the theory of discrete homoclinic attractors looks quite advanced, and it is, of course, richer than the corresponding theory of three-dimensional flows, see e.g. [25,26,29,31,32]. On the other hand, there is also a certain feedback here: a detailed study of discrete homoclinic attractors also has led to the discovery of new types of attractors of three-dimensional flows (for example, Lorenz attractors with several equilibria [32]). By the way, these studies are still ongoing.

We note that all spiral attractors of three-dimensional flows and maps, including the Shilnikov attractors, are quasiattractors: although they contain nontrivial hyperbolic subsets, arbitrarily small smooth perturbations also lead to the appearance of stable periodic orbits with extremely narrow absorbing domains. Usually, such orbits cannot be detected in standard numerics, and the attractor still seems to be chaotic. Nevertheless, some stable short-period orbits may appear to be visible – the corresponding parameter domains are called stability windows, and their existence in specific models is a well-known phenomenon in nonlinear dynamics [2,23].

However, spiral attractors of four-dimensional flows, as already shown by Turaev and Shilnikov [52], can be robustly chaotic. Actually, in the paper [52], the theory of pseudohyperbolic attractors, which remain chaotic for all small smooth perturbations, was laid down, and a geometric example of an attractor containing a saddle-focus with a three-dimensional stable and one-dimensional unstable manifold was constructed. Recently, an example of such an attractor in a system of four differential equations (in the generalized Lorenz model) was found in Gonchenko et al. [30]. In the case of four-dimensional maps, the corresponding examples of pseudohyperbolic spiral attractors are not yet known. But we are confident in their existence, including for Shilnikov discrete pseudohyperbolic attractors, and intend to consider the corresponding problems in the nearest future.

Notes

1. Their list is quite large; we will indicate only some of the most known three-dimensional models. Thus, spiral chaos was found in radio-electronic devices, such as the Chua circuit [17], the Anishchenko-Astakhov generator [3], in the optical laser systems [4,5,41,53], in chemical systems [6,11], in a certain class of models describing the behaviour of neurons [18], in biophysical experiments [40], in electromechanical systems [15,35], in electrochemical processes [13,39], in nonlinear convection in magnetic fields [42], in mechanical systems [33], etc.
2. Of course, when inside the whirlpool, there are no other attractors – for example, local bifurcations may lag behind global bifurcations and then, it can happen that both the limit cycle L_μ is stable and a homoclinic loop exists.
3. Such type of bifurcation of doubling of invariant curve is called a component-doubling bifurcation, see more details in Gonchenko et al. [28].

Disclosure statement

No potential conflict of interest was reported by the author(s).

Funding

This paper was carried out in the framework of the Russian Ministry of Science and Education [grant number 0729-2020-0036]. A. Gonchenko was supported by the RSciF [grant number 20-71-00079] (Section 4 and 5). Yu. Bakhanova and A. Kazakov were supported by the RSciF [grant number 19-71-10048] (Section 3). S. Gonchenko, A. Kazakov and E. Samylina thank the Theoretical Physics and Mathematics Advancement Foundation 'BASIS' [grant number 20-7-1-36-5], for support of scientific investigations.

References

[1] V.S. Afraimovich and L.P. Shilnikov, *Invariant two-dimensional tori, their breakdown and stochasticity*, in *Methods of qualitative theory of differential equations*, Gorky, (1983), pp. 3–26. [English translation in Am. Math. Soc. Transl., Ser. 2, 149, (1991), pp. 201–212].

[2] V.S. Afraimovich and L.P. Shilnikov, *Strange attractors and quasiattractors*. in *Nonlinear Dynamics and Turbulence*, G.I. Barenblatt, G. Iooss, and D.D. Joseph, eds., Pitman, New York, 1983. pp. 1–28.

[3] V.S. Anishchenko, *Complex oscillations in simple systems*, (1990), (in Russian).

[4] F.T. Arecchi, R. Meucci, and W. Gadomski, Laser dynamics with competing instabilities, Phys. Rev. Lett. 58(21) (1987), pp. 2205.

[5] F.T. Arecchi, A. Lapucci, R. Meucci, J.A. Roversi, and P.H. Coullet, Experimental characterization of Shil'nikov chaos by statistics of return times, EPL (Europhysics Letters) 6(8) (1988), pp. 677.

[6] F. Argoul, A. Arneodo, and P. Richetti, Experimental evidence for homoclinic chaos in the Belousov-Zhabotinskii reaction, Phys. Lett. A 120(6) (1987), pp. 269–275.

[7] A. Arneodo, P.H. Coullet, and C. Tresser, Occurence of strange attractors in three-dimensional Volterra equations, Phys. Lett. A 79(4) (1980), pp. 259–263.

[8] A. Arneodo, P.H. Coullet, and C. Tresser, Possible new strange attractors with spiral structure, Commun. Math. Phys. 79 (1981), pp. 573–579.

[9] A. Arneodo, P. Coullet, and C. Tresser, Oscillators with chaotic behavior: an illustration of a theorem by Shil'nikov, J. Stat. Phys. 27(1) (1982), pp. 171–182.

[10] A. Arneodo, P.H. Coullet, E.A. Spiegel, and C. Tresser, Asymptotic chaos, Phys. D 14 (1985), pp. 327–347.

[11] A. Arneodo, F. Argoula, J. Elezgarayab, and P. Richettia, Homoclinic chaos in chemical systems, Phys. D: Nonlinear Phenom. 62(1) (1993), pp. 134–169.

[12] Yu.V. Bakhanova, A.O. Kazakov, E.Yu. Karatetskaya, A.D. Kozlov, and K.A. Safonov, On homoclinic attractors of three-dimensional flows, Iz. VUZ. Appl. Nonlinear Dynam. 28(3) (2020), pp. 231–258.

[13] M.R. Bassett and J.L. Hudson, Shilnikov chaos during copper electrodissolution, J. Physical Chemistry 92(24) (1988), pp. 6963–6966.

[14] L.A. Belyakov, Bifurcation set in a system with homoclinic saddle curve, Math. Notes Acad. Sci. USSR 28(6) (1980), pp. 910–916.

[15] J.C. Chedjou, P. Woafo, and S. Domngang, Shilnikov chaos and dynamics of a self-sustained electromechanical transducer, J. Vib. Acoust. 123(2) (2001), pp. 170–174.

[16] L. Chua, M Komuro, and T. Matsumoto, The double scroll family, IEEE Trans. Circuits Syst. 33(11) (1986), pp. 1072–1118.

[17] L.O. Chua, M. Komuro, and T. Matsumoto, The double scroll family, Circuits Syst. IEEE Trans. 33(11) (1986), pp. 1072–1118.

[18] U. Feudel, A. Neiman, X. Pei, W. Wojtenek, H. Braun, M. Huber, and F. Moss, Homoclinic bifurcation in a Hodgkin-Huxley model of thermally sensitive neurons, Chaos: Interdiscip. J. Nonlinear Sci. 10(1) (2000), pp. 231–239.

[19] J.A. Gallas, The structure of infinite periodic and chaotic hub cascades in phase diagrams of simple autonomous flows, Int. J. Bifurcation Chaos 20(02) (2010), pp. 197–211.

[20] P. Gaspard, Generation of a countable set of homoclinic flows through bifurcation, Phys. Lett. A 97(1–2) (1983), pp. 1–4.

[21] P. Gaspard and G. Nicolis, What can we learn from homoclinic orbits in chaotic dynamics? J. Stat. Phys. 31(3) (1983), pp. 499–518.

[22] S.V. Gonchenko, D.V. Turaev, P. Gaspard, and G. Nicolis, Complexity in the bifurcation structure of homoclinic loops to a saddle-focus, Nonlinearity 10(2) (1997), pp. 409.
[23] S.V. Gonchenko, L.P. Shilnikov, and D.V. Turaev, Quasiattractors and homoclinic tangencies, Comput. Math. Appl. 34(2–4) (1997), pp. 195–227.
[24] A.S. Gonchenko, S.V. Gonchenko, and L.P. Shilnikov, Towards scenarios of chaos appearance in three-dimensional maps, Rus. J. Nonlinear Dyn. 8 (2012), pp. 3–28.
[25] S.V. Gonchenko, A.S. Gonchenko, I.I. Ovsyannikov, and D. Turaev, Examples of Lorenz-like attractors in henon-like maps, Math. Model. Nat. Phenom. 8(5) (2013), pp. 48–70.
[26] A.S. Gonchenko, S.V. Gonchenko, A.O. Kazakov, and D.V. Turaev, Simple scenarios of onset of chaos in three-dimensional maps, Int. J. Bifur. Chaos 24(08) (2014), pp. 1440005.
[27] S.V. Gonchenko, A.S. Gonchenko, A.O. Kazakov, A.D. Kozlov, and Yu.V. Bakhanova, Mathematical theory of dynamical chaos and its applications: review. part 2. spiral chaos of three-dimensional flows, Iz. VUZ. Appl. Nonlinear Dyn. 27(5) (2019), pp. 7–52.
[28] A.S. Gonchenko, S.V. Gonchenko, and D.V. Turaev, Doubling of invariant curves and chaos in three-dimensional diffeomorphisms, Chaos 31 (2021), pp. 113130.
[29] A.S. Gonchenko, M.S. Gonchenko, A.D. Kozlov, and E.A. Samylina, On scenarios of the onset of homoclinic attractors in three-dimensional non-orientable maps, Chaos: Interdiscip. J. Nonlinear Sci. 31(4) (2021), pp. 043122.
[30] S. Gonchenko, A. Kazakov, and D. Turaev, Wild pseudohyperbolic attractor in a four-dimensional Lorenz system, Nonlinearity 34(4) (2021), pp. 2018.
[31] A.S. Gonchenko and S.V. Gonchenko, Variety of strange pseudohyperbolic attractors in threedimensional generalized henon maps, Phys. D. 337 (2016), pp. 43–57.
[32] S.V. Gonchenko and A.S. Gonchenko, On discrete Lorenz-like attractors, Chaos: Interdiscip. J. Nonlinear Sci. 31(2) (2021), pp. 023117.
[33] M.E. Henderson, M. Levi, and F. Odeh, The geometry and computation of the dynamics of coupled pendula, Int. J. Bifur. Chaos 1(01) (1991), pp. 27–50.
[34] E. Karatetskaia, A. Shykhmamedov, and A. Kazakov, Shilnikov attractors in three-dimensional orientation-reversing maps, Chaos: Interdiscip. J. Nonlinear Sci. 31(1) (2021), pp. 011102.
[35] M.T.M. Koper, P. Gaspard, and J.H. Sluyters, Mixed-mode oscillations and incomplete homoclinic scenarios to a saddle focus in the indium/thiocyanate electrochemical oscillator, J. Chem. Phys. 97(11) (1992), pp. 8250–8260.
[36] C. Letellier, P. Dutertre, and B. Maheu, Unstable periodic orbits and templates of the Rössler system: toward a systematic topological characterization, Chaos: Interdiscip. J. Nonlinear Sci. 5(1) (1995), pp. 271–282.
[37] S. Malykh, Y. Bakhanova, A. Kazakov, K. Pusuluri, and A. Shilnikov, Homoclinic chaos in the Rössler model, Chaos: Interdiscip. J. Nonlinear Sci. 30(11) (2020), pp. 113126.
[38] T. Menacer, R. Lozi, and L.O. Chua, Hidden bifurcations in the multispiral Chua attractor, Int. J. Bifur. Chaos 26(14) (2016), pp. 1630039.
[39] T. Noh, Shilnikov chaos in the oxidation of formic acid with bismuth ion on Pt ring electrode, Electrochim. Acta. 54(13) (2009), pp. 3657–3661.
[40] D. Parthimos, D.H. Edwards, and T.M. Griffith, Shilnikov homoclinic chaos is intimately related to type-III intermittency in isolated rabbit arteries: role of nitric oxide, Phys. Rev. E. 67(5) (2003), pp. 051922.
[41] A.N. Pisarchik, R. Meucci, and F.T. Arecchi, Theoretical and experimental study of discrete behavior of Shilnikov chaos in a CO2 laser, Eur. Phys. J. D-At. Mol. Opt. Phys. 13(3) (2001), pp. 385–391.
[42] A.M. Rucklidge, Chaos in a low-order model of magnetoconvection, Phys. D: Nonlinear Phenom. 62(1) (1993), pp. 323–337.
[43] L.P. Shilnikov, Some cases of generation of periodic motions in n-space, Dokl. Akad. Nauk. Russ. Acad. Sci. 143(2) (1962), pp. 289–292.
[44] L.P. Shilnikov, Some cases of generation of period motions from singular trajectories, Mat. Sb. 103(4) (1963), pp. 443–466.
[45] L.P. Shilnikov, A case of the existence of a denumerable set of periodic motions, Dokl. Akad. Nauk. Russ. Acad. Sci. 160(3) (1965), pp. 558–561.

[46] L.P. Shilnikov, Existence of a countable set of periodic motions in a four-dimensional space in an extended neighborhood of a saddle-focus, Dokl. Akad. Nauk 172(1) (1967), pp. 54–57.
[47] L.P. Shilnikov, On the generation of a periodic motion from trajectories doubly asymptotic to an equilibrium state of saddle type, Mat. Sb. 119(3) (1968), pp. 461–472.
[48] L.P. Shilnikov, On the question of the structure of an extended neighborhood of a structurally stable state of equilibrium of saddle-focus type, Mat. Sb.(NS) 81(123) (1970), pp. 92–103.
[49] L.P. Shilnikov, Bifurcation theory and turbulence, Nonlinear and Turbulent Processes in Physics. In *Proceedings of the Second International Workshop*, 10–25 October, 1983, R. Z. Sagdeev, ed., Harwood Academic Publishers, 1984. p. 1627.
[50] L.P. Shilnikov, Bifurcation theory and turbulence I, in *Methods of Qualitative Theory of Differential Equations*, Gorki University, 1986, pp. 150–163.
[51] L.P. Shilnikov and A. Shilnikov, Shilnikov bifurcation, Scholarpedia 2(8) (2007), pp. 1891.
[52] D.V. Turaev and L.P. Shilnikov, An example of a wild strange attractor, Sb. Math. 189 (1998), pp. 291–314.
[53] C.S. Zhou, J. Kurths, E. Allaria, S. Boccaletti, R. Meucci, and F.T. Arecchi, Constructive effects of noise in homoclinic chaotic systems, Phys. Rev. E. 67(6) (2003), pp. 066220.

Chaotic attractors of discrete dynamical systems used in the core of evolutionary algorithms: state of art and perspectives

Ivan Zelinka and Roman Senkerik

ABSTRACT

This paper delves into the intersection between discrete dynamical systems and bio-inspired metaheuristic algorithms. Chaos, an inherent phenomenon manifested by both continuous and discrete dynamic nonlinear systems has become an essential component of engineering design. This study examines the relationship between chaos and bio-inspired metaheuristic algorithms from two distinct angles: the presence of chaos within the realm of algorithms, and the application of bio-inspired algorithms for the identification, control, or synthesis of complex systems. Furthermore, this paper offers an in-depth exploration of the interplay between metaheuristics and complex nonlinear dynamics, including potential avenues for future research. The presented simulation studies, along with the design of objective functions for optimization and the implementation of metaheuristics, serve as a valuable foundation for subsequent experiments involving discrete nonlinear systems and complex systems with time delays. Such systems may benefit from multiparametric or multicriteria optimization approaches, paving the way for novel advancements in the field.

1. Introduction

In modern science, difference and differential equations play an indispensable role in the mathematical apparatus. They can be used to solve many real engineering problems and define fundamental research theorems. Classical tools have long been used to solve such equations and systems of equations, but in the last several decades, computers and algorithms have come to play a very important role. The family of algorithms allowing us to solve problems based on difference equations is very rich. The metaheuristic algorithms (including evolutionary algorithms and swarm intelligence) can not only be used to optimize tasks defined by set of differential equations but also to design their structure given the measured data from a given problem. They can also be used to control physical systems studied and described using difference or differential equations, control and synthesize chaotic systems, and solve/optimize many other complex problems.

This paper represents a summary of selected areas with an interesting intersection of both classical mathematics in the field of difference equations and their solution and modern optimization approaches that belong to the family of metaheuristics, specifically evolutionary algorithms and swarm intelligence. The main goal of this paper is only to draw attention to the existence of the intersection – fusion of these two fields and their importance. The focus here is on discrete nonlinear systems producing chaotic behaviour and how such systems can be modeled, stabilized, identified, or possibly synthesized using modern algorithmic techniques.

The organization of this paper is the following. First, a brief background on deterministic chaos, metaheuristics (evolutionary computational techniques), and the motivation that has led to many research papers linking discrete or continuous chaos and metaheuristics are provided. Selected important cases are presented, such as the analysis of chaos in metaheuristics, the evolutionary synthesis of new chaotic systems, and especially the evolutionary control of discrete chaotic maps and real systems. These technical demonstrations include a literature review, references to related publications, and a brief discussion. Finally, all the findings are summarized in a brief conclusion.

1.1. Deterministic chaos

One of the important moments in the twentieth century is the emergence of the 'new science' popularly known as 'Chaos.' Many researchers were involved in its creation, and this gradually emerging discipline can be traced back to the 19th century. For example, so-called Lyapunov exponents are typical examples. These exponents, usually referred to as λ, characterize the rate at which nearby trajectories in a dynamical system diverge or converge over time. It provides a measure of the sensitivity of a system to initial conditions, which is a key feature of chaotic systems. The Lyapunov exponent quantifies the exponential separation rate between two initially close trajectories in phase space as they evolve in time. If the exponent is positive, it indicates that the trajectories diverge exponentially, suggesting chaos. Conversely, if the exponent is negative, the trajectories converge exponentially, indicating stability. A zero Lyapunov exponent implies that the trajectories neither converge nor diverge significantly, resulting in a neutral or marginally stable behaviour. More details can be found in survey paper [3].

Another example is the celestial mechanic's field, specifically the study of the motion of three bodies in gravitational interaction [39]. The three-body problem is one of the fundamental demonstrations of the deterministic chaos of so-called Hamiltonian systems.

Popularization and increased interest in chaos are linked to E. N. Lorenz, who became famous for discovering the so-called 'Butterfly Effect,' which he presented in the 1960s. The butterfly effect refers to the impossibility of long-term weather forecasting (max. 2–3 days) and, thus to its chaotic nature. The so-called catastrophe theory is closely related to chaos theory, developed by the French mathematician Rene Thom in the 1960s [52].

In recent decades new types of chaotic systems have been discovered [9, 53], which according to [55], are so-called dual systems. In parallel with these systems, other types have been described and intensively studied, such as (dissipative) Lozi, Tinkerbell, Ikeda, Sinai, Burger and Duffing systems, or (conservative) Chirikov, Arnold, Baker, Nose-Hoover, and Henon-Heiles systems. These systems are mainly derived from existing physical systems. Typical examples are the Lorenz system (weather), the Duffing system (steel belt in a magnetic field), and the logistic equation (biological hunter-prey system).

1.2. Evolutionary computational techniques

Optimization algorithms are a powerful tool for solving many problems in engineering practice. They are usually used where solving a problem by analytical methods is inappropriate or impracticable. 'Metaheuristics' is an established term for modern nature-inspired optimization methods, first used in 1986 [23]. Without any doubt, it can be said that nowadays, evolutionary algorithms (EAs) represent the most popular metaheuristics [17]. In computer science, EAs or so-called evolutionary computation techniques (ECTs) are a subfield of artificial intelligence (AI) that allows for solving complex optimization problems. ECTs are inspired by Darwinian principles of natural selection as a critical mechanism of evolution and Mendel's law of heredity [5, 21].

The operation of evolutionary algorithms can be generally characterized as the sequence of operations (Algorithm 1). To get a clear idea of the functionality of the typical sequence of operations, it is still necessary to define the relationship between the optimization problem itself and the EA/ECT, i.e. how the algorithm obtains feedback on the quality of the solution to control the evolutionary process using specific operators. This connecting element is the objective function (or fitness function). The best solution is the global extreme – the maximum or minimum of the objective function.

Algorithm 1 Evolution algorithm

```
(Randomly) generate the initial population of solutions;
Evaluate each solution in the population;
while the termination condition is not satisfied do
    Select the solutions which are highly suitable for reproduction (parents);
    Recombine pairs of parents (crossover);
    Mutate the resulting offspring;
    Evaluate (quality of) new solutions;
    Replace inferior solutions with new solutions;
end while
Return the best solution found;
```

As already mentioned, metaheuristics often find inspiration in biological phenomena. Over time, two large groups of algorithms have emerged, distinguished by different philosophies of function (mainly due to different 'inspirations'). Thus, in the literature and in the scientific community, it is possible to find the labelling of these two important groups as 'classical EA' and 'swarm' algorithms. Classical evolutionary algorithms generally use an iterative process representing the evolution of a population. This class of algorithms is characterized by the use of special operators to drive the evolutionary process to find the ideal solution, such as the genome, mutation, crossover (recombination), and others. The classical evolutionary algorithms mainly include evolutionary strategies [6], genetic algorithms (GA) [24] and differential evolution (DE) [11, 51].

GAs is a popular example of a bio-inspired optimization technique that mimics the process of natural selection to find optimal or near-optimal solutions to complex problems. GAs are widely used in various domains, including optimization, machine learning, and artificial intelligence, due to their ability to effectively explore large and complex search spaces. GAs works by iteratively evolving a population of candidate solutions using simple steps. At the end of the algorithm, the solution with the highest fitness value in the final population is considered the best-found solution.

(1) *Initialization*: Randomly create a population of candidate solutions (usually represented as binary strings or other data structures).
(2) *Fitness Evaluation*: Assess the quality of each candidate solution in the population by calculating its fitness value using a predefined objective function. The fitness function is the normalized objective function in the interval 0–1.
(3) *Selection*: Choose a subset of the population for reproduction based on their fitness values. Higher-fitness solutions are more likely to be selected, emulating the survival of the fittest principle. Usually, the ranking of solutions or tournament system is used.
(4) *Crossover*: Generate new offspring solutions by recombining the selected parent solutions. This is typically done by exchanging parts of their genetic information, such as swapping sections of their binary strings.
(5) *Mutation*: Introduce small random changes into the offspring solutions to promote diversity in the population and prevent premature convergence to suboptimal solutions. This process is analogous to genetic mutations in biological systems. Simple binary inversion with very low probability is common.
(6) *Termination*: Repeat steps 2 to 5 for a predefined number of generations until a termination criterion is met (e.g. reaching a maximum number of generations, spending the budget of objective/fitness function evaluation, finding a solution with a satisfactory fitness value, or observing no significant improvement over several generations).

Many metaheuristics find inspiration in the collective intelligence of animal societies. These are called swarm algorithms. Some species make intelligent decisions without central control. Based on the limited knowledge of an individual, swarms achieve their optimization goals, e.g.for foraging. Modeling these events has led to the development of popular algorithms such as Particle Swarm Optimization (PSO) [18], Ant Colony Optimization [14], Self-organizing Migration Algorithm (SOMA) [65], whose successes were then followed by Artificial Bee Colony [26], Grey Wolf Optimizer [35] and many others.

Apart from the above, algorithms using inspiration in physical laws [41] or cognitive science [7] are also popular. However, it is important not to forget another important subfield of ECT, namely symbolic regression. Symbolic regression (symbolic feature identification) is an approach to search for features to analyze data sets to synthesize data-generating systems (models) fully. Symbolic regression is the process of synthesizing a final model from the basic simple functions, operators, and terminals (variables and constants) that correspond to, for example, a given data set.

John Koza first introduced the symbolic regression approach as the well-known concept of genetic programming (GP) [27], later followed by Conor Ryan with grammatical evolution (GE) [44]. There are also many other methods and areas, such as Evolutionary Hardware and Cartesian Genetic Programming [56]. Still, above all, we should mention the symbolic regression method, namely Analytical Programming (AP), developed by Prof. Ivan Zelinka [69] in 2001, which became the basis of many of the experiments presented here.

1.3. Motivation

The apparent motivation behind this research survey is to show how, and especially why, these algorithms are so important and interesting in nonlinear systems and how they can be applied in an engineering way.

As such, the issue of optimization has been an essential topic of research with various application domains for many past years. The problems of optimization have been solved by the classical mathematical apparatus, numerical methods (descent or pattern searching), techniques based on (non)linear programming (including mixed integer versions), and many more.

The computational and algorithmic complexity increases not only with the complexity of the optimized problem given by the number of parameters (arguments, dimensions) but also with the number of constraints and the domain of arguments of the optimized function (real, integer, logical, linguistic). Thus, to avoid unbearable computational demands in many cases, the need to implement complex solvers and often use the domain knowledge of a large team of experts, it is advantageous to use computational methods based on metaheuristics.

Using these methods requires only a very good knowledge of the optimized problem, often the (external) relationship between the input and output data, i.e. correctly defining the objective function whose optimization should lead to the problem's solution. Another advantage is that these algorithms, by their nature, always aim at finding the global extreme. Many of them, as population-based algorithms, provide at the end of their run a large set of different quality solutions, which may differ from each other in the value of the objective function and the position in the search space. Moreover, these algorithms are suitable for parallelization and execution on modern platforms. The disadvantage of these algorithms is that they work with stochastic operations, and there is no guarantee of optimality.

Figure 1 shows the landscape of an objective function (input/output mapping) constructed for synchronizing chaotic systems represented by a set of differential equations. Each point of this space represents one possible control action on the system through coupling parameters. As can be seen, the landscape is very complex (multimodal containing many local extremes), and finding an optimal solution representing an optimal control intervention using classical methods can be extremely difficult, if not impossible, even in this limited area. In this case, metaheuristics (evolutionary and swarm algorithms) stand out for their performance and efficiency, as they can find a good solution quickly. They can quickly explore the space and target specific areas even in such a very complex problem. The details, limitations, advantages, and disadvantages of using metaheuristics and different constructions of the objective function are discussed in the following sections devoted to different case studies linking metaheuristics and deterministic chaos. For more examples and discussions, refer to [68].

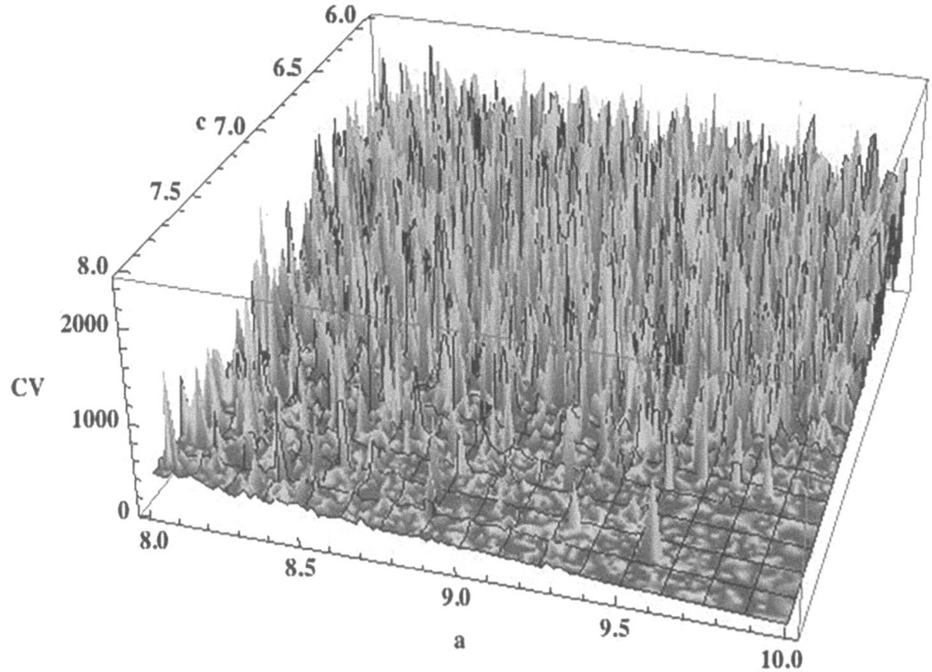

Figure 1. Objective function landscape for evolutionarily driven chaos synchronization problem. A typical example from [68], where the task was to find optimal values of coupling parameters *a* and *c*. *CV* stands for Cost function Value.

2. Chaos in the dynamics of artificial intelligence models and metaheuristics

For many reasons like analysis of population behaviour, tuning of parameters, or simply investigating the effectiveness and applicability of evolutionary algorithms to given tasks, it is very important to know how algorithms behave under certain conditions and whether there might be hidden problems in their dynamics. If we look at evolutionary algorithms, we find that they are actually discrete dynamical systems that can also be represented by differential or difference equations. Hence, we can assume that these algorithms can externally exhibit the behaviour that is proper to classical dynamical systems modeled by standard differential equations.

The first researchers to point this out were Agapi and Wright in 2001 [61] in their study of genetic algorithms. They demonstrated the presence of chaotic dynamics within the algorithm itself and discussed the impact on its operation. In [61], models of dynamic systems of genetic algorithms are considered when the population size goes to infinity. Their work is based on the research of [58–60]. The elegant theory of simple genetic algorithms is based on a random heuristic search on the idea of heuristic map G [61]. An important research point in [61] is that map G includes all the dynamics of a simple genetic algorithm. Sample bifurcation diagrams are shown in Figures 2–3. Ideas about chaos existence in a simple genetic algorithm are explained in detail in [61, 68]. In both publications was shown that chaos can be observed in heuristic algorithms, which certainly does not apply only to simple genetic algorithms.

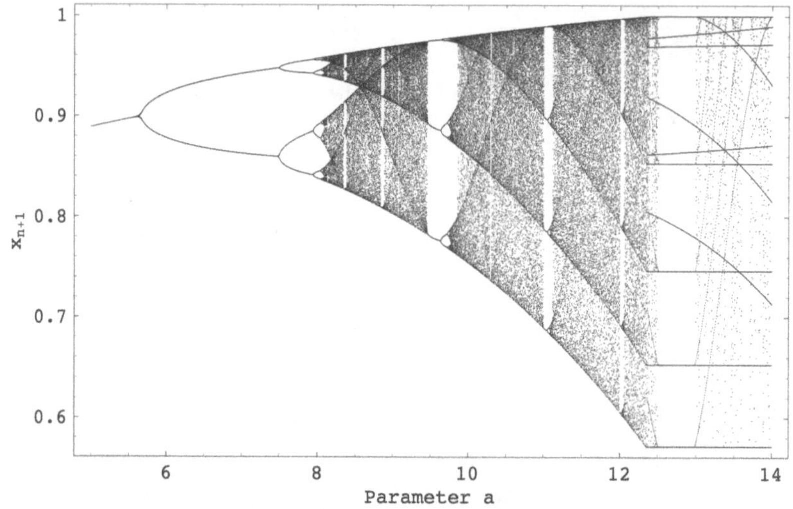

Figure 2. Bifurcation diagram of simple GA. The control parameter *a* belongs to the triplet of control parameters *a*, *b*, *T* of heuristic map G used to model the behaviour of GA. Reconstructed according to [61].

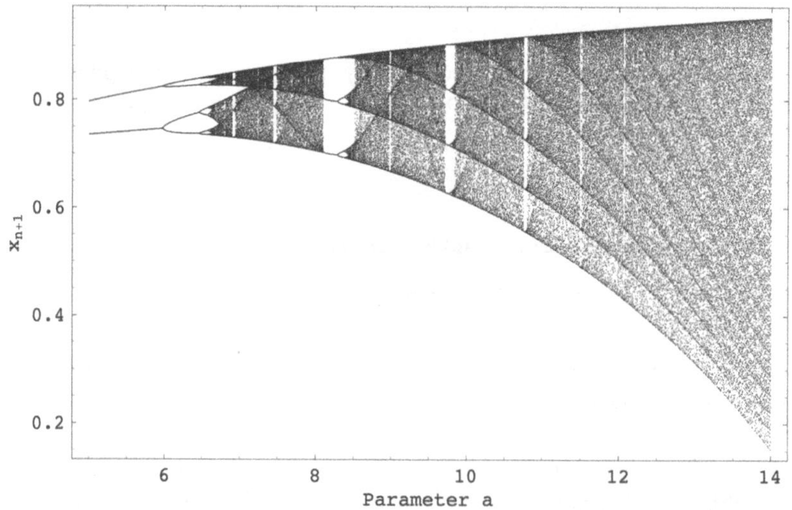

Figure 3. Bifurcation diagram of simple GA. The control parameter *a* belongs to the triplet of control parameters *a*, *b*, *T* of heuristic map G used to model the behaviour of GA. Reconstructed according to [61].

While AI algorithms and models are typically designed to be deterministic and predictable, some AI models can exhibit chaotic behaviour under certain conditions. Recurrent Neural Networks (RNNs) [33] are a class of neural networks that can process data sequences and maintain an internal state across time steps. RNNs, particularly when using nonlinear activation functions, can exhibit chaotic behaviour when their internal dynamics become sensitive to small changes in input or parameters. Another AI model exhibiting chaotic features is Echo State Networks (ESNs) [30], a reservoir computing model consisting of a large, fixed, and randomly connected recurrent neural network called the reservoir.

The reservoir can exhibit rich, complex, and even chaotic dynamics, depending on the choice of connection weights and activation functions. Finally, even if it is not directly an AI model, some Cellular Automata rules can lead to chaotic behaviour, as observed in the famous Game of Life by John Conway or Rule 30 by Stephen Wolfram.

The presence of chaos in AI models, metaheuristic, is often considered a challenge to be mitigated or controlled, as it can make the model's behaviour difficult to predict, analyze and interpret. However, in some cases, controlled chaos, or the state called 'the edge of chaos' can be harnessed to improve the exploration and learning capabilities of AI models.

3. Chaos synthesis and identification

The challenging task is not only to solve differential equations and their behaviour, but also to synthesize them based on existing data. In the experiments described in this section, we point out that evolutionary techniques can also synthesize a system of equations to satisfy predefined criteria (i.e. exhibits deterministic chaos).

In recent decades, there have been introduced many different approaches to the construction of chaotic systems, or also, in other words, the synthesis of chaos. Synthesizing chaotic systems can be done through various methods, often involving mathematical models, systematic approaches, or physical systems designed to exhibit chaotic behaviour [2, 10, 12, 13, 19, 54, 72].

An original approach towards synthesizing chaotic systems based on the principles of symbolic regression is proposed in [66]. It is a synthesis of dynamical systems that exhibit chaotic behaviour. The complex structure synthesis was based on symbolic objects (terminals, non-terminals) $\{x, A, +, -, \times, /\}$ contained in one of the simplest chaotic systems, the logistic equation, as in (1).

$$x_{n+1} = Ax_n(1 - x_n) \tag{1}$$

During these experiments, many new descriptions of chaotic systems in the form of differential equations were synthesized. In the Figures 4–8 are selected four examples of bifurcation diagrams of these artificially created chaotic systems, which demonstrate that evolution has been able to construct sets of equations to generate chaotic behaviour in specific interval. An example of the definition of such a synthesized system is given in (2) and Figure 4. More details can be found in [66, 68].

$$x_{n+1} = -\frac{A}{2x_n(-\frac{x_n^3}{A^3} - \frac{A^2}{x_n} + A)} \tag{2}$$

On the other hand, successful attempts have been made to identify the equations in the analytical description so that their behaviour fits the observed data exactly. The research, published in [70], is illustrated in Figure 9 which demonstrates finding an analytical description of a system such that the system generates the same bifurcation diagram as the original for which a description was searched. The figure shows essentially two bifurcation diagrams one in black and the other in red. The black bifurcation diagram is the original data to which the mathematical description was searched and the red bifurcation diagram is the diagram generated by the found mathematical description as constructed by

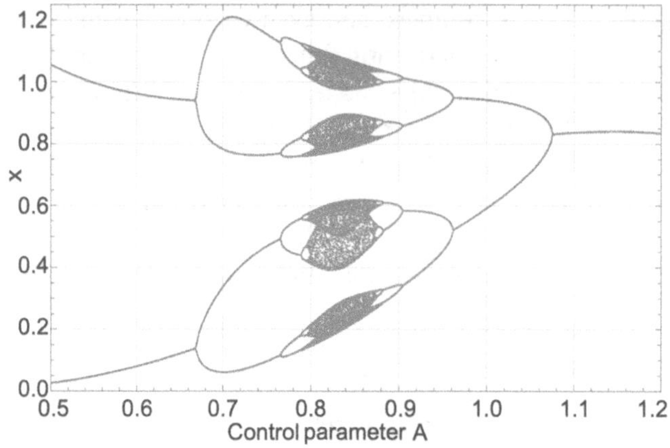

Figure 4. Bifurcation diagram of evolutionary synthesized chaotic system (2).

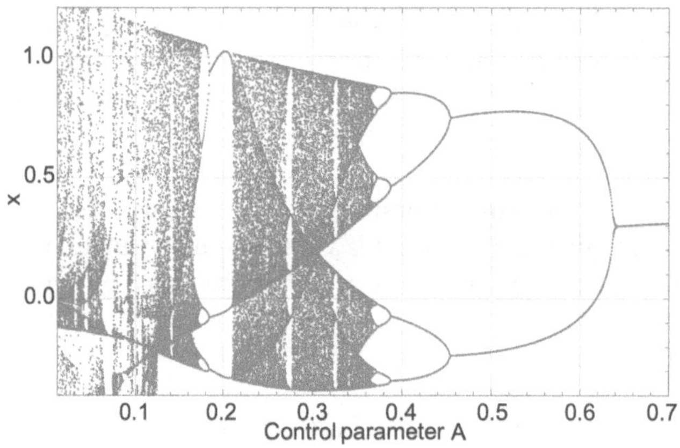

Figure 5. Bifurcation diagram of an evolutionary synthesized chaotic system, selected example from [66, 68].

the evolutionary algorithms. The difference between the two bifurcation diagrams is practically minimal. Finally, however, it should be noted that the above-mentioned approach of identifying chaos through bifurcation diagrams is based on the consideration of equilibrium states and therefore works well for self-excitation of chaos, but does not allow one to effectively reveal hidden chaotic attractors (for example, in systems without equilibrium states).

4. Control

This section aims to present possible approaches to control various chaotic systems using evolutionary (metaheuristic) algorithms. The focus here is on the design of an objective function that gives feedback to the metaheuristics on the quality of the process of searching the space for possible solutions. Furthermore, the influence of chaotic dynamics and the design of the objective function on the shape of the hyperplane and thus on the possibilities and limitations of the efficiency of metaheuristic algorithms is discussed. Individual case

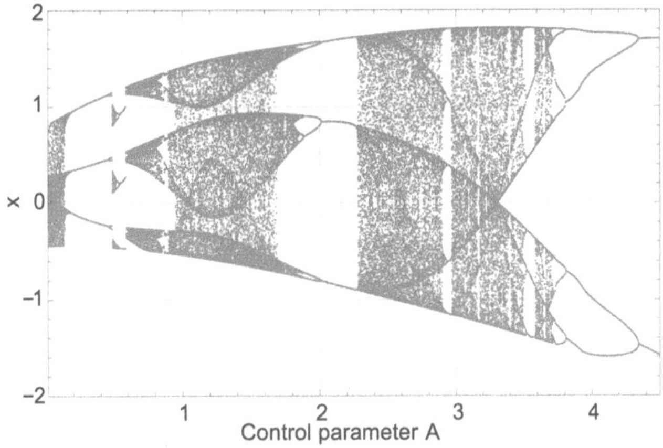

Figure 6. Bifurcation diagram of an evolutionary synthesized chaotic system, selected example from [66, 68].

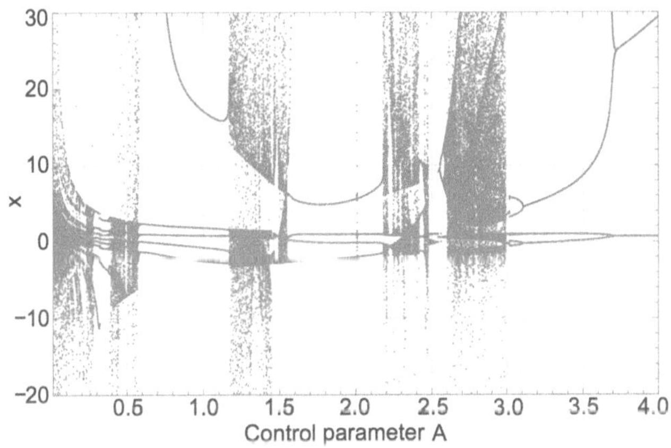

Figure 7. Bifurcation diagram of an evolutionary synthesized chaotic system, selected example from [66, 68].

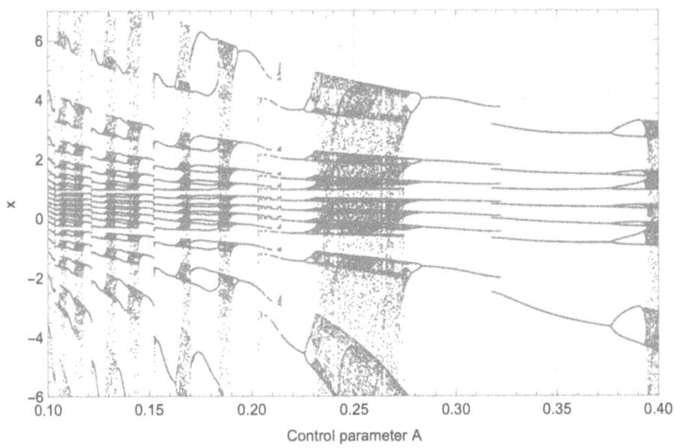

Figure 8. Bifurcation diagram of an evolutionary synthesized chaotic system, selected example from [66, 68]. A zoom of Figure 7.

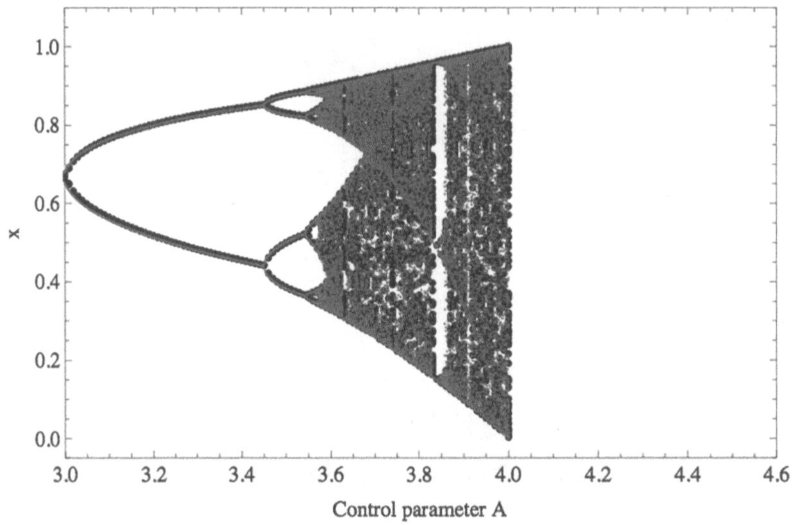

Figure 9. Bifurcation diagrams of discrete chaotic system identification. Black – original system, red – identified. Synthesis of a system based on the components of Equation (1).

studies are then presented in each subsection, including simple discrete chaotic maps with mention of coupled map lattices (CML) systems and an example of a real chaotic system.

4.1. Control of simple chaotic systems

Since the early 1990s, when OGY (Ott, Grebogi and Yorke) method [38] has been introduced, many chaos control methods have been developed like sliding mode [62], based on fuzzy systems [22] or local chaos control based on the Lyapunov approach [42]. Many targeting algorithms have been published with emphasis on a significant reduction of the stabilization time, simply to overcome several drawbacks of OGY [20] related to optimality and with the long initial chaotic waiting phase before the trajectory is stabilized. One of these techniques is the Pyragas' delayed feedback control method [40], which can be considered a 'targeting' algorithm. Despite the limitations of the Pyragas method identified in [28], the reason why this technique is suitable for metaheuristic-based optimization is the number of available control parameters, which have to be set otherwise by inaccurate estimation or demanding mathematical analysis. The application of evolutionary algorithms may lead to improved system behaviour and faster stabilization of the desired unstable periodic orbit (UPO) state – either a steady fixed point (p-1 orbit) or an arbitrarily stable periodic event (p-2 or higher orbit).

Many methods have also been developed to control the so-called space-time chaos that is generated, among other things, by CML systems. Control laws for CML have been derived based on a priori knowledge of the structure of CML systems, which usually remains hidden in the real world [45]. Methods such as external observer [8], evolutionary algorithms [66], etc. have also been used to control CMLs.

The connection between metaheuristics and deterministic chaos control has been explored and published in recent years in papers [4, 15, 32, 43, 63, 64, 67].

In the following two subsections, the control law is based on the above-mentioned method of Pyragas Extended Delayed Feedback – ETDAS. Therefore, visualizations of the objective (cost) function landscapes are shown for different combinations of ETDAS method parameters.

4.2. Objective function design

Nowadays, metaheuristics optimization algorithms are known to be an effective tool for almost any optimization problem with various levels of complexity and difficulty to solve. But the quality of optimization results usually depends on the design of the objective function, especially when evolutionary algorithms are used to control the chaos. Since we are dealing with the minimization problem for parametric optimization (minimizing the cost of stabilization), we are referring to the objective function as the cost function (*CF*). The cost function is used to transform the search space (optimized parameters) into a geometric problem by adding an additional dimension defining the quality of the solution, and the metaheuristic optimization algorithm then tries to find the global extreme of the function (minimum or maximum) by searching the neighbourhood, cooperations between solutions, or operators such as crossover and mutation.

It is well known that deterministic chaos and chaos control techniques are very sensitive to parameter settings and initial conditions. In the case of parametric optimization, they are also extremely sensitive to the design of the cost function used, which is used for evaluating quality for a certain combination of optimized parameters for the control/synchronization method. Regarding sensitivity to initial conditions, this can be prevented in more ways than the design of the optimization cost function discussed here. The dependence of chaos on the initial data can be overcome by the requirement to have one connected global attractor. This means all trajectories, except for a set of measure zero, are attracted to one chaotic set (a self-excited attractor with some mixing properties, e.g. in the classical Lorenz system). Regarding the reported results here for the Henon attractor, it is important to avoid multistability and the presence of hidden attractors (for details, please refer to [16]).

There are several approaches to designing an objective/cost function to control or synchronize chaos. The most critical factor for calculating the quality of the solution is to choose appropriate metrics or indicators describing that the system has been stabilized, and then to appropriately transform their values, add any penalties and other optimality criteria (number of steps, robustness for different initial conditions, etc.) The authors in [31] use the ITAE criterion (Integral of the Time weighted Absolute Error). The paper [1] uses complex definitions of cost functions implementing calculations with Jacobi matrices and spectral radius.

Below is a summary of proposals from the authors of this research survey that have been used in several published papers [37, 46, 47]. The various proposals include simple functions, more complex ones with multiple criteria (but still intended for single-criteria metaheuristic algorithms), and so-called black-box functions that do not require explicit prior knowledge of the periodic orbit for stabilization. A summary of the advantages and disadvantages of the various designs of the cost functions also complements the overall survey.

The proposal of the basic cost function (CF_{Simple}) is based on minimizing the area created by the sum of the differences in discrete times t between the required (target) state

TS_t and the real system output (actual state) AS_t on the whole simulation interval with a total of n discrete time steps (3). This CF design is very convenient for the evolutionary searching process due to the relatively favourable CF landscape. (See Figure 11.).

$$CF_{Simple} = \sum_{t=0}^{n} |TS_t - AS_t| \tag{3}$$

But this simple approach has one major drawback: the inclusion of the initial chaotic transient behaviour of the not-stabilized system in the value of the cost function. Consequently, the parameter search cannot be targeted for precise stabilization. The slight numerical difference in stabilization quality is suppressed by including the initial chaotic transient behaviour mentioned above. On the other hand, this proposal should aim the metaheuristic algorithm to search for the shortest possible overall initial chaotic behaviour, hence to time optimality. Still, it depends on the amplitude of the chaotic oscillations. Of course, this design is only suitable for stabilizing the system to a steady state (p-1 orbit) since the phase shift between TS_t and AS_t is not known in advance.

Different designs of the cost functions should secure the successful stabilization of either p-1 orbit (stable state) or any other higher periodic orbit anywise phase shifted. The so-called universal cost function is based on searching for the desired stabilized periodic orbit and thereafter, calculation of the difference between desired target state (value) and found actual periodic orbit on the short time interval - τ_s (20 iterations) from the point, where the first minimal (pre-set small) value of the difference between desired and actual system output is found (i.e. floating window for minimization – see Figure 10). This value is fixed and was set based on experimental results. It was purposely chosen so that there is no mismatch between the system's sensitivity to changes in adjustable parameters and changes in the values of the cost function. Furthermore, due to final cost values converging towards zero (in case of successful stabilization), this CF also allows the using of a stopping criterion, avoiding very time-demanding simulations. The CF$_{UNI}$ has the form (4).

$$CF_{UNI} = p + \sum_{t=\tau 1}^{\tau 2} |TS_t - AS_t| \tag{4}$$

Where:

τ_1 – the first min value of difference between TS and AS

τ_2 – the end of optimization interval ($\tau_1 + \tau_s$),

$p = 0$ if $\tau_i - \tau_2 \geq \tau_s$

$p = 10 * (\tau_i - \tau_2)$ if $\tau_i - \tau_2 < \tau_s$ (i.e. late stabilization).

In order to include the time optimality of the stabilization, it was necessary to modify the definition of the universal CF$_{UNI}$ to convert the newly defined multi-criteria problem into a single-criteria one. The simplest, but at the same time the most problematic from the point of view of motion on the search space, is that the entire CF is multiplied by the number of iterations (*NI*) of the first found the minimum value of the difference between the desired and the actual output of the system (the origin of the fully stabilized UPO). However, given the extreme sensitivity of the chaotic system to the control method parameter settings, this could override the emphasis on stabilization accuracy (i.e. multiplying part of the cost function by the order of tens to hundreds of iterations). At the same time, it is still necessary to

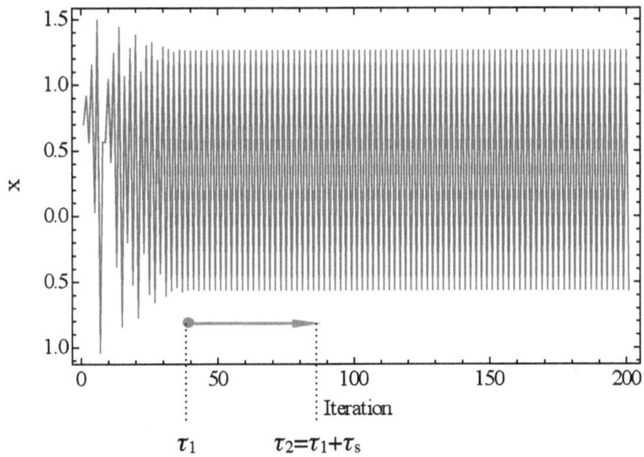

Figure 10. 'Floating window' for minimization.

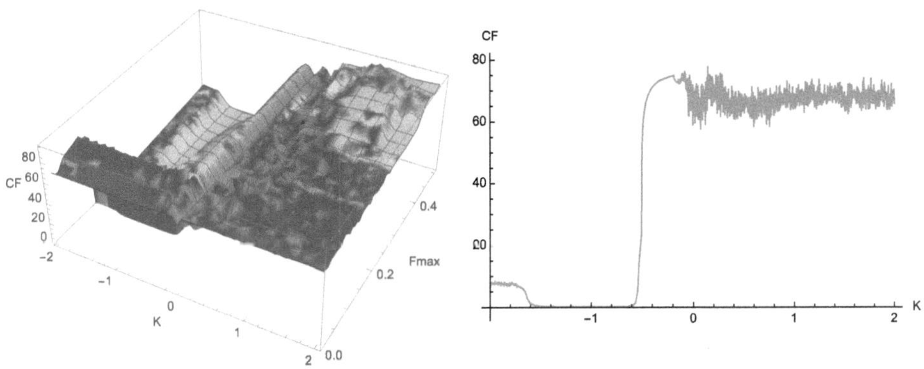

Figure 11. Examples of a CF_{Simple} landscapes, stabilization of p-1, parametric optimization of K and F_{max} for Pyragas control method.

work separately with the part of the cost function for stabilization quality and the part for time optimality. Indeed, it is not possible to simply perform the above-mentioned multiplication of CF_{UNI} and (*NI*), since in the case of full stabilization the result would be zero, regardless of the number of steps. It is, therefore, necessary to define a constant *SC* which, with respect to common techniques for converting a multi-criteria problem to a single-criteria problem, will provide approximately the same numerical value (in order) for both parts of the complex cost function. The value of *SC* (6) is computed using *ExpCF*, which is the value of the exponent of the power of 10 of the unpenalized basis part of the cost function (5). To avoid uncertainties associated with the stabilization value returning 0, a small constant is added to the value of the cost function defining the stabilization quality. Such a design also allows the introduction of preferential weights – either for quality or speed.

$$ExpCF = \log_{10}\left(\sum_{t=\tau 1}^{\tau 2} |TS_t - AS_t| + 10^{-15}\right) \tag{5}$$

$$SC = 10^{ExpCF} \tag{6}$$

The final design of targeting CF (CF_{MULTI}) has the form (7). The metaheuristics should find the solutions to secure the fast targeting of the desired behaviour of the system.

$$CF_{MULTI} = (NI \cdot SC) + p + \sum_{t=\tau 1}^{\tau 2} |TS_t - AS_t| \tag{7}$$

The last possible modification is to increase the robustness for a different range of initial conditions to ensure that the found control parameters are not optimal only for the initial conditions used in the optimization. The final CF value is computed as a sum of m repeated simulations for different initial conditions, where (for example) $x_{initial}$ is from the range 0.05–0.95 and uses step 0.1 (i.e. $m = 10$). Consequently, the metaheuristics should find robust solutions securing the fast targeting of the desired behaviour of the system for almost any initial conditions. The CF used for the evolutionary approach is given in (8):

$$CF_{MULTI} = \sum_{1}^{m} \left((NI \cdot SC) + p + \sum_{t=\tau 1}^{\tau 2} |TS_t - AS_t| \right), \tag{8}$$

The issue of pure searching for periodic orbits causes very chaotic, erratic, and discrete-type CF landscapes (see Figure 14). As a sort of compromise between the simple design of a cost function for the p-1 orbit only and the above universal designs requiring knowledge of the desired state (periodic orbit), experiments were performed with functions that should allow the defined order of the periodic orbit to be found. The trade-off is more from a user's point of view since transient chaotic oscillations are still included in the calculation of the cost function. In recent publications, this type of function has been referred to as a 'black box.' Based on the choice of the target orbit, a specific cost function is selected.

In this case, it is not possible to use the simple rule of minimizing the area created by the difference between the required and actual state on the whole simulation interval. It means that this cost function design does not take any numerical target state into consideration, but the selected target behaviour of the system. The proposal is based on the following simple rule. In discrete systems, the iteration $y(n)$ and $y(n + k)$ of output value, where k is the order of unstable periodic orbit, must be the same. The idea is then to minimize the area created by the difference between the n and $n + k$ output iteration on the whole simulation interval, thus at the same time, this proposal of CF should secure fast targeting into the close neighbourhood of desired periodic orbit. Also, it is crucial to include the penalization, which should avoid the finding of solutions, where the stabilization occurs on boundary values (here, for experiments with Henon map $\{-1.0, 1.5\}$) or oscillation between them occurs. This penalization was calculated as the sum of the number of iterations, where the system output reaches the saturation boundary value. The CF_1 for stabilization on p-1 orbit has form (9).

$$CF_{BB1} = p_{BB} + \sum_{t=0}^{n} |diff| \tag{9}$$

where: $p_{BB} = penalization$, $diff = y_{t+k} - y_t$, and $k =$ order of desired orbit.

Table 1. Simulation results characteristics for different CF types.

CF	Advantages	Disadvantages
CF_{Simple} (3)	Wide global extreme area, simple and fast evolutionary process.	Suitable only for p-1 orbit optimization case, no constraints or penalizations could be added.
CF_{UNI} (4)	Universal CF suitable for any UPO, arbitrary constraints or penalizations could be simply added.	High nonlinearity and chaotic nature of CF.
CF_{MULTI} (7)	Universal CF suitable for any UPO with optimization of time required for full stabilization	Extremely high nonlinearity and chaotic (discrete) nature of CF.
CF_{BB} (9)	Favorable CF landscape, no required knowledge of periodic orbit position	Naive approach only for periodic orbits of $k = 1$–5, since p-6 already consists possible rules for p-2 and p-3

In the case of higher periodic orbits, other simple numerical criteria had to be implemented. It defines that in the case of p-2 orbit, there must be some (bigger) difference between the selected output iterations. Considering the fact of minimizing the *CF* value, the numerical criterion in (9) had to be rewritten into the suitable form (10), which is an example for p-2 orbit, and c – is a small constant 1.10^{-16} which was added to prevent the evolutionary optimization from crashing, since upon finding the possible solution stabilized at p-1 orbit it returns the division by zero.

$$\frac{1}{\left|y_{t+1} - y_t\right| + c} \tag{10}$$

The example of cost function for p-2 orbit CF_2 has the form (11), and p-4 orbit has the form (12).

$$CF_{BB2} = p_{BB} + \sum_{t=0}^{n} \left|diff\right| + \frac{1}{\left|y_{t+1} - y_t\right| + c} \tag{11}$$

$$CF_{BB4} = p_{BB} + \sum_{t=0}^{n} \left|diff\right| + \frac{1}{\left|y_{t+3} - y_t\right| + c} \tag{12}$$

The brief overview of the simulation results characteristics given as the advantages/disadvantages for each single cost function design is given in Table 1. The cost functions landscapes (and their 2D or 3D cross-sections) are depicted in Figures 11–16. These figures show the dependence of the cost function landscape on the selected optimized parameters of Pyragas control technique (gain constant K and limitation of perturbation F_{max}. CF_{Simple} is shown in Figures 11 and 12 (for p-1 and p-2 orbit), CF_{UNI} in Figure 13, CF_{MULTI} in Figure 14, and finally two Figures 15 and 16 for black-box type cost functions (CF_{BB}), showing the differences between the task of finding the periodic orbit of two different types (p-1 and p-2).

4.3. Example of metaheuristic based control of discrete chaotic maps

This section contains illustrative examples of chaos control optimization results for the Henon map and Pyragas ETDAS technique. Figures 18–21 always show pairs of best-found results and re-runs (100 runs) for different initial conditions.

The results of the experimental simulations confirm the characteristics of the landscapes of the cost functions and their respective definitions. The optimization of the control

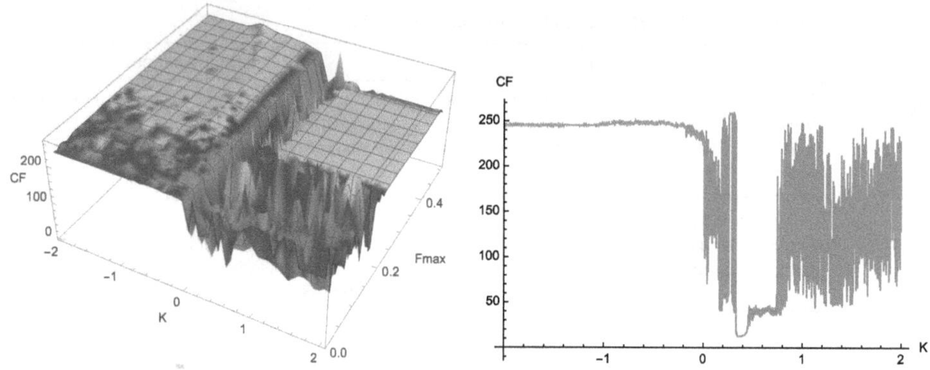

Figure 12. Examples of a CF_{Simple} landscapes, stabilization of p-2, parametric optimization of K and F_{max} for Pyragas control method.

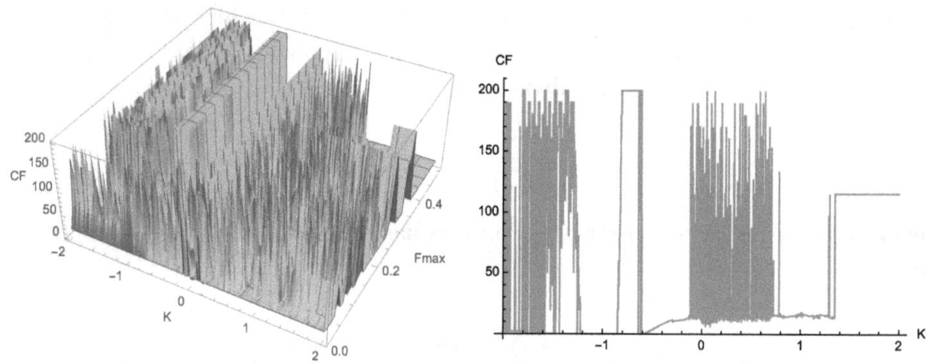

Figure 13. Examples of a CF_{UNI} landscapes, stabilization of p-1, parametric optimization of K and F_{max} for Pyragas control method.

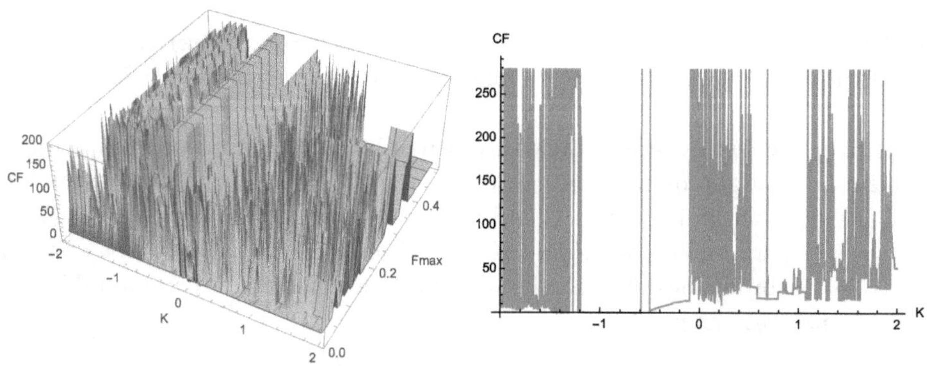

Figure 14. Examples of a CF_{MULTI} landscapes, stabilization of p-1, parametric optimization of K and F_{max} for Pyragas control method.

method parameters using CF_{Simple} (Figure 17) and CF_{UNI} (Figure 18) provides fast stabilization for the same initial conditions that were used for running the metaheuristics. For a broader range of initial conditions, then, both approaches do not always guarantee

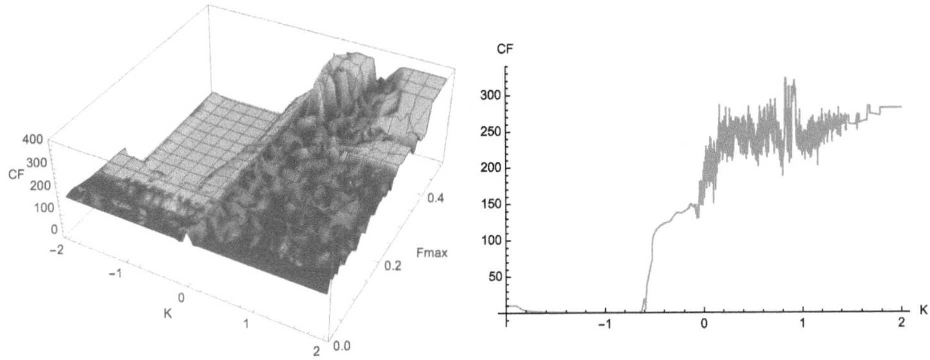

Figure 15. Examples of a CF_{BB1} landscapes, stabilization of p-1, parametric optimization of K and F_{max} for Pyragas control method.

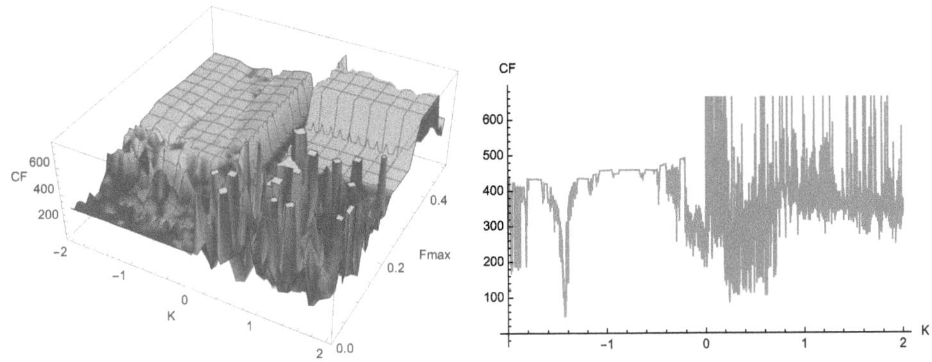

Figure 16. Examples of a CF_{BB2} landscapes, stabilization of p-2, parametric optimization of K and F_{max} for Pyragas control method.

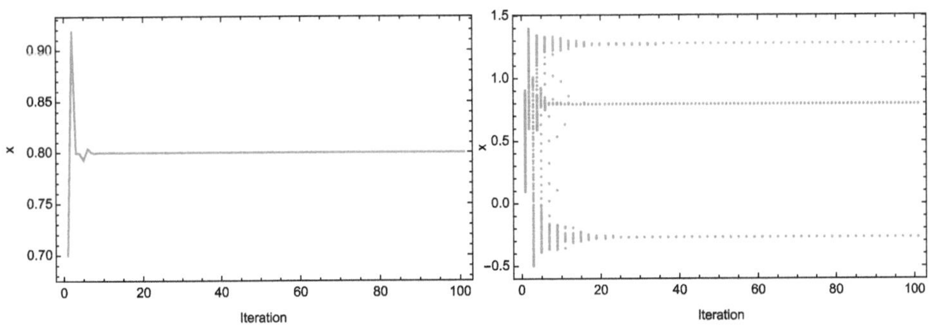

Figure 17. Stabilization of Henon map with evolutionary optimized ETDAS technique, CF_{Simple}.

stabilization, and in the case of CF_{UNI}, the highly chaotic landscape of the cost function and possible overfitting only to specific initial conditions (local optimum in an otherwise complex *CF* surface) are also manifest. The more computationally demanding CF_{MULTI}, then, ensures finding a robust combination of control parameters (Figure 19). The results of repeated simulations for different initial conditions and black box-type cost functions

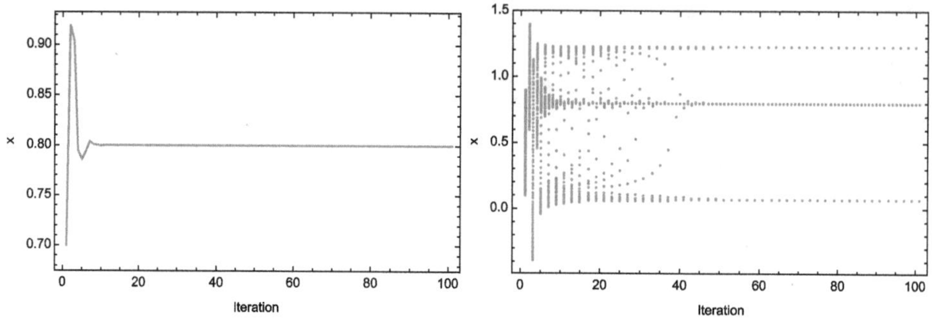

Figure 18. Stabilization of Henon map with evolutionary optimized ETDAS technique, CF_{UNI}.

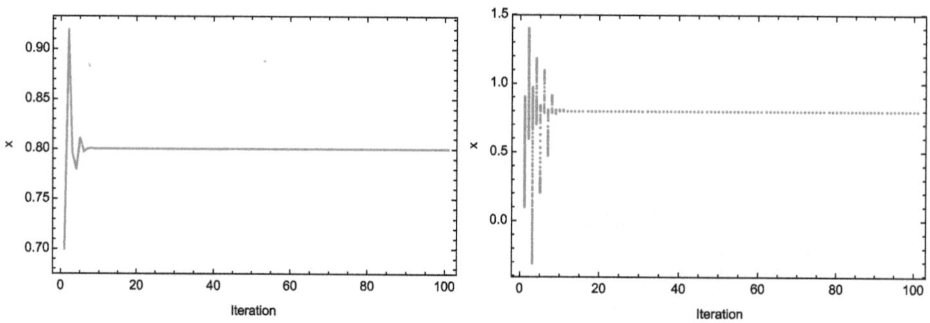

Figure 19. Stabilization of Henon map with evolutionary optimized ETDAS technique, CF_{MULTI}.

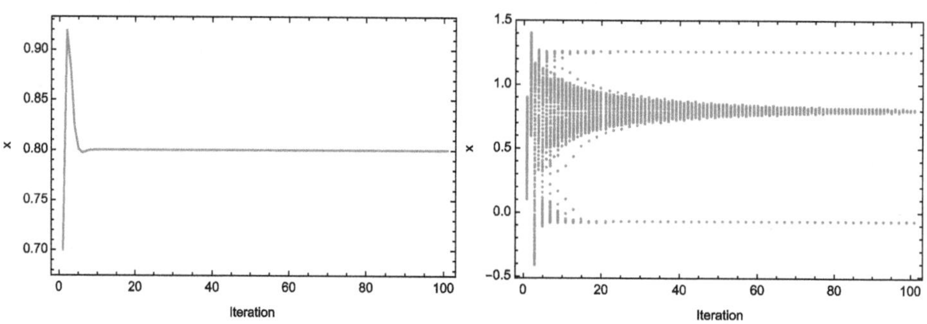

Figure 20. Stabilization of Henon map with evolutionary optimized ETDAS technique, CF_{BB1}.

depicted in Figures 20–21, in turn, then show that simply designed to find a particular type of system behaviour may not be the best solution for this particular case, despite the favourable cost function landscape for metaheuristic optimization. In essence, the results confirm that when metaheuristics are used to control complex (nonlinear) systems, there must be a mutually appropriate fit between the definition of the cost function and the properties of its landscape.

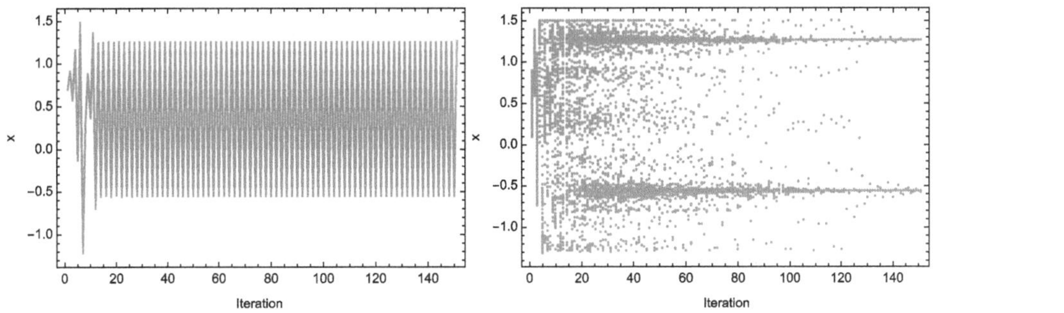

Figure 21. Stabilization of Henon map with evolutionary optimized ETDAS technique, CF_{BB2}.

4.4. Control of a real chaotic system – plasma reactor control

The evolutionary algorithms were also used to control the plasma reactor in real time. Specifically, this involved active noise compensation from the signal using evolutionary techniques. This signal is generated by a Langmuir probe that was inserted in the plasma reactor. It is a diagnostic probe used to analyze the energy distribution in plasma processes. Plasmas in general, especially plasmas driven by radio frequencies, are intrinsically nonlinear. Strong nonlinear potential waveforms around the probe embedded in the plasma, strongly distort the useful signal sensed by the probe. The cost of the research was to use evolutionary computation techniques to synthesize seven harmonic signals, with two parameters (frequency and amplitude). Thus, a total of 14 parameters have been optimized. The aforementioned seven harmonic signals were used to synthesize the resulting signal, which was intended to compensate for interference in the optimal case. The results from the experiments were published in [36].

4.5. Control of spatiotemporal chaos

Evolutionary computation techniques can handle more complex problems, such as even the stabilization of spatiotemporal chaos [64], as demonstrated in Figures 23–25. The cost function design was very similar to CF_{Simple} (3) described in section 4.2, where the feedback pinning value has been optimized based on the difference between an actual spatiotemporal chaotic system and desired state. In Figure 22 we can see an example of the landscape of the cost function, where each point of this curve represents the quality of the control intervention. The goal of the evolutionary techniques was to find the lowest point on this landscape. Of course, this is a very trivial demonstration, but the search space is still very complex. The research for example in [45] describes in detail the background and results in analysis.

4.6. Symbolic regression for chaos control

Another challenging task for evolutionary computation is undoubtedly symbolic regression in the feedback control rule synthesis problem. In recent publications, this approach is referred to as 'metaevolution.' The control technique (rules) can be viewed as a symbolic structure that can be synthesized according to the stabilization requirements of the chaotic

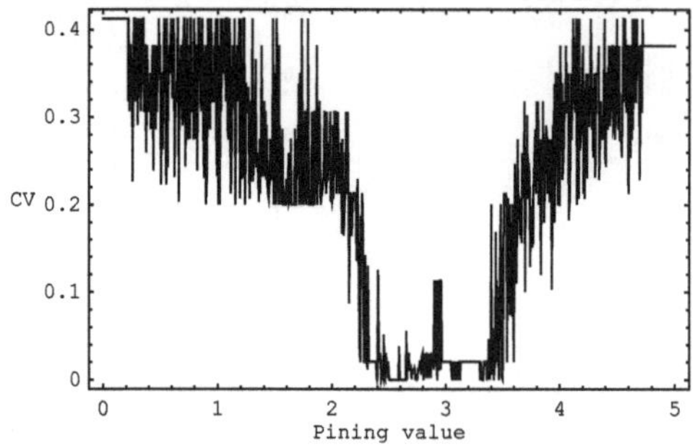

Figure 22. The CF landscape of CML chaos control by EA. The best control value (pinning) is located in the deepest position.

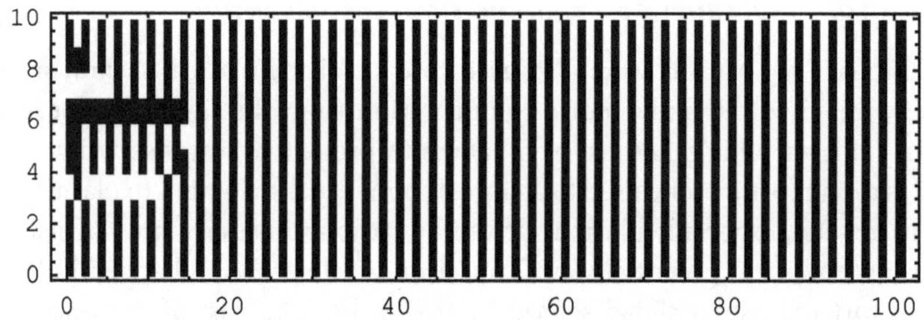

Figure 23. Stabilization and control of the spatiotemporal – CML chaotic system by EA. The *x* axis are iterations (time), *y* axis number of input sites. Two time and one spatial period stabilization are visible.

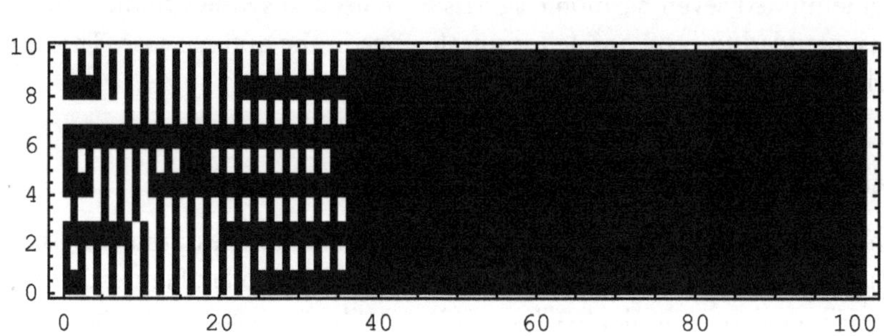

Figure 24. Stabilization and control of the spatiotemporal – CML chaotic system by EA. The *x* axis are iterations, *y* axis number of input sites. A stable state in time and space is achieved here (the black part).

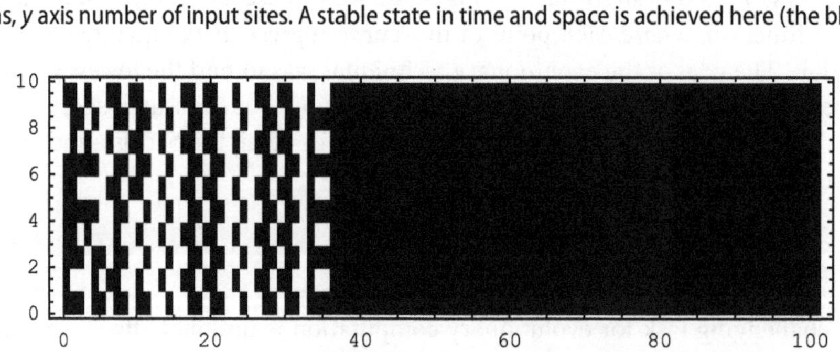

Figure 25. Stabilization and control of the spatiotemporal – CML chaotic system by EA. The *x* axis are iterations, *y* axis number of input sites. A stable state in time and space is achieved here (the black part).

system. The advantage is that it is not necessary to focus only on the 'common' state-of-the-art technique and estimate its optimal parameters. Using symbolic regression methods, the entire complex structure of the control method is generated, along with appropriate values of possible parameters. In addition, very specific constraints can be chosen, such as robust stabilization for selected types of chaotic systems, their regimes and initial conditions, or preferences for time, quality, and conditions for achieving stabilization. The disadvantage is of course the increase in computational complexity. More detail on this is written in the following papers [37, 47, 48].

4.7. Future challenges and possibilities

Research opportunities in the field of metaheuristically based control optimization, synthesis, or synchronization of chaotic systems are certainly open for the future. Experiments with real-world problems have shown that simpler and more strongly stochastic algorithm strategies are preferable for the aforementioned tasks. The available high-performance computing capabilities then allow the use of symbolic regression optimization methods and overall more complex cost function designs. The full potential of multi-criteria optimization can also be exploited thanks to the available multi-criteria evolutionary frameworks. For the metaheuristic algorithms themselves, it is then possible to use a number of new autoconfiguration frameworks like iRace [29] or SMAC [25], which makes it unnecessary to laboriously search for metaheuristic hyperparameter settings.

For a more detailed description and more results regarding the interconnection between deterministic chaos and evolutionary algorithms (or any general metaheuristics), we recommend as a starting reference [48, 68], where a large number of references to other works of similar type can be found.

5. Chaos for evolutionary computation

However, it is also necessary to focus on the evolutionary computational techniques (metaheuristics) themselves. In the last decade, it has become very popular to hybridize chaos and metaheuristics through the use of unconventional randomization schemes. Since stochastic processes are the basis of functionality in heuristic algorithms, chaotic systems are used as pseudo-random generators influencing, through their complex dynamics, e.g. the selection of individuals for the crossover process, the size of the mutation, the probabilities of performing other specific operations, and the actual movement on the hyperplane of possible solutions and the decision for the next steps. Examples of successful applications can be found [34, 49], benchmarking studies [57, 71], and exploring the impact of chaotic dynamics on population diversity or from the perspective of different types of chaotic systems [50].

6. Conclusion

This paper summarizes the intersection of two areas, namely deterministic chaos and so-called biologically inspired algorithms, or evolutionary computational techniques, which fall into the group of metaheuristic algorithms, and are a very important part of today's modern computer science and computational intelligence. This paper summarized two different aspects of the fusion of chaos and evolutionary computational techniques. First, the occurrence of chaos in evolutionary algorithms and swarm intelligence itself was discussed, and research opportunities are found there. In the second, the possibilities of using evolutionary computational techniques to synthesize, identify, or control chaotic systems were described, both as simulations of discrete chaotic maps and as control of real chaotic systems. Individual case studies are always briefly described, supported by graphical outputs, and possibilities for further research are discussed. In particular, in the area of evolutionary control of chaotic systems, the design of cost functions for optimization, the implications of their construction on the landscape, and the possibilities of search by evolutionary algorithms are described in detail. Given the continuous increase in computing power, the availability of parallel computing architectures, self-configuration frameworks for the hyperparameters of the metaheuristics used, and implementations of multi-criteria algorithms, it can be argued without a doubt that this survey paper can provide a solid foundation for further significant advances and significantly better results in the effective deployment of evolutionary computational techniques in research questions and applications related to deterministic chaos.

Disclosure statement

No potential conflict of interest was reported by the author(s).

Funding

Supported by a grant of SGS No. SP2023/050, VSB-Technical University of Ostrava, Czech Republic, by the Grant Agency of Czech Republic (GACR) project No. GF21-45465L, and further by the resources of A.I.Lab at the Faculty of Applied Informatics, Tomas Bata University in Zlin.

References

[1] T. Alexeeva, Q. Diep, N. Kuznetsov, T. Mokaev, and I. Zelinka, *Forecasting and control in overlapping generations model: Chaos stabilization via artificial intelligence*, preprint (2022). Available at arXiv, 2208.06345.

[2] J. Alvarez-Ramirez, H. Puebla, and I. Cervantes, *Stability of observer-based chaotic communications for a class of lur'e systems*, Int. J. Bifurcat. Chaos 12 (2002), pp. 1605–1618.

[3] L. Arnold and V. Wihstutz, *Lyapunov exponents: A survey*, in *Lyapunov Exponents: Proceedings of a Workshop held in Bremen, November 12–15, 1984*, Springer, 2006, pp. 1–26.

[4] A.T. Azar and S. Vaidyanathan, *Advances in Chaos Theory and Intelligent Control*, Vol. 337, Springer, 2016.

[5] T. Bäck and H.P. Schwefel, *An overview of evolutionary algorithms for parameter optimization*, Evol. Comput. 1 (1993), pp. 1–23.

[6] H.G. Beyer and H.P. Schwefel, *Evolution strategies–a comprehensive introduction*, Nat. Comput. 1 (2002), pp. 3–52.

[7] A. Byrski, E. Świderska, J. Łasisz, M. Kisiel-Dorohinicki, T. Lenaerts, D. Samson, B. Indurkhya, and A. Nowé, *Socio-cognitively inspired ant colony optimization*, J. Comput. Sci. 21 (2017), pp. 397–406.

[8] G. Chen, *Controlling Chaos and Bifurcations in Engineering Systems*, CRC press, 1999.

[9] G. Chen and T. Ueta, *Yet another chaotic attractor*, Int. J. Bifurcat. Chaos 9 (1999), pp. 1465–1466.

[10] K.M. Cuomo, *Synthesizing self-synchronizing chaotic systems*, Int. J. Bifurcat. Chaos 3 (1993), pp. 1327–1337.

[11] S. Das, S.S. Mullick, and P.N. Suganthan, *Recent advances in differential evolution–an updated survey*, Swarm. Evol. Comput. 27 (2016), pp. 1–30.
[12] A. Dmitriev, A. Panas, and S. Starkov, *Ring oscillating systems and their application to the synthesis of chaos generators*, Int. J. Bifurcat. Chaos 6 (1996), pp. 851–865.
[13] A. Dmitriev, E. Efremova, L. Kuzmin, and A. Anagnostopoulos, *High dimensional RC–oscillators of chaos*, in *Proc. of Int. Symp. NOLTA*, Miyagi, Japan, Oct 28–Nov 1, 2001, pp. 139–142.
[14] M. Dorigo, V. Maniezzo, and A. Colorni, *Ant system: Optimization by a colony of cooperating agents*, IEEE Trans. Syst. Man Cybern. Part B (Cybern.) 26 (1996), pp. 29–41.
[15] L. dos Santos Coelho and D.L. de Andrade Bernert, *An improved harmony search algorithm for synchronization of discrete-time chaotic systems*, Chaos Solitons Fractals 41 (2009), pp. 2526–2532.
[16] D. Dudkowski, S. Jafari, T. Kapitaniak, N.V. Kuznetsov, G.A. Leonov, and A. Prasad, *Hidden attractors in dynamical systems*, Phys. Rep. 637 (2016), pp. 1–50.
[17] D. Dumitrescu, B. Lazzerini, L.C. Jain, and A. Dumitrescu, *Evolutionary Computation*, CRC press, 2000.
[18] R. Eberhart and J. Kennedy, *A new optimizer using particle swarm theory*, in *MHS'95. Proceedings of the Sixth International Symposium on Micro Machine and Human Science*, IEEE, 1995, pp. 39–43.
[19] K. Eguchi, T. Inoue, and A. Tsuneda, *Synthesis and analysis of a digital chaos circuit generating multiple-scroll strange attractors*, IEICE T. Fund. Electr. Commun. Comput. Sci. 82 (1999), pp. 965–972.
[20] B.I. Epureanu and E.H. Dowell, *On the optimality of the ott-grebogi-yorke control scheme*, Phys. D: Nonlinear Phenomena 116 (1998), pp. 1–7.
[21] A.E. Ezugwu, A.K. Shukla, R. Nath, A.A. Akinyelu, J.O. Agushaka, H. Chiroma, and P.K. Muhuri, *Metaheuristics: A comprehensive overview and classification along with bibliometric analysis*, Artif. Intell. Rev. 54 (2021), pp. 4237–4316.
[22] G. Feng and G. Chen, *Adaptive control of discrete-time chaotic systems: A fuzzy control approach*, Chaos Solitons Fractals 23 (2005), pp. 459–467.
[23] F. Glover, *Future paths for integer programming and links to artificial intelligence*, Comput. Oper. Res. 13 (1986), pp. 533–549.
[24] J.H. Holland, *Adaptation in Natural and Artificial Systems: An Introductory Analysis with Applications to Biology, Control, and Artificial Intelligence*, MIT press, 1992.
[25] F. Hutter, H.H. Hoos, and K. Leyton-Brown, *Sequential model-based optimization for general algorithm configuration*, in *International Conference on Learning and Intelligent optimization*, Springer, 2011, pp. 507–523.
[26] D. Karaboga, *An idea based on honey bee swarm for numerical optimization*, Tech. Rep., Technical report-tr06, Erciyes university, engineering faculty, computer ..., 2005
[27] J.R. Koza, *Genetic programming as a means for programming computers by natural selection*, Stat. Comput. 4 (1994), pp. 87–112.
[28] N. Kuznetsov, G. Leonov, and M. Shumafov, *A short survey on pyragas time-delay feedback stabilization and odd number limitation*, IFAC-PapersOnLine 48 (2015), pp. 706–709.
[29] M. López-Ibáñez, J. Dubois-Lacoste, L.P. Cáceres, M. Birattari, and T. Stützle, *The irace package: Iterated racing for automatic algorithm configuration*, Oper. Res. Perspect. 3 (2016), pp. 43–58.
[30] M. Lukoševičius, *A practical guide to applying echo state networks*, in *Neural Networks: Tricks of the Trade*, 2nd ed., Springer, Berlin, 2012, pp. 659–686.
[31] R. Matousek and T. Hulka, *Stabilization of higher periodic orbits of the chaotic logistic and hénon maps using meta-evolutionary approaches*, in *2019 IEEE Congress on Evolutionary Computation (CEC)*, IEEE, 2019, pp. 1758–1765.
[32] R. Matousek, L. Dobrovsky, P. Minar, and K. Mouralova, *A note about robust stabilization of chaotic hénon system using grammatical evolution*, in *Proceedings of the 2014 International Nostradamus conference on modern methods of prediction, modeling and analysis of nonlinear systems*, June 23–25, 2014, Ostrava, Czech Republic.
[33] L. Medsker and L.C. Jain, *Recurrent Neural Networks: Design and Applications*, CRC press, 1999.

[34] M. Metlicka and D. Davendra, *Chaos driven discrete artificial bee algorithm for location and assignment optimisation problems*, Swarm. Evol. Comput. 25 (2015), pp. 15–28.

[35] S. Mirjalili, S.M. Mirjalili, and A. Lewis, *Grey wolf optimizer*, Adv. Eng. Softw. 69 (2014), pp. 46–61.

[36] L. Nolle, I. Zelinka, A.A. Hopgood, and A. Goodyear, *Comparison of an self-organizing migration algorithm with simulated annealing and differential evolution for automated waveform tuning*, Adv. Eng. Softw. 36 (2005), pp. 645–653.

[37] Z.K. Oplatkova, R. Senkerik, I. Zelinka, and M. Pluhacek, *Analytic programming in the task of evolutionary synthesis of a controller for high order oscillations stabilization of discrete chaotic systems*, Comput. Math. Appl. 66 (2013), pp. 177–189.

[38] E. Ott, C. Grebogi, and J.A. Yorke, *Controlling chaos*, Phys. Rev. Lett. 64 (1990), pp. 1196.

[39] H. Poincaré, *On the problem of three bodies and equations of dynamics*, Acta Math. 13 (1890), pp. A3–A270.

[40] K. Pyragas, *Delayed feedback control of chaos*, Philos. Trans. R. Soc. A: Math. Phys. Eng. Sci. 364 (2006), pp. 2309–2334.

[41] E. Rashedi, H. Nezamabadi-Pour, and S. Saryazdi, *Gsa: A gravitational search algorithm*, Inf. Sci. (Ny) 179 (2009), pp. 2232–2248.

[42] H. Richter and K.J. Reinschke, *Optimization of local control of chaos by an evolutionary algorithm*, Phys. D: Nonlinear Phenom. 144 (2000), pp. 309–334.

[43] H. Richter, *An evolutionary algorithm for controlling chaos: The use of multi-objective fitness functions*, in *International Conference on Parallel Problem Solving from Nature*, Springer, 2002, pp. 308–317.

[44] C. Ryan, J.J. Collins, and M.O. Neill, *Grammatical evolution: Evolving programs for an arbitrary language*, in *European Conference on Genetic Programming*, Springer, 1998, pp. 83–96.

[45] E. Schöll and H.G. Schuster (eds.), *Handbook of Chaos Control*, 2nd ed., WILEY-VCH, Weinheim, 2008.

[46] R. Senkerik, I. Zelinka, D. Davendra, and Z. Oplatkova, *Evolutionary optimisation of hénon map control: A black box approach*, Int. J. Oper. Res. 13 (2012), pp. 129–146.

[47] R. Senkerik, Z. Oplatkova, I. Zelinka, and D. Davendra, *Synthesis of feedback controller for three selected chaotic systems by means of evolutionary techniques: Analytic programming*, Math. Comput. Model. 57 (2013), pp. 57–67.

[48] R. Senkerik, Z. Kominkova Oplatkova, I. Zelinka, B. Chramcov, D.D. Davendra, and M. Pluhacek, *Utilization of analytic programming for the evolutionary synthesis of the robust multi-chaotic controller for selected sets of discrete chaotic systems*, Soft. Comput. 18 (2014), pp. 651–668.

[49] R. Senkerik, M. Pluhacek, I. Zelinka, D. Davendra, and Z. Kominkova Oplatkova, *Comparison of chaos driven PSO and differential evolution on the selected PID tuning problem*, in *IFIP International Conference on Computer Information Systems and Industrial Management*, Springer, 2015, pp. 67–76.

[50] R. Senkerik, A. Viktorin, M. Pluhacek, and T. Kadavy, *On the population diversity for the chaotic differential evolution*, in *2018 IEEE Congress on Evolutionary Computation (CEC)*, IEEE, 2018, pp. 1–8.

[51] R. Storn and K. Price, *Differential evolution–a simple and efficient heuristic for global optimization over continuous spaces*, J. Global Optim. 11 (1997), pp. 341–359.

[52] R. Thom, *tabilité structurelle et morphogénèse: essai d'une théorie générale des modèles*, Mathematical physics monograph series 17 (1972), p. 362.

[53] T. Ueta and G. Chen, *Bifurcation analysis of chen's equation*, Int. J. Bifurcat. Chaos 10 (2000), pp. 1917–1931.

[54] T. Ushio, *Synthesis of synchronized chaotic systems based on observers*, Int. J. Bifurcat. Chaos 9 (1999), pp. 541–546.

[55] A. Vaněček and S. Čelikovský, *Control Systems: From Linear Analysis to Synthesis of Chaos*, Prentice Hall International (UK) Ltd., 1996.

[56] Z. Vasicek and L. Sekanina, *On area minimization of complex combinational circuits using cartesian genetic programming*, in *2012 IEEE Congress on Evolutionary Computation*, IEEE, 2012, pp. 1–8.
[57] A. Viktorin, M. Pluhacek, and R. Senkerik, *Success-history based adaptive differential evolution algorithm with multi-chaotic framework for parent selection performance on CEC2014 benchmark set*, in *2016 IEEE Congress on Evolutionary Computation (CEC)*, IEEE, 2016, pp. 4797–4803.
[58] M.D. Vose, The Simple Genetic Algorithm: Foundations and Theory, MIT press, 1999.
[59] M.D. Vose and G.E. Liepins, *Punctuated equilibria in genetic search*, Complex Syst. 5 (1991), pp. 31–44.
[60] M.D. Vose and A.H. Wright, *Simple genetic algorithms with linear fitness*, Evol. Comput. 2 (1994), pp. 347–368.
[61] A.H. Wright and A. Agapie, *Cyclic and chaotic behavior in genetic algorithms*, in *Proceedings of the 3rd Annual Conference on Genetic and Evolutionary Computation*, Citeseer, 2001, pp. 718–724.
[62] S.K. Yang, C.L. Chen, and H.T. Yau, *Control of chaos in lorenz system*, Chaos Solitons Fractals 13 (2002), pp. 767–780.
[63] I. Zelinka, *Investigation on evolutionary deterministic chaos control*, IFAC Proc. Vol. 38 (2005), pp. 1101–1106.
[64] I. Zelinka, *Real-time deterministic chaos control by means of selected evolutionary techniques*, Eng. Appl. Artif. Intell. 22 (2009), pp. 283–297.
[65] I. Zelinka, *Soma-self-organizing migrating algorithm*, in *Self-Organizing Migrating Algorithm*, Springer, Cham, 2016, pp. 3–49.
[66] I. Zelinka, G. Chen, and S. Celikovsky, *Chaos synthesis by means of evolutionary algorithms*, Int. J. Bifurcat. Chaos 18 (2008), pp. 911–942.
[67] I. Zelinka, R. Senkerik, and E. Navratil, *Investigation on evolutionary optimization of chaos control*, Chaos Solitons Fractals 40 (2009), pp. 111–129.
[68] I. Zelinka, S. Celikovský, H. Richter, and G. Chen, *Evolutionary Algorithms and Chaotic Systems*, Vol. 267, Springer, 2010.
[69] I. Zelinka, D. Davendra, R. Senkerik, R. Jasek, and Z. Oplatkova,*Analytical Programming - a Novel Approach for Evolutionary Synthesis of Symbolic Structures*, in *Evolutionary Algorithms*, E. Kita (ed.), InTech, 2011. ISBN: 978-953-307-171-8.
[70] I. Zelinka, M. Chadli, D. Davendra, R. Senkerik, and R. Jasek, *An investigation on evolutionary reconstruction of continuous chaotic systems*, Math. Comput. Model. 57 (2013), pp. 2–15.
[71] I. Zelinka, Q.B. Diep, V. Snášel, S. Das, G. Innocenti, A. Tesi, F. Schoen, and N.V. Kuznetsov, *Impact of chaotic dynamics on the performance of metaheuristic optimization algorithms: An experimental analysis*, Inf. Sci. (Ny) 587 (2022), pp. 692–719.
[72] T. Zhou, G. Chen, and S. Čelikovský, *An algorithm for computing heteroclinic orbits and its application to chaos synthesis in the generalized lorenz system*, IFAC Proc. Vol. 38 (2005), pp. 1079–1084.

Chaos in popular metaheuristic optimizers – a bibliographic analysis

Michal Pluhacek, Anezka Kazikova, Adam Viktorin, Tomas Kadavy and Roman Senkerik

ABSTRACT
This paper presents an overview of the history and recent efforts in combining chaos theory and evolutionary computation techniques. Various algorithms from the evolutionary computation domain, also known as metaheuristic algorithms, have been successfully enhanced with chaotic components in the past. Numerous ways to incorporate chaos have been examined, and many impressive results have been reported. Implementations of discrete chaotic maps such as Lozi, Hénon, and logistic map as generators of chaotic pseudo-random sequences for controlling evolution operators in metaheuristics have achieved significant popularity. In this survey, we focus on the research field itself and perform a bibliographical analysis to show how broad and active is nowadays the research field of chaos-enhanced metaheuristics and what are some of the most recent works published.

1. Introduction

Since its inception, the theory of chaos [19,29] has inspired and influenced researchers in many distant research fields. In this brief survey, we aim to show evidence that the field of evolutionary computing is no exception. We show that, albeit our understanding of reasons is limited, the chaotic sequences empower the evolutionary computation techniques to better performance.

In the past decades, the research and application of metaheuristics, a commonly used designation for nature-inspired heuristic optimizers (also frequently noted as evolutionary algorithms (EAs)), expanded dramatically. Nowadays, finding an engineering optimization problem that some metaheuristic optimizer has not yet been applied to is nearly impossible. In the quest to develop more powerful optimizers, the researchers are exploring various methods of further algorithm performance improvements, e.g. self-adaptivity, learning mechanics, or hybridization.

Among the less conventional approaches for enhancing the performance of metaheuristics is the incorporation of chaos as a source of stochasticity. As a key operation of metaheuristics, stochasticism is typically used for three basic operators – selection, crossover, and mutation. The authors of individual implementations of chaos-driven metaheuristics have the choice to implement chaotic dynamics in any of them, in different combinations, or all of them. The chaos-driven selection operator affects either the selection of the solution to the next steps in the algorithm or the probability of selection. Thanks to self-organization, patterns, and a certain periodicity in the chaotic signal, the formation of subpopulations and self-organization can occur. Crossover and mutation operators can then use the properties mentioned above to transform them into the selection of specific solution dimensions (genes) or their mutation. A particular case is, in a certain sense, influencing the communication between solutions in swarm algorithms, using patterns in a chaotic signal to control the social and cognitive properties of the swarm. In addition, specific modifications using chaos, such as local search algorithms, can also be used to enhance metaheuristics

1.1. A brief history of chaos implementation into metaheuristics

In their pioneering work 'Using Chaos in Genetic Algorithms,' Determan and Foster [8] speculate that chaos might be present in natural evolutionary processes. They design experiments to investigate if chaos implementation into the genetic algorithm (in the form of a mutation operator based on the logistic map) will affect the algorithm's performance in solving several test cases, e.g. the Iterated Prisoner's Dilemma problem. For their test cases, they conclude that 'the use of chaos is beneficial.' A significant performance gain is observed compared to a similar but non-chaotic method.

While the work of Determan and Foster might be considered the first step for chaos and metaheuristics, the article published in 2003 by Caponetto et al. [3] is by many considered the origin of general interest in chaos enhanced metaheuristics. In their work, the authors show that by simply switching the source of pseudo-random numbers for a chaotic sequence, the performance of the genetic algorithm can be improved without any intervention in the algorithm itself. This 'plug-in' property of chaotic sequences is very convenient for practitioners. The authors conclude that 'numerous examples and statistical tests showed an improvement of the EAs when chaotic sequences were used instead of random processes.' While there are many chaotic systems to choose from [29], many researchers typically use only the logistic map as the pseudo-random number generator (PRNG); it is, therefore, worth noting that the authors of [3] compared several additional chaotic systems, namely the tent map, sinusoidal iterator, Gauss map, Lozi map, and Chua's Oscillator.

In 2005, Liu et al. [18] chose perhaps the most popular metaheuristic, Particle Swarm Optimization (PSO), for their experiment with chaos enhancement. In this case, the chaotic sequences were not directly used in the PSO, but an additional component noted, 'chaotic local search,' was designed to improve the algorithm's performance. The logistic map was used as the source of the chaotic sequences, and the authors showed not only improved performance of the PSO (in comparison with several other methods) but, for probably the first time, drew attention to the fact that using chaotic sequences often leads to speedup of the metaheuristics.

Coelho and Lopes used the logistic map in 2006 [9] to improve the performance of the Differential Evolution (DE) algorithm with chaos-based mutation. DE is recognized as one of the best-performing general-use metaheuristics. For the given scenario, the newly developed chaotic DE showed superior performance and inspired several follow-up studies. In 2009, Coelho and Mariani used the normalized output of the Hénon map to alter the convergence behaviour of the PSO algorithm [10]. In their approach, the control parameters affecting the particle trajectory toward points of interest are generated based on the output of the chaotic map.

Davendra et al. in 2010 [7] further confirmed the observations of Caponetto et al. [3] by using chaotic sequences as a source of pseudo-random numbers for DE and Self-organizing Migrating Algorithm (SOMA), When applied to the task of PID (proportional–integral–derivative) controller design, the chaotic variants of SOMA and DE outperformed the compared methods. The main contribution of this work lies in further cementing the fact that performance improvement can be achieved without introducing any special operators or additional modifications but by simply changing the canonical PRNG for a chaotic sequence. In this case, the Lozi map served as the chaotic PRNG (CPRNG).

The work of Davendra et al. was further expanded by Pluhacek et al. in [26]. Firstly, six different CPRNGs based on six two-dimensional chaotic maps (Lozi, burgers, dissipative, Arnold's cat, Sinai, and Tinkerbell) were tested as CPRNGs for the PSO algorithm with global topology. The authors conclude that various CPRNGs lead to different convergence behaviour, with some CPRNGs promoting fast exploitation while others promoting exploratory behaviour of the algorithm. These observations led to the idea that combining two different CPRNGs within one run of the algorithm might bring further performance improvement. In this case, the CPRNGs were switched when the algorithm failed to improve over a certain number of iterations. The results showed that switching the CPRNGs leads to renewed algorithm convergence and helps it escape local optima. In 2015 Pluhacek et al. further generalized this approach using the ensemble learning technique to let each particle decide which CPRNG to use at which optimisation phase [25].

The above-presented brief introduction to the main research steps in the field of chaos-enhanced metaheuristics is summarized in the following Table 1. It can be observed that in a mere 15 years, the field of metaheuristics has evolved from the first experiments with chaos in GA towards various chaos implementations and operators in DE, PSO, or SOMA and further developing into simultaneous use of multiple chaotic systems. Each of these iterative research steps has inspired the subsequent steps and numerous other follow-up studies and re-implementations of proposed methods into other metaheuristics.

Of course, the question arises as to what advantages and disadvantages chaotic dynamics in metaheuristics can actually bring. However, our goal is not to present a technical study mapping the plethora of implementation possibilities. The advantages and disadvantages themselves also would have to be classified according to the contribution of chaos to the performance of metaheuristics, its efficiency, impact on the observed features of the solutions population, and adaptivity of the hyperparameters. Last but not least, different algorithmic and programming languages implementations, strategies and the dependence of the performance and observed features of the metaheuristic on the instance of the problem being solved itself should be considered (i.e. famous No Free Lunch Theorem – NFL [33]).

Table 1. Chaos-enhanced metaheuristics – main research steps timeline.

Year	Title	Chaos implementation	Chaotic maps
1999	Using chaos in genetic algorithm	chaotic mutation	logistic
2003	Chaotic sequences to improve the performance of evolutionary algorithms	CPRNG	Lozi, tent, sinusoidal, Gauss, Chua's
2005	Improved particle swarm optimization combined with chaos	chaotic local search	logistic
2006	Supply Chain Optimization Using Chaotic Differential Evolution Method	chaotic mutation	logistic
2009	A novel chaotic particle swarm optimization approach using Hénon map and implicit filtering local search for economic load dispatch	CPRNG	Hénon
2010	Chaos driven evolutionary algorithms for the task of PID control	CPRNG	Lozi
2014	Particle swarm optimization algorithm driven by multichaotic number generator	CPRNG	Lozi, burgers, dissipative, Arnold's cat, Sinai, Tinkerbell

1.2. The scope of this survey

Nowadays, there are more than 300 metaheuristic algorithms [22], many belonging to the class of swarm-based methods. A number of them have been in recent years subject to chaos implementation, resulting in the introduction of many algorithms such as the chaotic whale optimization [16], firefly algorithm with chaos [12], chaotic bat algorithm [13], butterfly optimization algorithm with chaos [1], chaotic dragonfly algorithm [28], or chaotic vortex search algorithm [15], to name a few. Many of the novel metaheuristics have been criticized for lack of novelty and other fundamental issues [30]. Due to that and mainly to keep this survey in reasonable length, we elect to investigate further only three algorithms, probably the most popular and well-established metaheuristics: the genetic algorithm, differential evolution, and particle swarm optimization.

The rest of the paper is structured as follows: The next section presents an extensive bibliographical analysis, focusing on the quantitative indicators of research interest regarding the chaos-enhanced metaheuristics and community activity and outreach. The following section details some of the most recent works in the field, highlighting the sustained interest of the scientific community in combining chaos and metaheuristics. The performance of the chaos-enhanced metaheuristics is briefly discussed in the following section.

2. Bibliographical analysis

As presented in the previous section, the beginnings of chaos implementation into metaheuristics are tied to the genetic algorithm. Performing a simple combined search for the terms 'genetic algorithm' and 'chaos' in the ISI Web of Science database[1] returns numerous results, as exampled in Figure 1 that covers the years 1999–2020. Given the time required for the indexation of various publications, the year 2020 is (at the time of writing this article) the last with a conclusive number of publications indexed in the database. The dotted line in the graph represents the moving average. Similar graphs for two other well-known

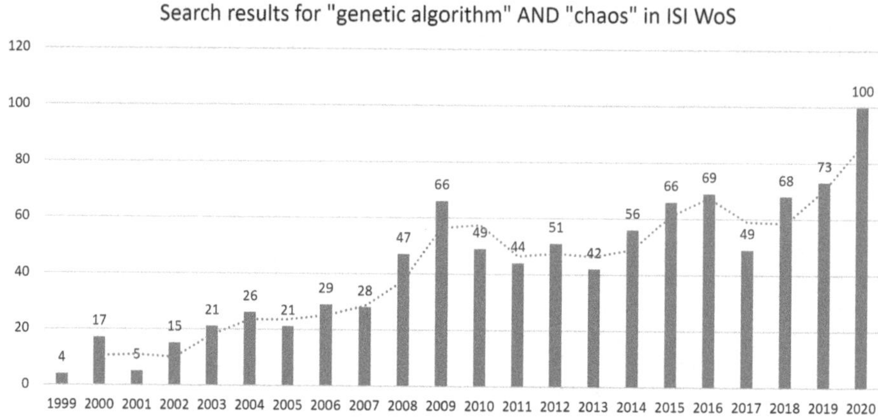

Figure 1. Search results for 'genetic algorithm' AND 'chaos' in ISI WoS.

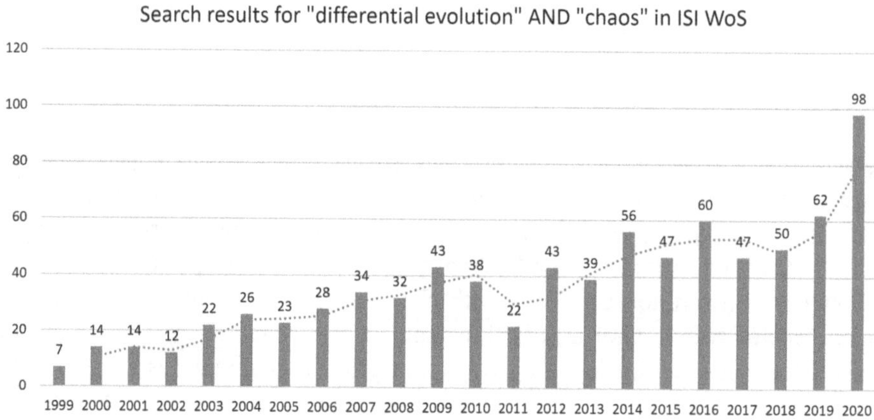

Figure 2. Search results for 'differential evolution' AND 'chaos' in ISI WoS.

and popular metaheuristics, the Differential evolution and Particle swarm optimization, are given in Figures 2 and 3.

It is easy to spot that the research community's interest in chaos and various metaheuristics peaked for the first time around 2009, and another increase in interest was noticeable around 2014–2016. Finally, 2020 shows by far the highest number of published papers. Interestingly enough, the dynamic of new publications seems to be algorithm-dependent. As highlighted by direct comparison in Figure 4, the GA and DE show quite similar publication activity and increased publication rate. On the other hand, PSO shows a more dynamic increase, and while the interest in combining PSO and chaos started several years later, the PSO boasts the highest number of publications in 2020.

One explanation for the results reported above might be that the basic search is too broad and reflects more the general increase in the number of scientific papers published rather than increasing interest in chaos implementation. To investigate further, we performed additional searches, focusing on 'Authors keywords' as defined in ISI WoS. The results of this altered search are given in Figure 5 for GA, Figure 6 for DE, and Figure 7 for

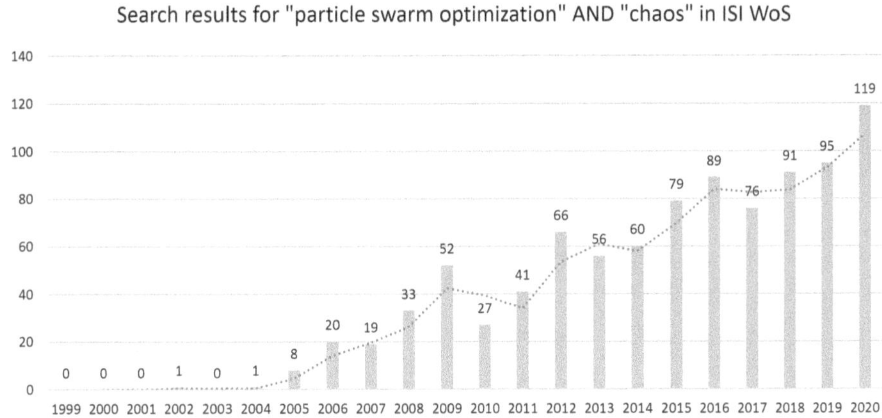

Figure 3. Search results for 'particle swarm optimization' AND 'chaos' in ISI WoS.

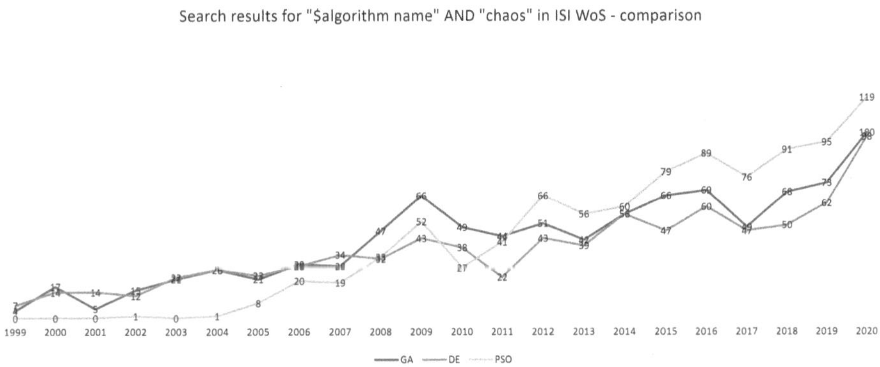

Figure 4. Search results comparison.

PSO. The comparison is shown in Figure 8. Please note that non-empty search results for the keywords are available only from 2004.

At first glance, it is clear that the volume of published works is much lower according to the keywords search. While some trends persist, e.g. the peak of 'interest' around 2009, some are surprisingly inconsistent. Namely, the number of search results for DE is diminishing steeply towards the year 2020. Further, the results for PSO are at a maximum in 2012 rather than 2020, where the interest is stagnating rather than increasing. The peak of interest in chaos and GA around 2009 is much more profound in this search than in the first basic search. The year 2017 seems to represent an important milestone for all three investigated algorithms, with a significant drop in search results. However, while GA and PSO seem to recover from this drop, the interest in chaos and DE seems to drop altogether. These observations are particularly noticeable in the comparative Figure 8 that illustrates the different trajectories of publication activity regarding different algorithms. In addition, in Table 2, we present the document type overview, as summed up for the keywords search results. Somewhat surprisingly, the PSO shows an unprecedented number of journal articles in comparison with DE and GA, where only about one-third of papers concerning DE and chaos (according to the keyword search) are published in journals; it is more than

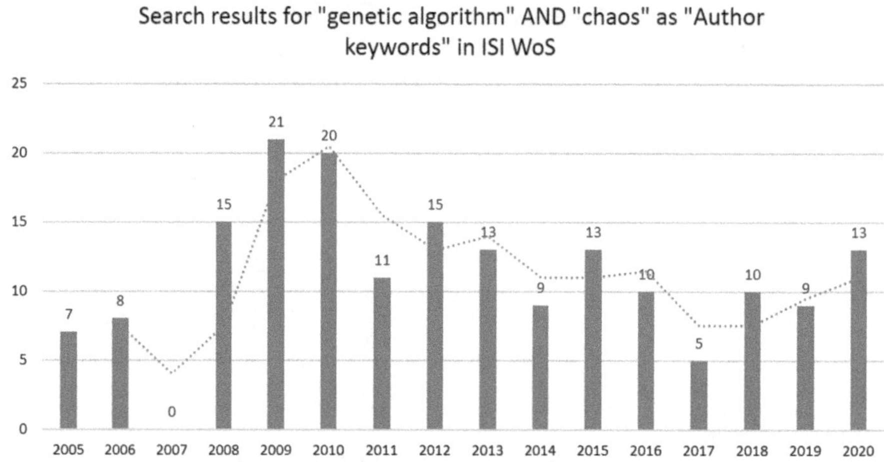

Figure 5. Search results for 'genetic algorithm' AND 'chaos' as 'Author keywords' in ISI WoS.

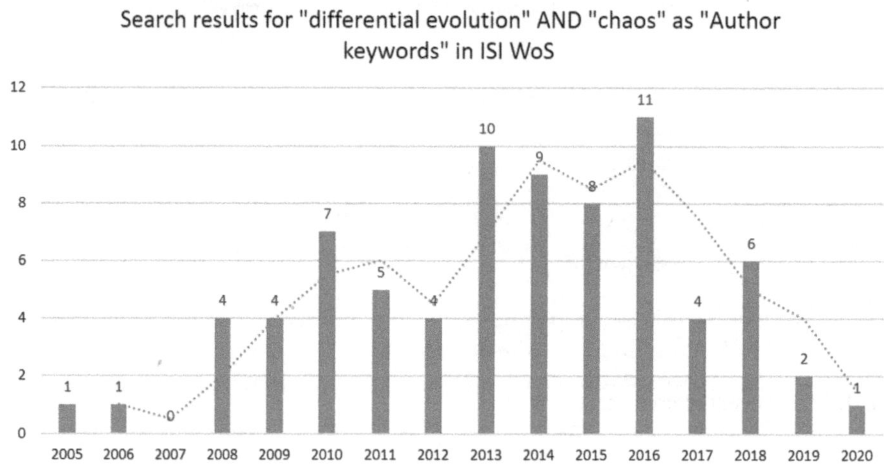

Figure 6. Search results for 'differential evolution' AND 'chaos' as 'Author keywords' in ISI WoS.

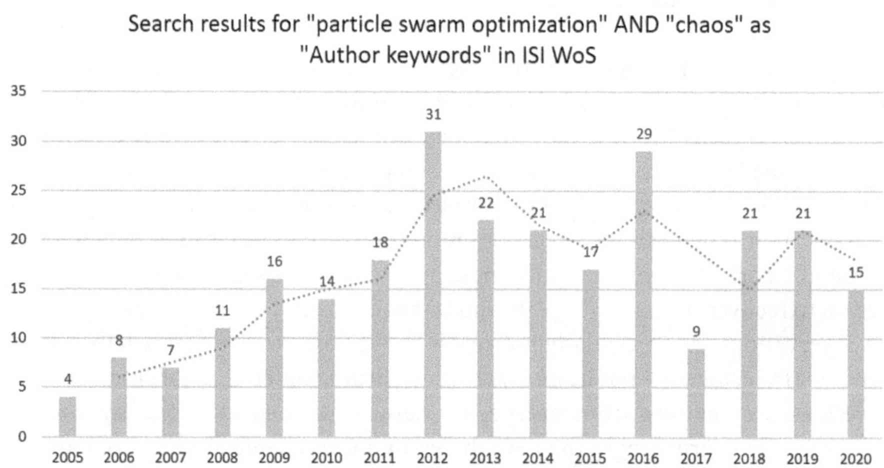

Figure 7. Search results for 'particle swarm optimization' AND 'chaos' as 'Author keywords' in ISI WoS.

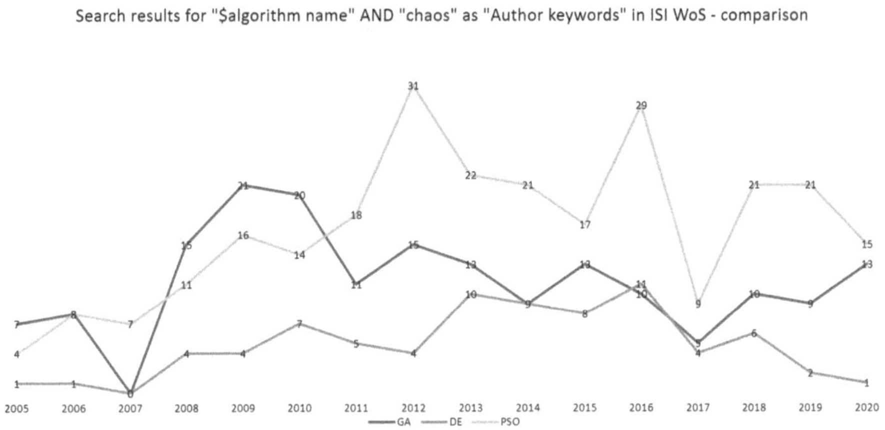

Figure 8. Search results as 'Author keywords' comparison.

Table 2. Document type in keywords search results in ISI WoS.

	No. results			% of total	
Algorithm	Article	Proceeding paper	Total[1]	Article	Proceeding paper
GA	75	105	179[2]	41.899	58.659
DE	28	53	77	36.364	68.831
PSO	144	125	264[3]	54.545	47.348

[1]Some publications have both article and proceeding paper tags assigned to them in the database.[2]The search for GA returned two book chapters in addition.[3]The search for PSO returned two additional book chapters, one early access article and one retracted publication.

half for PSO, Given that, (in general) journal publications tend to take more time to publish over conference papers, we might see this as one of the reasons for the slight delay in publications quantity increase for PSO in comparison to GA and to some extent DE.

It is important to stress that neither the basic search nor the keywords search can fully represent the number of papers concerning the implementation of chaos into a particular metaheuristic. The only way to acquire such exact results would be by reading and manually cataloguing tens of thousands of research papers, which is well beyond the scope of this survey. The data presented in this study are meant to highlight specific general trends and give the reader a broad idea about the significance of the topic in question.

2.1. Community outreach

Following the quantitative analysis of the published works, we look more closely at the scientific community authoring the publications concerning the chaos and selected metaheuristics. Recognizing that the keyword search might better reflect reality, we use only the data from the keywords search in 2004–2020 in this analysis. Firstly, we look at the number of unique authors for each of the three examined algorithms, as recognized by the ISI WoS database (Table 3). The PSO exhibits the highest number of unique authors (715), and despite also having the highest number of publications (264), it boasts the highest unique author per publication ratio. The GA comes second in all the metrics, with DE having the

Table 3. The total number of publications, unique authors and unique authors per publication ratio.

Algorithm	Authors	Publications	A/P Ratio
GA	445	179	2.486
DE	125	77	1.623
PSO	715	264	2.708

Table 4. Number of publications authored or co-authored by the top 10 authors for each algorithm.

Algorithm:	GA		DE		PSO	
Authors order	Publications	% of total	Publications	% of total	Publications	% of total
Author 1	10	5.587	34	44.156	12	4.545
Author 2	5	2.793	30	38.961	12	4.545
Author 3	4	2.235	30	38.961	12	4.545
Author 4	3	1.676	15	19.481	8	3.03
Author 5	3	1.676	14	18.182	7	2.652
Author 6	3	1.676	13	16.883	6	2.273
Author 7	3	1.676	6	7.792	5	1.894
Author 8	3	1.676	5	6.494	4	1.515
Author 9	3	1.676	3	3.896	4	1.515
Author 10	3	1.676	3	3.896	4	1.515

lowest number of unique authors, producing the lowest number of publications with the lowest ratio of unique authors per publication.

In Table 4, the publication output of each algorithm's top ten most productive authors is presented (the data has been anonymized). Please note that as multiple authors co-author most publications, the percentage adds up to over 100%. The DE field seems to be dominated by a small group of authors, with the most productive author producing almost half of all publications in the database. As for GA, the most productive author has produced ten publications, which account for approx. 5.5% and the percentage drops fast for other authors. In the case of the PSO algorithm, three authors share the top spot (co-authoring all twelve publications) with just about 4.5% of all publications. The percentage for other authors drops similarly to GA. The above-presented data hint that the GA and PSO with chaos components are attracting broad interest from numerous authors. At the same time, DE seems much more restricted to a relatively small community.

Finally, in Tables 5 and 6, we present the distribution of publications by the country of authors (as identified by ISI WoS). It should be no surprise that the countries list for DE is much shorter than for PSO and GA. Interestingly, the top 5 countries in the lists for GA and PSO are the same, albeit with slight variations in order.

3. Recent works

In this section, we would like to highlight some of the most recent works in the field of chaos-enhanced metaheuristics, with a focus on GA, DE and PSO.

3.1. Genetic algorithm

In [34], a chaotic GA variant using the logistic map is successfully applied to help with intelligent information recommendation based on user preference. Another variant of chaotic

Table 5. Number of publications per country for DE.

DE		
CZECH REPUBLIC	44	57.143
PEOPLES R CHINA	26	33.766
INDIA	4	5.195
USA	3	3.896
AUSTRALIA	1	1.299
FRANCE	1	1.299
IRAN	1	1.299
NORWAY	1	1.299
SOUTH KOREA	1	1.299
TURKEY	1	1.299
VIETNAM	1	1.299

Table 6. Number of publications per country for GA and PSO.

GA[1]			PSO[2]		
Country	Publications	% of total	Country	Publications	% of total
PEOPLES R CHINA	121	67.598	PEOPLES R CHINA	189	71.591
IRAN	12	6.704	TAIWAN	19	7.197
TAIWAN	11	6.145	CZECH REPUBLIC	15	5.682
USA	10	5.587	IRAN	11	4.167
CZECH REPUBLIC	9	5.028	USA	11	4.167
INDIA	7	3.911	ENGLAND	9	3.409
SINGAPORE	4	2.235	JAPAN	7	2.652
ENGLAND	3	1.676	INDIA	5	1.894
BELGIUM	2	1.117	BRAZIL	3	1.136
CANADA	2	1.117	SERBIA	3	1.136
EGYPT	2	1.117	SOUTH AFRICA	3	1.136
ALGERIA	1	0.559	TURKEY	3	1.136
AUSTRALIA	1	0.559	ALGERIA	2	0.758
GERMANY	1	0.559	AUSTRALIA	2	0.758
GREECE	1	0.559	EGYPT	2	0.758
INDONESIA	1	0.559	FRANCE	2	0.758
ITALY	1	0.559	MALAYSIA	2	0.758
MALAYSIA	1	0.559	VIETNAM	2	0.758
NETHERLANDS	1	0.559	GERMANY	1	0.379
PORTUGAL	1	0.559	ITALY	1	0.379
SAUDI ARABIA	1	0.559	MEXICO	1	0.379
SCOTLAND	1	0.559	PORTUGAL	1	0.379
SLOVAKIA	1	0.559	SINGAPORE	1	0.379
SOUTH KOREA	1	0.559	SOUTH KOREA	1	0.379
SWITZERLAND	1	0.559	SPAIN	1	0.379
TUNISIA	1	0.559	SWEDEN	1	0.379
TURKEY	1	0.559	TUNISIA	1	0.379

[1] 2 records (1.117%) did not contain country data.[2] 1 record (0.379%) did not contain country data.

GA called the Chaos Directed GA (CDGA) is proposed in [27] and is used to tackle the challenge of water distribution network design. The authors propose three variants of CDGA (CDGAI – CDGAIII) built around chaotic systems with different dimensionality. CDGA I uses the one-dimensional logistic map, CDGA II is built around the two-dimensional Hénon map, and CDGA III uses the three-dimensional Lorenz map. The authors conclude that CDGA II and III excel in consistently converging into the optima.

Another example of the recent use of chaos-enhanced GA is presented in [2] to solve the relay node placement for multifaceted heterogeneous wireless sensor networks. The authors propose combining two chaotic modifications to the GA, a tent map-based generation of the initial population and a logistic map-based chaotic crossover. The authors report significant improvement in performance thanks to proposed modifications.

3.2. *Differential evolution*

Adding chaotic local search (CLS) is a popular modification in metaheuristics. Recently, a DE variant using the CLS has been proposed in [14]. In this study, the authors show that multiple (twelve to be exact) chaotic maps are typically more effective than just one. The proposed algorithm has been tested on IEEE CEC 2005, 2013, and 2018 benchmark sets with superior performance reported. Another variant of DE using local search based on the chaotic logistic map is presented in [20]. This variant is designed for multi-objective optimization ad applied to the unbalanced transportation problem considering fuel consumption. The algorithm is found effective and able to outperform the state-of-the-art methods.

3.3. *Particle swarm optimization*

As will be presented in this section, the recent works combining chaos and PSO seem to focus mainly on solving specific real-world challenges. One recent exception is the Chaos Adaptive Particle Swarm Optimization Algorithm (CAPSO) proposed in [11]. The authors use a logistic map for chaotic population initialization and further propose an adaptive chaotic search mechanic to improve the algorithm's performance. Comparing the algorithm with several modern PSO variants on IEEE CEC 2013, the authors report promising performance, unfortunately not supported by statistical evaluation for significance.

Among the many recent applications of chaos-enhanced PSO, optimizers are, e.g. tracking of mobile objects [31], or optimal parameter estimation of polymer electrolyte membrane fuel cells model [23] (using the original CEPSO from 2009). A one-dimensional chaotic mapping method is used in PSO to solve a hybrid scheduling task for multi-equipment at an automated terminal in [17]. The CEPSO algorithm is also applied in [35] to solve motion planning and tracking for position-posture control of a planar underactuated Manipulator. In [6], the authors propose an updated dynamic inertia weight and velocity calculation formulas for PSO in combination with chaotic population initialization (using the ever-popular logistic map). The algorithm is then applied to an unmanned aerial vehicle (UAV) safe path planning with promising performance, reporting significant improvements in solution quality and convergence speed.

Various chaotic neighbourhood search strategies are employed in recent works by Pan et al. [24] and Chen et al. [5]. The first in a multi-objective design of an axial permanent magnet eddy current coupler, the latter to solve fault-tolerant control of a five-phase permanent magnet synchronous motor.

4. Performance of chaos-enhanced metaheuristics

According to the above-presented data, the chaos enhancement of metaheuristics seems to be an established research field with an active research community. Therefore, it is

Table 7. Score values of TOP 5 algorithms at the IEEE CEC 2016 competition – algorithms ranked based on overall fitness summation for all the dimensions[2].

	LSHADE_EpSin	UMOEASII	SSEABC	iLSHADE	MCSHADE
$D = 10$	1.5056E+03	1.4440E+03	2.1054E+03	1.9776E+03	1.9588E+03
$D = 30$	3.1753E+03	4.3771E+03	7.6832E+03	5.3241E+03	1.0612E+04
$D = 50$	5.8812E+03	1.5920E+04	1.9148E+04	1.8028E+04	4.5461E+04
$D = 100$	3.3286E+04	2.9613E+04	3.0626E+04	2.2337E+05	1.9618E+05
Total score	4.3848E+04	5.1354E+04	5.9563E+04	2.4870E+05	2.5421E+05
Rank	1	2	3	4	5

quite surprising that chaos-enhanced algorithms only very rarely participate in the regular benchmarking competitions organized by the metaheuristics community. Such a rare event happened at the IEEE CEC 2016 Single objective numerical optimization competition.[2] The algorithm in question was a variant of differential evolution called Success History Based Adaptive Differential Evolution Algorithm with Multi Chaotic Framework for Parent Selection (MC-SHADE) [32]. The MC-SHADE algorithm has ranked fifth among the nine competitors. For illustration, the final score values and rankings of the top 5 algorithms of the competition, as presented by the authors of the competition, are given in Table 7. The winner of the competition was a different DE variant, as has been the case for the majority of recent IEEE CEC competitions. It is, therefore, somewhat surprising that more chaos-enhanced DE algorithms are not entering the competition.

Despite our best efforts, we were unable to find any other example of chaos-enhanced metaheuristics entering the IEEE CEC competitions, nor the BBOB workshop series.[3] As a result, the performance of various chaotic metaheuristics can only be judged from the reported results in the corresponding research papers. However, a thorough performance investigation and comparison, e.g. as reported in [14], are quite rare. Typically the scope of papers is limited to specific application or performance comparison on a small set of benchmark problems with an often questionable selection of competitors.

However, the following list of general observations and conclusions from our own experience and the works referenced in previous sections can be derived:

- There are many various ways to incorporate chaos into a general metaheuristic algorithm, from using it as a source of pseudo-random numbers to specific modifications such as chaotic local search or chaotic crossover. Any (or all) of these might lead to performance improvement.
- Chaos implementation typically affects the speed of convergence of a particular metaheuristic and its ability to escape local optima.
- Combining multiple chaotic components or multiple chaotic systems has yielded even better results.
- While chaos is shown to improve the performance of many algorithms' basic variants significantly, its positive effect might be mitigated in more complex algorithm variants, especially in self-adaptive algorithms, algorithms with a high number of control parameters, and algorithms using advanced learning methods.
- Swarm algorithms tend to benefit more from chaotic dynamics than 'classical' metaheuristics such as genetic algorithm or differential evolution. Testing on artificial test

problems showed mixed results, but excellent results have been achieved in real-world optimization (especially in the discrete domain).

4.1. Computational complexity

It is important to note the computational complexity of implementing chaotic PRNG. While accessing an element in a vector or data structure may be faster than using the built-in Random function of a programming language, both methods have linear time complexity. However, using chaos requires initializing the system and generating a sufficient chaotic sequence, which may result in longer response times compared to the built-in function. Empirical evidence suggests that after about 2500 pseudo-random value requests, the benefits of using chaos in terms of time complexity outweigh the initial investment. The exact point at which this occurs depends on various factors, such as the type of algorithm, metaheuristic strategy, and whether integer or real numbers are being generated. The expression of the exact O notation relation is then tied to the particular implementation.

5. Conclusion

Albeit the combination of chaos theory and evolutionary computing is still deemed by many as an unconventional approach, it has been presented in this survey that numerous successful examples of such interconnection exist in the literature and that the research field is active and new works are published annually. Based on the literature overview, the main focus of the community currently seems to be on using chaos to improve the performance of a particular metaheuristic facing a specific optimization challenge rather than generalized theoretical studies on the effect of chaotic sequences on the inner workings of metaheuristics.

The researchers of the genetic algorithm show a surprisingly stable level of interest in the chaotic enhancement with a broad author community participating, with PSO seemingly following the pattern and making up for its later start with a higher quantity of papers in the last years. The interest of the DE community in chaos implementation seems to be much less stable and influenced mainly by a few authors.

It is well documented in the literature that chaos implementation into metaheuristics often leads to improved performance, faster convergence, and better ability to avoid local optima. While the logistic map seems (likely for its simplicity) to be the most popular chaotic sequence generator for metaheuristics, other chaotic systems, namely the Hénon and Lozi maps, are also popular and frequently used.

Unfortunately, the main question, 'Why does chaos improve the performance of metaheuristics?' remains unanswered and is not discussed in most papers. There are currently several open questions in this field, as many published studies have used chaotic dynamics to enhance algorithm performance but without providing sufficient theoretical explanations or justifications. However, future research and challenges in this area align with current trends in bioinspired metaheuristics, specifically in the area of automated algorithm configuration. These methods aim to find the optimal algorithm instance or hyperparameters. Additionally, predicting the performance of algorithms in a specific context is also a relevant issue. Research into multi-chaotic generators, the parametric setup of chaotic systems, and the implementation of chaotic systems in algorithm strategies all fall within the realm of automated algorithm configuration. Another important area of study is the behaviour of solution populations in metaheuristic algorithms when incorporating another form of randomization.

In addition to the utilization of chaos in metaheuristics, soft-computing and other optimization tools, there exists a contrasting approach [4,21] that is not covered in this study.

Undoubtedly the early pioneers of chaos theory could not foresee that their discoveries would decades later inspire and influence research in such a distant discipline as evolutionary computing and subsequently allow solving various complex tasks of engineering optimization. According to this survey, there is significant research activity in chaos-enhanced metaheuristics. It is likely, that new and even more elaborated concepts of chaos and metaheuristics interconnections will emerge in the future.

Notes

1. https://www.webofscience.com, Search date: October 14th, 2022.
2. https://github.com/P-N-Suganthan/CEC2016.
3. http://numbbo.github.io/workshops/.

Disclosure statement

No potential conflict of interest was reported by the author(s).

Funding

This work was supported by the Internal Grant Agency of Tomas Bata University under the Projects no. IGA/CebiaTech/2023/004, and further by the resources of A.I.Lab at the Faculty of Applied Informatics, Tomas Bata University in Zlin (ailab.fai.utb.cz).

References

[1] S. Arora and S. Singh, *An improved butterfly optimization algorithm with chaos*, J. Intell. Fuzzy Syst.32 (2017), pp. 1079–1088.

[2] P.S. Banerjee, S.N. Mandal, D. De and B. Maiti, *CGARP: Chaos genetic algorithm-based relay node placement for multifaceted heterogeneous wireless sensor networks*, Innov. Syst. Softw. Eng. (2022), pp. 1–16. https://doi.org/10.1007/s11334-022-00439-5

[3] R. Caponetto, L. Fortuna, S. Fazzino and M.G. Xibilia, *Chaotic sequences to improve the performance of evolutionary algorithms*, IEEE Trans. Evol. Comput. 7 (2003), pp. 289–304.

[4] M. Chadli and I. Zelinka, *Chaos synchronization of unknown inputs Takagi–Sugeno fuzzy: Application to secure communications*, Comput. Math. Appl. 68 (2014), pp. 2142–2147.

[5] C. Chen, Y. Song, Y. Zhang, J. Tian, S. Gao and B. Lang, *Adaptive fault-tolerant control of five-phase permanent magnet synchronous motor current using chaotic-particle swarm optimization*, Front. Energy Res. 10 (2022), pp. 1331.

[6] H. Chu, J. Yi and F. Yang, *Chaos particle swarm optimization enhancement algorithm for UAV safe path planning*, Appl. Sci. 12 (2022), p. 8977.

[7] D. Davendra, I. Zelinka and R. Senkerik, *Chaos driven evolutionary algorithms for the task of PID control*, Comput. Math. Appl. 60 (2010), pp. 1088–1104.

[8] J. Determan and J.A. Foster, *Using chaos in genetic algorithms*, in *Proceedings of the 1999 Congress on Evolutionary Computation-CEC99 (Cat. No. 99TH8406)*, Vol. 3, IEEE, 1999, pp. 2094–2101.

[9] L. dos Santos Coelho and H.S. Lopes, *Supply chain optimization using chaotic differential evolution method*, in *2006 IEEE International Conference on Systems, Man and Cybernetics*, Vol. 4, 2006, pp. 3114–3119.
[10] L. dos Santos Coelho and V.C. Mariani, *A novel chaotic particle swarm optimization approach using Hénon map and implicit filtering local search for economic load dispatch*, Chaos Solitons Fract. 39 (2009), pp. 510–518.
[11] Y. Duan, N. Chen, L. Chang, Y. Ni, S.V.N.S. Kumar and P. Zhang, *CAPSO: Chaos adaptive particle swarm optimization algorithm*, IEEE. Access 10 (2022), pp. 29393–29405.
[12] A.H. Gandomi, X.S. Yang, S. Talatahari and A.H. Alavi, *Firefly algorithm with chaos*, Commun. Nonlinear Sci. Numer. Simul. 18 (2013), pp. 89–98.
[13] A.H. Gandomi and X.S. Yang, *Chaotic bat algorithm*, J. Comput. Sci. 5 (2014), pp. 224–232.
[14] S. Gao, Y. Yu, Y. Wang, J. Wang, J. Cheng and M. Zhou, *Chaotic local search-based differential evolution algorithms for optimization*, IEEE Trans. Syst. Man Cybern. Syst. 51 (2021), pp. 3954–3967.
[15] F.S. Gharehchopogh, I. Maleki and Z.A. Dizaji, *Chaotic vortex search algorithm: metaheuristic algorithm for feature selection*, Evol. Intell. 15 (2022), pp. 1777–1808.
[16] G. Kaur and S. Arora, *Chaotic whale optimization algorithm*, J. Comput. Des. Eng. 5 (2018), pp. 275–284.
[17] J. Li, J. Yang, B. Xu, Y. Yang, F. Wen and H. Song, *Hybrid scheduling for multi-equipment at U-shape trafficked automated terminal based on chaos particle swarm optimization*, J. Mar. Sci. Eng. 9 (2021), p. 1080.
[18] B. Liu, L. Wang, Y.H. Jin, F. Tang and D.X. Huang, *Improved particle swarm optimization combined with chaos*, Chaos Solitons Fract. 25 (2005), pp. 1261–1271.
[19] E.N. Lorenz, *Deterministic nonperiodic flow*, J. Atmosph. Sci. 20 (1963), pp. 130–141.
[20] R. Luo, S. Ji and T. Ji, *An effective chaos-driven differential evolution for multi-objective unbalanced transportation problem considering fuel consumption*, Appl. Soft Comput. 101 (2021), Article ID 107058.
[21] K. Mohamed, M. Chadli and M. Chaabane, *Unknown inputs observer for a class of nonlinear uncertain systems: An LMI approach*, Int. J. Autom. Comput. 9 (2012), pp. 331–336.
[22] D. Molina, J. Poyatos, J. Del Ser, S. García, A. Hussain and F. Herrera, *Comprehensive taxonomies of nature-and bio-inspired optimization: Inspiration versus algorithmic behaviour, critical analysis recommendations*, Cogn. Comput. 12 (2020), pp. 897–939.
[23] M.T. Özdemir, *Optimal parameter estimation of polymer electrolyte membrane fuel cells model with chaos embedded particle swarm optimization*, Int. J. Hydrog. Energy 46 (2021), pp. 16465–16480.
[24] P. Pan, D. Wang and B. Niu, *Design optimization of APMEC using chaos multi-objective particle swarm optimization algorithm*, Energy Rep. 7 (2021), pp. 531–537.
[25] M. Pluhacek, R. Senkerik and D. Davendra, *Chaos particle swarm optimization with Eensemble of chaotic systems*, Swarm Evol. Comput. 25 (2015), pp. 29–35.
[26] M. Pluhacek, R. Senkerik and I. Zelinka, *Particle swarm optimization algorithm driven by multichaotic number generator*, Soft Comput. 18 (2014), pp. 631–639.
[27] S.N. Poojitha, V. Jothiprakash and B. Sivakumar, *Chaos-directed genetic algorithms for water distribution network design: An enhanced search method*, Stoch. Environ. Res. Risk Assess. 36 (2022), pp. 3377–3393. https://doi.org/10.1007/s00477-022-02200-7
[28] G.I. Sayed, A. Tharwat and A.E. Hassanien, *Chaotic dragonfly algorithm: An improved metaheuristic algorithm for feature selection*, Appl. Intell. 49 (2019), pp. 188–205.
[29] J.C. Sprott, *Chaos and Time-Series Analysis*, Vol. 69, Oxford University Press, Oxford, 2003.
[30] K. Sörensen, *Metaheuristics—The metaphor exposed*, Int. Trans. Oper. Res. 22 (2015), pp. 3–18.
[31] S. Teymori and P. Babaei, *A chaos-enhanced accelerated PSO algorithm in reliable tracking of mobile objects*, Int. J. Comput. Appl. Technol. 67 (2021), pp. 263–274. https://doi.org/10.1504/IJCAT.2021.121537
[32] A. Viktorin, M. Pluhacek and R. Senkerik, *Success-history based adaptive differential evolution algorithm with multi-chaotic framework for parent selection performance on CEC2014*

benchmark set, in *2016 IEEE Congress on Evolutionary Computation (CEC)*, IEEE, 2016, pp. 4797–4803.

[33] D.H. Wolpert and W.G. Macready, *No free lunch theorems for optimization*, IEEE Trans. Evol. Comput. 1 (1997), pp. 67–82.

[34] L. Yin and P. Zhao, *User preference intelligent information recommendation system based on chaos genetic algorithm*, J. Ambient Intell. Hum. Comput. (2021), pp. 1–10. https://doi.org/10.1007/s12652-020-02611-w

[35] P. Zhang, X. Lai, Y. Wang and M. Wu, *Chaos-PSO-based motion planning and accurate tracking for position-posture control of a planar underactuated manipulator with disturbance*, Int. J. Control Autom. Syst. 19 (2021), pp. 3511–3521. https://doi.org/10.1007/s12555-020-0553-z

Ramified continua as global attractors of C^1-smooth self-maps of a cylinder close to skew products

Lyudmila Efremova

ABSTRACT
In the paper we consider C^1-smooth self-maps of a cylinder close to C^1-smooth skew products (and satisfying some additional conditions). We study such geometric property of the maps, as existence of C^1-smooth invariant local lamination, and apply this geometric property to the proof of the geometric integrability of maps under consideration. Using obtained results we construct the example of the family of C^1-smooth maps close to skew products so that each map from this family admits the global attractor, which is a one-dimensional ramified continuum with a complicated topological structure. The global attractor of every map from the family under consideration consists of arcs of two types. On the unique circle (which is the arc of first type) the map is mixing; on arcs of second type of different lengths homeomorphic to a closed interval (the family of such arcs has continuum cardinality) the map is not mixing. The topological structure of the global attractor and dynamical properties of trajectories on the attractor lead to the property of dense intermittency (in the complement to the attractor) of attraction sets of different ω-limit sets, the union of which coincides with the global attractor.

1. Introduction

(1) In the current time the theory of dynamical systems on complicated continua, in particular, on ramified continua, is intensively developing (see, e.g. [1, 12,33,34,38,39,44,46]). There are dynamical systems (in particular, ordinary differential equations) that admit continua with a complicated topological structure as their attractors (see, e.g. [42]). For example, Plykin attractor [49,50] is an indecomposable continuum. Indecomposable continua also arise in the study of inverse limits of maps with homoclinic tangencies [7]. Dendrites appear not only in considerations of inverse limits of maps with attractors of Hénon [35] and Lozi [43] types (see, e.g. [8,13]), but also in investigations of limit sets of some Kleinian groups on hyperbolic 3-manifolds [14,45]. General pseudo-arcs are realized as inverse limits of maps connected with attractors of disc homeomorphisms [17].

Let a discrete dynamical system with the phase space M be given by a map $F : M \to M$. Let $B \subseteq M$ be *an absorbing domain* in M, i.e. the inclusion $F(\overline{B}) \subset B$ holds, where $\overline{(\cdot)}$ is the closure of a set.

Definition 1.1: *The maximal attractor* $A_{\max}$ in the absorbing domain B is said to be the set

$$A_{\max} = \bigcap_{n=1}^{+\infty} F^n(B).$$

An invariant set A is said to be *an attractor* of F if there exists an absorbing domain, for which A is the maximal attractor [4, Ch. 1, Section 1.8].

If the whole phase space M is an absorbing domain, for which A is the maximal attractor, then we say that A is the *global attractor.*

In the case, when the topological entropy $h(F_{|A})$ (see [2] for the definition of the topological entropy) of the restriction of the map F on the attractor A is positive, we say that A is a *chaotic attractor.*

A connected compact Hausdorff space A is called *a continuum* [41, Ch. 5, Section 47, I].

A connected subset of a continuum A whose closure is homeomorphic to the interval $[0, 1]$ is called *an arc*. The preimages of 0 and 1 in A under this homeomorphism are called *the endpoints of the arc* (see, e.g. [34]).

Let, as usually, $\mathbb{N}$, $\mathbb{R}^1$, $\mathbb{C}$ be the sets of natural numbers, real numbers and complex numbers respectively. Although there are continua that admit points of infinite order, in this paper we consider continua with points of only finite order.

Definition 1.2 ([16]): Let A be a continuum, z be a point in A. We say that z is a point of the order $n \in \mathbb{N}$ if z is a unique common endpoint of every two of exactly n arcs contained in A.

By *a ramified continuum* we mean a continuum admitting points of an order $n \geq 3$. Such points are called *ramification points.*

(2) Let a self-map F of the phase space M (equipped with x and y coordinates) be presented in the form

$$F(x, y) = (f(x) + \mu(x, y), g_x(y)), \quad \text{where } g_x(y) = g(x, y), \quad (x, y) \in M. \tag{1}$$

If M is the plane, then we obtain Hénon map for

$$f(x) + \mu(x, y) = 1 - ax^2 + y, \quad g_x(y) = by \quad [35]$$

and Lozi map for

$$f(x) + \mu(x, y) = 1 - a|x| + y, \quad g_x(y) = by \quad [43].$$

Here a, b are parameters. If M is a cylinder (or a torus), then we obtain a Belykh map for

$$f(x) + \mu(x, y) = x + h(x) + y, \quad g_x(y) = \lambda(h(x) + y) \quad [10],$$

where λ is a parameter, $0 < \lambda \leq 1$, and a function h satisfies some conditions described in [10]. Smooth maps of type (1) (satisfying some additional conditions) in the case, when M is a compact plane rectangle, are studied in [27–29].

Let M be a compact two-dimensional cylinder $M = S^1 \times I_2$. Here S^1 is a circle, I_2 is a compact interval of the real line. In this paper, we consider C^1-smooth maps (1) close (in the C^1-norm) to skew products on the cylinder M and so that a function μ satisfies zero boundary conditions.

Give necessary explanations. Let $C^1(S^1)$ be the class of C^1-smooth self-maps of the circle S^1 endowed with the standard C^1-norm $|| \cdot ||_{C^1(S^1)}$. Let the family of ε-neighbourhoods $B^1_{1,\varepsilon}(\varphi)$ for every $\varphi \in C^1(S^1)$ and $\varepsilon > 0$ form the base of the C^1-topology in $C^1(S^1)$ generated by the C^1-norm $|| \cdot ||_{C^1(S^1)}$.

Definition 1.3 ([5]): A map $\varphi \in C^1(S^1)$ is said to be *structurally stable* (in the C^1-norm) if for every $\delta > 0$ there exists $\varepsilon > 0$ such that for a map $\psi \in B^1_{1,\varepsilon}(\varphi)$ one can find δ-close in the C^0-norm (of the uniform convergence) to the identity map homeomorphism $h : S^1 \to S^1$ satisfying the equality

$$h \circ \varphi = \psi \circ h.$$

Denote by $\Omega(\cdot)$ the nonwandering set of a map (for the definition of the nonwandering set see, e.g. [48, Ch. 0, Section 0.2]).

A map $\varphi \in C^1(S^1)$ is said to be Ω-*stable* (in the C^1-norm) if for every $\delta > 0$ there exists $\varepsilon > 0$ such that for a map $\psi \in B^1_{1,\varepsilon}(\varphi)$ one can find δ-close in the C^0-norm to the identity map homeomorphism $h : \Omega(\varphi) \to \Omega(\psi)$ satisfying the equality

$$h \circ \varphi_{|\Omega(\varphi)} = \psi_{|\Omega(\psi)} \circ h.$$

Remark 1.1: Definition 1.3 shows that the Ω-stable map $\varphi \in C^1(S^1)$ satisfying $\Omega(\varphi) = S^1$, is C^1-structurally stable.

Recall the concept of the degree of a self-map of S^1. For this goal we use the multiplicative presentation of the circle

$$S^1 = \{x \in \mathbb{C} : |x| = 1\} = \{e^{2\pi i t} : t \in \mathbb{R}^1\}.$$

Then for every continuous map $\varphi : S^1 \to S^1$ there is a continuous map $\widehat{\varphi} : \mathbb{R}^1 \to \mathbb{R}^1$ satisfying

$$\varphi \circ \exp = \exp \circ \widehat{\varphi},$$

where $\exp t = e^{2\pi i t}$ for every $t \in \mathbb{R}^1$.

The above continuous map $\widehat{\varphi}$ is said to be *a lifting* of the circle map φ [48, Ch. 1, Section 1.3].

Definition 1.4 ([48, Ch. 1, Section 1.3]): *The degree* of a map φ is said to be an integer number

$$\deg \varphi = \frac{1}{n}(\widehat{\varphi}(t+n) - \widehat{\varphi}(t)),$$

that does not depend on $t \in \mathbb{R}^1$; moreover, it is defined for an integer number n for any lifting $\widehat{\varphi} : \mathbb{R}^1 \to \mathbb{R}^1$ of the circle map φ.

To formulate conditions for the map of type (1) we need the main result of the paper [36].

Theorem 1.5 ([36]): *In the space $C^1(S^1)$ there is an open everywhere dense set* $\mathbf{L}$ *of maps that equals the union of two subsets* $\mathbf{L}_1$ *and* $\mathbf{L}_2$*, where* $\mathbf{L}_1$ *is the set of* Ω*-stable maps with a completely disconnected nonwandering set, and* $\mathbf{L}_2$ *is the set of structurally stable maps such that every* $\varphi \in \mathbf{L}_2$ *is topologically conjugate with an expanding map*[1] *of the same degree* $deg\, \varphi$*, where* $|deg\, \varphi| > 1$.

Further, we interpret a circle S^1 as the unit interval $[0, 1]^*$ with identified points 0 and 1. Let $I_2 = [a, b]$.

In this paper we consider the class $C^1_L(M)$ of C^1-smooth maps (1), where $f \in \mathbf{L}$, and μ satisfies the condition:

(i_μ) $\mu(0, y) = \mu(1, y) = \mu(x, a) = \mu(x, b) = 0$ for $x \in [0, 1]^*$, $y \in [a, b]$.

Let the base of the C^1-topology in $C^1_L(M)$ generated by the standard C^1-norm $|| \cdot ||_{C^1_L(M)}$, be given by the family of ε-balls $B^1_\varepsilon(F)$ for every $F \in C^1_L(M)$ and $\varepsilon > 0$.

We also need the class $C^1(M, S^1)$ of C^1-smooth maps of the cylinder M into the circle S^1 endowed with the standard C^1-norm $|| \cdot ||_{C^1(M,S^1)}$. Then for every $y \in I_2$ we have:

$$||\mu||_{C^1(S^1)} \leq ||\mu||_{C^1(M,S^1)}. \tag{2}$$

Let the family of ε-balls $B^1_{(2,1),\varepsilon}(\psi)$ for every $\psi \in C^1(M, S^1)$ and $\varepsilon > 0$ form the base of the C^1-topology in $C^1(M, S^1)$ induced by the C^1-norm $|| \cdot ||_{C^1(M,S^1)}$.

Let $\delta > 0$. Since $f \in \mathbf{L}$ (see Theorem 1.5) then we find $\varepsilon > 0$, using Definition 1.3. We suppose in this paper that $\mu \in C^1(M, S^1)$ satisfies the 'condition of smallness' in the C^1-norm (see also [28,29]). It means that the following inequality is valid:

$(ii_{\mu,\varepsilon})$ $||\mu||_{C^1(M,S^1)} < \varepsilon$.

We use first and second natural projections $pr_1 : M \to S^1$ and $pr_2 : M \to I_2$ of the cylinder M on the circle S^1 and on the interval I_2 respectively. Consider $f \in \mathbf{L}$ as the function $f \circ pr_1$ of two variables $(x, y) \in M$. By the above condition $(ii_{\mu,\varepsilon})$ we have $(f + \mu) \in B^1_{(2,1),\varepsilon}(f \circ pr_1)$. By inequality (2) we have for every $y \in I_2$:

$$(f + \mu) \in B^1_{1,\varepsilon}(f). \tag{3}$$

(3) Analogously [28], we give the coordinate presentation of iterations of map (1). We use notations

$$\varphi_y(x) = f(x) + \mu_y(x), \quad \text{where } \mu_y(x) = \mu(x, y). \tag{4}$$

For every point (x^0, y^0) and $n = 1$ we set

$$\varphi_{y^0,n}(x^0) = \varphi_{y^0}(x^0), \quad g_{(x^0,y^0),n}(y^0) = g_{x^0}(y^0); \tag{5}$$

for $n \geq 2$

$$\begin{aligned} \varphi_{y^0,n}(x^0) &= \varphi_{y^{n-1}} \circ \ldots \circ \varphi_{y^0}(x^0), \\ g_{(x^0,y^0),n}(y^0) &= g_{\varphi_{y^{n-2}}(x_{n-2})} \circ \ldots \circ g_{x^0}(y^0). \end{aligned} \tag{6}$$

By (1), (5) and (6) the following equality holds:

$$F^n(x^0, y^0) = (\varphi_{y^0,n}(x^0), \quad g_{(x^0,y^0),n}(y^0)), \quad n \geq 1. \tag{7}$$

(4) The survey [25] contains results on the dynamics of skew products of interval maps. As opposed to [25], we consider here a skew product Φ on a cylinder M, where

$$\Phi(x, y) = (\varphi(x), g_x(y)) \quad \text{for every } (x, y) \in M. \tag{8}$$

Here $\varphi : S^1 \to S^1$ is a map, which is the quotient of the skew product Φ, and $g_x : I_2 \to I_2$ is the fibre map for each $x \in S^1$. By (8) we obtain for a natural number $n > 1$

$$\Phi^n(x, y) = (\varphi^n(x), g_{x,n}(y)), \quad \text{where } g_{x,n}(y) = g_{\varphi^{n-1}(x)} \circ \ldots \circ g_x(y). \tag{9}$$

There is a deep correlation between maps (1) and skew products. In fact, similar to [28], introduce the one-parameter family of skew products $\Phi_t : M \to M$

$$\Phi_t(x, y) = (\varphi_t(x), g_x(y)).$$

with variable quotients. Here $t \in I_2$, $\varphi_t(x) = f(x) + \mu_t(x)$. Then for every horizontal fibre $y = t$ the equality holds:

$$F(x, y)_{|y=t} = \Phi_t(x, y)_{|y=t}.$$

This implies the equality

$$F^n(x^0, y^0) = \Phi_{y^{n-1}} \circ \ldots \circ \Phi_{y^0}(x^0, y^0) \tag{10}$$

for a point $(x^0, y^0) \in M$ (detailed explanations independent of the topology of the phase space are given in [28]).

By formulas (7) and (10) we have for each $n \geq 1$:

$$(\varphi_{y^0,n}(x^0), g_{(x^0,y^0),n}(y^0)) = \Phi_{y^{n-1}} \circ \ldots \circ \Phi_{y^0}(x^0, y^0).$$

(5) In this paper we study such geometric property, as existence of the C^1-smooth invariant local lamination, which is the invariant lamination in the case $f \in \mathbf{L}_1$ and the invariant foliation in the case $f \in \mathbf{L}_2$, for maps from the set $C^1_L(M)$ such that for every $\delta > 0$ there is $\varepsilon > 0$, for which the condition $(ii_{\mu,\varepsilon})$ is fulfilled. We also prove the geometric integrability of these maps (for definitions see Section 2).

The great attention is paid to the consideration of the family of C^1-smooth maps $F_k : M \to M$ $(k > 1, k \in \mathbb{N}, I_2 = [0, 1])$ satisfying

$$F_k(x, y) = (f_k(x) + \mu(x, y), g(x, y)) \quad \text{for } (x, y) \in M, \tag{11}$$

where

$$f_k(x) = kx \,(\text{mod}\, 1) \tag{12}$$

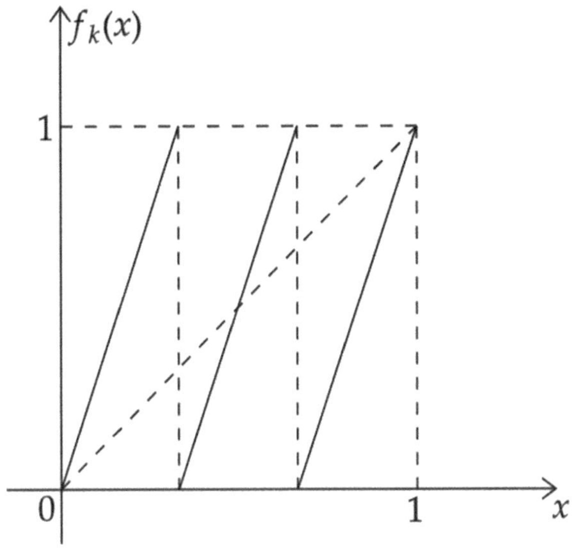

Figure 1. The graph of f_k for $k = 3$.

Since $k > 1$ then (12) means that f_k is the canonical expanding map of the circle [53], [48, Ch.1, Section 1.3], $\deg f_k = k, f_k \in \mathbf{L}_2$. Moreover, using conditions (i_μ) and $(ii_{\mu,\varepsilon})$, we also obtain the correct equality $\deg(f_k + \mu) = k$ for a map $(f_k + \mu) : S^1 \to S^1$ for every $y \subset [0, 1]$.

We choose the map $g : M \to [0, 1]$ (the choice of g will be described in Section 4) so that the constructed map (11) with $f_k : S^1 \to S^1$ satisfying (12) has only nonhyperbolic periodic points and possesses a global one-dimensional attractor, which is a ramified continuum with a complicated topological structure. This attractor consists of arcs of two types. On the circle $S^1 \times \{0\}$ (which is the unique arc of first type) the map is mixing; on arcs of different lengths of second type homeomorphic to a closed interval (the family of such arcs has continuum cardinality) the map is not mixing (compare with [21]). The topological structure of the global attractor and dynamical properties of trajectories on the attractor lead to the property of dense intermittency (in the complement to the attractor) of attraction sets of different ω-limit sets, the union of which coincides with the global attractor (for details, see Section 4). These considerations are possible due to geometric properties of maps (1).

Finishing this part of the paper, point out that Belykh two-dimensional maps (in particular, the Belykh map on a cylinder) have only hyperbolic periodic points and admit chaotic attractors [10].

2. Preliminaries

In the paper, we study not only the topological structure of the attractor, but also the dynamic properties of the trajectories belonging to it. Therefore, this section contains a survey of the main properties we need, first, of C^1-smooth Ω-stable circle maps from the set $\mathbf{L}_1$ and C^1-smooth structurally stable circle maps from the set $\mathbf{L}_2$ and, second, of skew products and geometrically integrable self-maps of a cylinder.

(1) Denote by $\tau(f)$ the set of (least) periods of periodic points of a map $f \in C^1(S^1)$. Set $q = \inf \tau(f)$.

Add to Theorem 1.5 the information on the structure of the nonwandering set of a map $f \in \mathbf{L}$.

Proposition 2.1 ([18,34, Ch. 2.b, c]): *If $f \in \mathbf{L}_1$ then either*

(2.1.1) *for every $p \in \tau(f)$ there is i, where $0 \le i \le \nu$ for some $\nu < +\infty$, such that $p = 2^i q$; the nonwandering set $\Omega(f)$ is finite and consists of hyperbolic periodic points, or*

(2.1.2) *there is $p \in \tau(f)$ such that $p/q \notin \{2^i\}_{i \ge 0}$, and*

$$\Omega(f) = \Omega_r(f) \bigcup \Omega_p(f),$$

where $\Omega_r(f)$ is the nonempty rarefied part of the nonwandering set that consists of a finite number of hyperbolic periodic points; $\Omega_p(f)$ is the nonempty perfect part of the set $\Omega(f)$ that consists of a finite number of locally maximal quasiminimal [2] *(i.e. maximal quasiminimal in some neighbourhood) hyperbolic perfect nowhere dense sets.*

(2.1.3) *For $f \in \mathbf{L}_2$ equalities $\Omega(f) = \Omega_p(f) = S^1$ hold.*

If $f \in \mathbf{L}_1$, and there is $p \in \tau(f)$ such that $p/q \notin \{2^i\}_{i \ge 0}$; or $f \in \mathbf{L}_2$, then

(2.1.4) *$\Omega_p(f) = \overline{Per(f)}$, where $Per(\cdot)$ is the set of periodic points of a map;*

(2.1.5) *for every number $\theta > 0$ there is a periodic orbit, which approximates $\Omega_p(f)$ up to θ, it means that for a finite cover of $\Omega_p(f)$ by intervals of the length $\le \theta$ there is a periodic orbit that intersects every interval of this cover, at list, in one point;*

(2.1.6) *there are numbers $\alpha = \alpha(f) > 0$ and $c = c(f) > 1$ such that the inequality*

$$|(f^n(x))'| > \alpha c^n$$

holds for every $x \in \Omega_p(f)$ and $n \ge 1$.

(2) Let $T^0(M)$ ($T^1(M)$) be the space of continuous (C^1-smooth) skew products (8) on the cylinder M. We also use the set $T^1_L(M)$ of maps (8) such that $\varphi \in \mathbf{L}$. Then $T^1_L(M) \subset T^1(M)$. We need the information on the nonwandering set of a skew product on a cylinder and methods of this set description.

The following claim is the partial case of the general property of 'projection on the subspace of the codimension 1' of the nonwandering set of continuous skew products proved in the paper [31] for skew products on n-dimensional ($n \ge 2$) cells, cylinders and tori.

Lemma 2.2: *Let $\Phi \in T^0(M)$, $M = S^1 \times I_2$. Then the equality holds*

$$\Omega(\varphi) = pr_1(\Omega(\Phi)).$$

The Ω-function introduced in [19] for skew products of interval maps (see also [25]) gives a powerful functional tool for the study of the nonwandering set. In [31] the concept of the Ω-function is generalized to the skew products on n-dimensional ($n \ge 2$) cells, cylinders and tori. Here we give the definition of the Ω-function for skew products on a compact two-dimensional cylinder.

We use the topological space 2^{I_2} of all closed subsets of the segment I_2 endowed with the exponential topology, i. e. the weakest topology in which the sets 2^A are open in 2^{I_2} for open sets A, and closed in 2^{I_2} for closed sets A (see [41, Ch.1, Section 17, I]). Here 2^A means the set of all closed subsets lying in $A \subset I_2$.

Definition 2.3 ([31]): *The Ω-function* of a skew product $\Phi \in T^0(M)$ is defined to be the function $\Omega^{\Phi} : \Omega(\varphi) \to 2^{I_2}$ such that for any $x \in \Omega(\varphi)$ the equality holds:

$$\Omega^{\Phi}(x) = (\Omega(\Phi))(x),$$

where $(\Omega(\Phi))(x)$ is the slice of the nonwandering set $\Omega(\Phi)$ by the fibre over x.

The slice $(A')(x)$ of a set $A' \subseteq M$ by the vertical fibre over a point $x \in S^1$ is the following set:

$$\left(A'\right)(x) = \{y \in I_2 : (x, y) \in A'\}.$$

As it follows from Definition 2.3, the Ω-function of a skew product $\Phi \in T^0(M)$ has the real dynamical sense: its graph in the phase space M coincides with the nonwandering set $\Omega(\Phi)$.

Analogously [25], we present Φ^n for every $n \geq 1$ in the form

$$\Phi^n = \Phi_{n,1} \circ \Phi_n, \tag{13}$$

where

$$\Phi_n(x, y) = (id(x), g_{x,n}(y)); \tag{14}$$

$$\Phi_{n,1}(x, y) = (\varphi^n(x), id(y)). \tag{15}$$

Here $id(x)$, $id(y)$ are values of the identity maps of the circle S^1 in a point x and the interval I_2 in a point y, respectively.

Introducing maps (14) and (15), we obtain an opportunity to define the following dynamical multifunctions, first, auxiliary functions and, second, suitable functions (definitions for skew products of interval maps see in [19,25]).

Definition 2.4: By *auxiliary functions for the Ω-function* of a continuous skew product (8), satisfying $\varphi(\Omega(\varphi)) = \Omega(\varphi)$, we mean multifunctions $\Omega_n^{\Phi} : \Omega(\varphi) \to 2^{I_2}$, $n \geq 1$ such that

$$\Omega_n^{\Phi}(x) = \Omega(\varphi_{x,n}) \quad \text{for any } x \in \Omega(\varphi). \tag{16}$$

Multifunctions $\overline{\Omega}_n^{\Phi} : \Omega(\varphi) \to 2^{I_2}$ whose graphs in M are closures $\overline{(\cdot)}$ of graphs of auxiliary functions Ω_n^{Φ} are said to be *suitable for the Ω-function* of a continuous skew product (8), satisfying $\varphi(\Omega(\varphi)) = \Omega(\varphi)$, and for these functions equalities hold:

$$\overline{\Omega}_n^{\Phi}(x) = \left(\overline{\Omega}_n^{\Phi}\right)(x) \quad \text{for each } x \in \Omega(\varphi).$$

Here $(\cdot)(x)$ is the slice of a set by the vertical fibre over x.

By Definition 2.4, formula (14) and [25] suitable functions for the Ω-function have a real dynamical sense: for every $n \geq 1$ the graph of $\overline{\Omega}_n^{\Phi}$ in the phase space M coincides with the graph of the Ω-function of the restriction of the map Φ_n on the set $\Omega(\varphi) \times I_2$, i.e. with the nonwandering set $\Omega(\Phi_{n|\Omega(\varphi)\times I_2})$.

After we have defined the auxiliary functions Ω_n^{Φ} (the suitable functions $\overline{\Omega}_n^{\Phi}$) we should move every point (x, y) on the graph of Ω_n^{Φ} (of $\overline{\Omega}_n^{\Phi}$) to the point $(\varphi^n(x), y)$, using the direct product $\Phi_{n,1}$ $(n \geq 1)$ (see the formula (15)). Then we obtain two types of *approximating multifunctions* $\Omega^{\Phi}_{n,1} : \Omega(\varphi) \to 2^{I_2}$ and $\overline{\Omega}^{\Phi}_{n,1} : \Omega(\varphi) \to 2^{I_2}$ satisfying:

$$\Omega^{\Phi}_{n,1}(x) = \left(\Phi_{n,1}(\Omega^{\Phi}_n)\right)(x), \quad \overline{\Omega}^{\Phi}_{n,1}(x) = \left(\Phi_{n,1}(\overline{\Omega}^{\Phi}_n)\right)(x) \tag{17}$$

respectively for any $x \in \Omega(\varphi)$. Here Ω^{Φ}_n ($\overline{\Omega}^{\Phi}_n$) is the graph in M of the auxiliary (suitable) multifunction. By the above we arrive to a point $(x, y) \in \Omega^{\Phi}_{n,1}$ (to a point $(x, y) \in \overline{\Omega}^{\Phi}_{n,1}$) from points $(\overline{x}, y)$, where $\overline{x}$ is an arbitrary point of n-th full preimage of x with respect to the map $\varphi_{|\Omega(\varphi)}$. Therefore, the equalities hold:

$$\Omega^{\Phi}_{n,1}(x) = \bigcup_{\overline{x} \in \{(\varphi_{|\Omega(\varphi)})^{-n}(x)\}} \Omega^{\Phi}_n(\overline{x}), \quad \text{and} \quad \overline{\Omega}^{\Phi}_{n,1}(x) = \bigcup_{\overline{x} \in \{(\varphi_{|\Omega(\varphi)})^{-n}(x)\}} \overline{\Omega}^{\Phi}_n(\overline{x}), \tag{18}$$

where $(\varphi_{|\Omega(\varphi)})^{-n}(x)$ is n-th full preimage of x with respect to $\varphi_{|\Omega(\varphi)}$.

Equalities (18) demonstrate the memory effect in noninvertable skew products. This effect significantly complicates the description of nonwandering sets of skew products.

In the case of skew products with a C^1- Ω-stable (structurally stable) quotient approximating multifunctions inherit such properties of auxiliary or suitable multifunctions as low and upper semicontinuity and, hence, continuity (for skew products of interval maps see [25]).

Finishing this part of Section 2, devoted to skew products, we formulate the result we need in future, in Section 4. Denote by $\omega_{(\cdot)}(z)$ the ω-limit set of the trajectory of a point z with respect to a map $(\cdot)$ (for the definition of an ω-limit set see, e.g. [48, Ch. 0, Section 0.2]).

Theorem 2.5: *Let* $\Phi \in T^1_L(M)$, *and there be a sequence of natural numbers* $\{n_i\}_{i\geq 1}$ *such that auxiliary functions* $\Omega^{\Phi}_{n_i}$ *are continuous for all* $i \geq i_0$ *for some* $i_0 \geq 1$. *Here natural numbers* n_i *(return times) are so that for every* $x \in \Omega(\varphi)$ *and every its neighbourhood* $U_1(x)$ *one can find an* n_i *in the above number sequence satisfying:*

$$U_1(x) \bigcap \varphi^{n_i}(U_1(x)) \neq \emptyset.$$

Then $\bigcup_{(x,y)\in M} \omega_{\Phi}((x, y))$ *is a closed set, and for any neighbourhood* U *of the set* $\bigcup_{(x,y)\in M} \omega_{\Phi}((x, y))$ *there exists an* $m = m(U)$ *such that the sojourn time outside* U *of the trajectory of an arbitrary point in* M *does not exceed* m.

Theorem 2.5 extends the main result of the paper [51] on the case of skew products from $T^1_L(M)$ satisfying some additional conditions. The proof of Theorem 2.5 is analogous to the proof of this statement for skew products of interval maps, which is available in the survey [25].

(3) Formulate the definition of the geometric integrability of a map on a cylinder (for a map on a compact rectangle in the plane see [9,25,28,30]).

Definition 2.6: A map $F : M \to M$ $(M = S^1 \times I_2)$ is said to be *geometrically integrable on a nonempty F-invariant set* $A \subseteq M$ if there exist a self-map ψ of an arc $J \subseteq S^1$ and ψ-invariant set $B(\psi) \subseteq J$ such that the restriction $F_{|A(F)}$ is semiconjugate with the restriction

$\psi_{|B(\psi)}$ by means of a continuous surjection $H : A(F) \to B(\psi)$, i.e. the following equality holds:

$$H \circ F_{|A(F)} = \psi_{|B(\psi)} \circ H. \tag{19}$$

The map $\psi_{|B(\psi)}$ is said to be *the quotient* of $F_{|A(F)}$.

Maps F and ψ in Definition 2.6 can be discontinuous. Moreover, in [26] the geometric integrability definition is extended in the case of some multifunctions with noncompact domains in the plane, and in [32] the generalization of the geometric integrability concept on the case of maps with phase spaces of high dimensions is obtained.

To formulate the geometric criterion of the integrability of a self-map of a cylinder recall the concept of one-dimensional C^r- local lamination, $r \geq 0$, (see [6, Ch.1, Section 1.2] for the definition of d-dimensional local lamination, $d \geq 1$) that generalizes concepts of one-dimensional lamination and one-dimensional foliation.

Let L_α be a C^r-curve[3], $L_\alpha \subset M$, L_α be C^r-regularly embedded to M.

Definition 2.7: Let A be a subset of M satisfying $A = \bigcup_\alpha L_\alpha$. Here α belongs to an index set; C^r-curves $\{L_\alpha\}_\alpha$ are pairwise disjoint. The family of curves $\{L_\alpha\}_\alpha$ is said to be *one-dimensional C^r- local lamination (without singularities)*, if for every point $z \in A$, $z = z(x, y)$, there exist a neighbourhood $U(z) \subset M$ and a C^r-diffeomorphism for $r \geq 1$ or homeomorphism for $r = 0$ $\chi : U(z) \to \mathbb{R}^2$ (here $\mathbb{R}^2$ is the plane) such that every connected component of the intersection $U(z) \bigcap L_\alpha$ (if it is not empty) is mapping by means of χ into a straight line such that

$$\chi_{|U(z)\bigcap L_\alpha} : U(z) \bigcap L_\alpha \to \chi(U(z) \bigcap L_\alpha)$$

is a C^r-diffeomorphism for $r \geq 1$ or homeomorphism for $r = 0$ on the image.

The set A satisfying the above equality is said to be *a support of the local lamination* $L(A)$, curves L_α are said to be *fibres*. If A is a closed set, $A \neq M$, then we say about *one-dimensional C^r-lamination*; if $A = M$ then we say about *one-dimensional C^r-foliation*.

Formulate the geometric criterion of the (geometric) integrability of self-maps of a cylinder (the paper [30] contains analogous results for plane maps).

Theorem 2.8: *Let F be a self-map of M, (F may not be set using equality* (1)*), $A(F)$ be a nonempty closed F-invariant subset of the cylinder M satisfying*

$$pr_2(A(F)) = I_2. \tag{20}$$

Let $J \subseteq S^1$ be an arc, ψ be a self-map of J, $B(\psi) \subseteq J$ be a closed ψ-invariant set.

Then $F_{|A(F)}$ is the geometrically integrable map with the quotient $\psi_{|B(\psi)}$ by means of a continuous surjection $H : A(F) \to B(\psi)$ such that for every $y \in I_2$ the map H is an injection on x, if and only if $A(F)$ is the support of a continuous invariant one-dimensional lamination for $A(F) \neq M$ (of a continuous invariant one-dimensional foliation for $A(F) = M$) with fibres $\{\gamma_{x'}\}_{x' \in B(\psi)}$ that are pairwise disjoint graphs of continuous functions $x = x_{x'}(y)$ for every $y \in I_2$. Moreover, the inclusion

$$F(\gamma_{x'}) \subseteq \gamma_{\psi(x')} \tag{21}$$

holds[4].

The following Theorem 2.9 contains the analytic criterion for the geometric integrability of a cylinder map in the sense of Definition 2.6.

Theorem 2.9: *Let F be a self-map of M (F may not be set using equality* (1)*), $A(F)$ be a nonempty closed F-invariant subset of M satisfying the equality* (20)*. Let $J \subseteq S^1$ be an arc, ψ be a self-map of J, $B(\psi) \subseteq J$ be a closed ψ-invariant set.*

Then $F_{|A(F)}$ is geometrically integrable with the quotient $\psi_{|B(\psi)}$ by means of a continuous surjection $H : A(F) \to B(\psi)$ such that for every $y \in I_2$ the map H is an injection on x, if and only if there is a homeomorphism $\widetilde{H}$ that maps the set $A(F)$ on the set $B(\psi) \times I_2$ and reduces the restriction $F_{|A(F)}$ to the skew product $F_{|B(\psi)\times I_2}$ satisfying the equality

$$F_{|B(\psi)\times I_2}(u, v) = (\psi_{|B(\psi)}(u), g_{x'}(v)), \quad \text{where } x' = pr_1 \circ \widetilde{H}^{-1}(u, v). \tag{22}$$

Here $\widetilde{H}^{-1} : B(\psi) \times J \to A(F)$ is the inverse homeomorphism for $\widetilde{H}$, and

$$\widetilde{H}(x, y) = (H(x, y), y), \quad \text{for all } (x, y) \in A(F). \tag{23}$$

Proofs of Theorems 2.8 and 2.9 repeat proofs of the corresponding claims given in [30].

3. Existence of a C^1-smooth local lamination, geometric integrability of maps (1)

The existence of a continuous invariant foliation for two-dimensional Lorenz type maps is proved in [3], and existence of a C^1-smooth invariant foliation (with C^2-smooth fibres) for these maps is proved in [52]. The proof of a continuous invariant lamination existence for C^1-smooth maps, obtained by small perturbations of skew products of interval maps with Ω-stable quotient, one can find in [28]. The sketch of the proof of the existence of C^1-smooth invariant lamination for these maps is given in [29].

The main result of this part of the paper is the existence theorem of C^1-smooth local lamination for C^1-smooth cylinder maps of the type (1).

Theorem 3.1: *Let $\Phi \in T^1_L(M)$ be a map of the form* (8)*. Let $\delta > 0$. Then there is an ε-neighbourhood $B^1_\varepsilon(\Phi)$ of the map Φ in the space $C^1_L(M)$ such that every map $F \in B^1_\varepsilon(\Phi)$ obtained from Φ by means of the C^1-smooth perturbation $\mu = \mu(x, y)$, where μ satisfies the condition ($ii_{\mu,\varepsilon}$), has an invariant C^1-smooth local lamination $L^1_{loc}(F)$, which is a lamination for $f \in \mathbf{L}_1$, and a foliation for $f \in \mathbf{L}_2$. Fibres of this local lamination start from the points of the set $\Omega(f) \times \{a_2\}$ and are pairwise disjoint graphs of C^1-smooth functions $x = x(y)$ on the interval I_2. Moreover, every curvilinear fibre is ε'-close in the C^1-norm (for some $\varepsilon' > 0$, $\varepsilon' = \varepsilon'(\delta)$) to the vertical closed interval that starts from the same initial point of the set $\Omega(f) \times \{a_2\}$ just as the curvilinear fibre.*

Take an arbitrary $\delta > 0$ and fix it. Using Definition 1.3, we find a number $\varepsilon > 0$ for δ. Then a map $\varphi \in B^1_{1,\varepsilon}(f) \bigcap \mathbf{L}_1$ ($\varphi \in B^1_{1,\varepsilon}(f) \bigcap \mathbf{L}_2$) is Ω-conjugate (topologically conjugate) to f by means of a homeomorphism that is δ-close to the identity map of the set $\Omega(f)$ (of the circle S^1) in the C^0-norm of the uniform convergence.

We use the ε-neighbourhood $B^1_\varepsilon(\Phi)$ of the map Φ in the space $C^1_L(M)$. Then by formulas (1), (8) and the property ($ii_{\mu,\varepsilon}$) the inequality

$$||F-\Phi||_{C^1_L(M)} = ||\mu||_{C^1(M,S^1)} < \varepsilon$$

holds for a map $F \in B^1_\varepsilon(\Phi)$.

Important conclusion follows from the above inequality: for every $y \in I_2$ we have $(f+\mu) \in B^1_{1,\varepsilon}(f)$. Hence, $(f+\mu)$ is Ω-conjugate to f by means of a homeomorphism that is δ-close in the C^0-norm to the identity map of the set $\Omega(f)$ if $f \in \mathbf{L}_1$; and $(f+\mu)$ is topologically conjugate to f by means of a homeomorphism that is δ-close in the C^0-norm to the identity map of the circle S^1 if $f \in \mathbf{L}_2$.

Let ε_m-neighbourhoods $B^1_{\varepsilon_m}(\Phi^m)$ of iterations Φ^m for every $m > 1$ in the space $C^1_L(M)$ correspond to the chosen ε-neighbourhood $B^1_\varepsilon(\Phi)$ of the skew product Φ. By $F \in B^1_\varepsilon(\Phi)$ we have: $F^m \in B^1_{\varepsilon_m}(\Phi^m)$. Using formulas (6)–(8), we obtain the inequalities

$$||\varphi_{y,m} - f^m||_{C^1(M,S^1)} \le ||F^m - \Phi^m||_{C^1_L(M)} < \varepsilon_m;$$

moreover, for every $y \in I_2$ the inequality holds:

$$||\varphi_{y,m} - f^m||_{C^1(S^1)} < ||\varphi_{y,m} - f^m||_{C^1(M,S^1)}.$$

Therefore,

$$\varphi_{y,m} \in B^1_{(2,1),\varepsilon_m}(f^m \circ pr_1) \quad \text{and,} \quad \text{the more so,} \quad \varphi_{y,m} \in B^1_{1,\varepsilon_m}(f^m); \tag{24}$$

moreover, by (3) and (7) –(9) we also have

$$\varphi^m_y \in B^1_{(2,1),\varepsilon_m}(f^m \circ pr_1) \quad \text{and,} \quad \text{the more so,} \quad \varphi^m_y \in B^1_{1,\varepsilon_m}(f^m) \tag{25}$$

for every $y \in I_2$. Here $B^1_{(2,1),\varepsilon_m}(f^m \circ pr_1)$ is the ε_m-neighbourhood of $(f^m \circ pr_1)$ in the space $C^1(M, S^1)$, and $B^1_{1,\varepsilon_m}(f^m)$ is the ε_m-neighbourhood of f^m in the space $C^1(S^1)$. The neighbourhood $B^1_{1,\varepsilon_m}(f^m)$ consists of maps that are Ω-conjugate to f^m by means of homeomorphisms δ-close in the C^0-norm to the identity map on the nonwandering set $\Omega(f^m) - \Omega(f)$ (see Proposition 2.1) if $f \in \mathbf{L}_1$, and of maps that are topologically conjugate to f^m by means of homeomorphisms δ-close in the C^0-norm to the identity map of the circle S^1 if $f \in \mathbf{L}_2$.

At first, repeating arguments of the paper [28] and using the above properties (24)–(25), we prove existence of a continuous one-dimensional local lamination.

Proposition 3.2: *Let $\Phi \in T^1_L(M)$ be a map of the form (8). Let $\delta > 0$. Then there is an ε-neighbourhood $B^1_\varepsilon(\Phi)$ of the map Φ in the space $C^1_L(M)$ such that every map $F \in B^1_\varepsilon(\Phi)$ obtained from Φ by means of the C^1-smooth perturbation $\mu = \mu(x,y)$, where μ satisfies the condition $(ii_{\mu,\varepsilon})$, has an invariant continuous one-dimensional local lamination $L_{loc}(F)$, which is a continuous one-dimensional lamination for $f \in \mathbf{L}_1$, and a continuous one-dimensional foliation for $f \in \mathbf{L}_2$. Fibres of this local lamination start from the points of the set $\Omega(f) \times \{a_2\}$ and are pairwise disjoint graphs of continuous functions $x = x(y)$ on the interval I_2. Moreover, every curvilinear fibre is δ-close in the C^0-norm to the vertical segment that starts from the same initial point of the set $\Omega(f) \times \{a_2\}$ just as the curvilinear fibre; fibres that start from the points (x, a_2), where x is f-periodic point, are C^1-smooth.*

Recall that we prove existence of C^1-smooth invariant one-dimensional fibres with respect to F^m for a natural number m, where $m \in \tau(f)$, using the nonlocal existence theorem of C^1-smooth implicit function $x = x(y)$ satisfying the initial condition $x(a_2) = x^0$ (see the condition (i_μ)), where x^0 is an arbitrary f-periodic point, and the equation

$$\varphi_{y,m}(x) = x \tag{26}$$

on the interval I_2.

Move on to the proof of Theorem 3.1.

Proof: (1) At first, prove that each C^1-smooth fibre started from a point of the set $Per(f) \times \{a_2\}$ is ε'-close in the C^1-norm to the vertical segment that starts from the same initial point just as the curvilinear fibre (the choice of $\varepsilon' > 0$ is described below).

By Proposition 2.1 the equality holds:

$$Per(f) = Per_s(f) \bigcup Per_u(f), \tag{27}$$

where $Per_s(f) \subset \Omega_r(f)$ is a nonempty finite set of sinks if $f \in \mathbf{L}_1$, and $Per_s(f) = \emptyset$ if $f \in \mathbf{L}_2$; $Per_u(f)$ is a set of sources that is empty if and only if $f \in \mathbf{L}_1$, and $Per(f)$ is a single-point set. Then there is a natural number n_* such that the equalities hold (for details of the proof see [29]):

$$\begin{aligned} \inf_{(x;y)\in L_{loc,u}(F)} \left| \frac{\partial}{\partial x}\varphi_{y,n_*}(x) \right| &= M^*; \quad \text{where } M^* > 1; \\ \sup_{(x;y)\in L_{loc,s}(F)} \left| \frac{\partial}{\partial x}\varphi_{y,n_*}(x) \right| &= M_*; \quad \text{where } M_* < 1. \end{aligned} \tag{28}$$

Here $L_{loc,u}(F)$ is the closure (in M) of the set of fibres that start from points of the set $Per_u(f) \times \{a_2\}$, and $L_{loc,s}$ is the (finite) set of fibres that start from points of the set $Per_s(f) \times \{a_2\}$. Suppose further that $n_* = 1$. If $n_* \neq 1$ then we get over consideration of the map F^{n_*}.

Using the nonlocal existence theorem of a C^1-smooth implicit function and Equation (26), we have

$$\frac{\mathrm{d}}{\mathrm{d}y}x(y) = -\frac{\frac{\partial}{\partial y}\varphi_{y,m}(x)}{\frac{\partial}{\partial x}\varphi_{y,m}(x) - 1} \tag{29}$$

for every point $(x, y) \in \gamma_{x^0}$ of a curvilinear fibre γ_{x^0} started from a point (x^0, a_2), where $x^0 \in Per(f)$.

Give an estimate for the absolute value of the derivative $\frac{\mathrm{d}}{\mathrm{d}y}x(y)$ in every point of a curvilinear fibre γ_{x^0}. For this goal we construct the special δ-trajectory of the real trajectory $\{F^j(x; y)\}_{j\geq 0}$. Let I_x be the vertical closed interval that starts from the point $(x;\, a_2)$. Set

$$z_0 = (x^0;\, y); \quad z_1 = (x^1; y^1); \ldots;\, z_i = (x^i; y^i); \ldots; z_{m-1} = (x^{m-1};\, y^{m-1}) \ldots;$$

$$z_{rm+i} = (x^{rm+i};\, y^{rm+i}) \ (r \geq 0;\ 0 \leq i \leq m-1);$$

where

$$\begin{aligned} x^{rm+i} &= f^{rm+i}(x^0) = f^i(x^0); \\ y^{rm+i} &= g_{(x;y);\, rm+i}(y) \quad (\text{here } 1 \leq i \leq m-1;\ m \geq 2) \end{aligned}$$

Then $(x^{rm+i}; y^{rm+i}) \in I_{x^i}$. Since

$$F^{rm+i}(x;y) = (\varphi_{y;\,rm+i}(x);\, g_{(x;y);\,rm+i}(y)); \quad \text{and} \quad F^{rm+i}(x;y) \in \gamma_{x^i},$$

then $\{z_{rm+i}\}_{r\geq 0;\,0\leq i\leq m-1}$ is δ-trajectory for the real trajectory of the point $(x; y)$. Moreover,

$$|f^j(x^0) - \varphi_{y;\,j}(x)| < \delta.$$

Since $f^j(x) \equiv f^j(x^0)$, then for every $j \geq 1$ we obtain the inequality

$$\left|\frac{\partial}{\partial y}\varphi_{y,j}(x)\right| < \varepsilon.$$

Therefore, using (29), we have

$$\left|\frac{\mathrm{d}}{\mathrm{d}y}x(y)\right| < \frac{\varepsilon}{|\frac{\partial}{\partial x}\varphi_{y,m}(x) - 1|} \leq \frac{\varepsilon}{||\frac{\partial}{\partial x}\varphi_{y,m}(x)| - 1|}. \tag{30}$$

If $(x, y) \in \gamma_{x^0}$, where $\gamma_{x^0} \in L_{loc,u}$, then

$$\left|\frac{\mathrm{d}}{\mathrm{d}y}x(y)\right| < \frac{\varepsilon}{(M^*)^m - 1} < \frac{\varepsilon}{M^* - 1}. \tag{31}$$

In this case we set $\varepsilon' = \max\{\delta, \varepsilon/(M^* - 1)\}$.

Let $(x, y) \in \gamma_{x^0}$, where $\gamma_{x^0} \in L_{loc,s}$. Then

$$\left|\frac{\mathrm{d}}{\mathrm{d}y}x(y)\right| < \frac{\varepsilon}{1 - (M_*)^m} < \frac{\varepsilon}{1 - M_*}. \tag{32}$$

In this case we set $\varepsilon' = \max\{\delta, \varepsilon/(1 - M_*)\}$.

(2) Further, we use notations $\{x_{x^0}\}_{x^0\in Per(f)}$ for C^1-smooth functions with graphs $\{\gamma_{x^0}\}_{x^0\in Per(f)}$. By Lagrange mean value theorem and formulas (31)–(32) the family $\{x_{x^0}\}_{x^0\in Per(f)}$ is precompact in the C^0-norm on the closed interval I_2. Moreover, by (31)–(32) for all $x^0 \in Per(f)$, $y', y'' \in I_2$ we have the following inequality for derivatives

$$\left|\left(\frac{\mathrm{d}}{\mathrm{d}y}x_{x^0}(y)\right)_{|y=y'} - \left(\frac{\mathrm{d}}{\mathrm{d}y}x_{x^0}(y)\right)_{y=y''}\right| < 2\varepsilon'.$$

Thus, families of functions $\{x_{x^0}\}_{x^0\in Per(f)}$ and their continuous derivatives $\{\frac{d}{dy}x_{x^0}\}_{x^0\in Per(f)}$ defined on the compact interval I_2 are precompact in the C^0-norm (for the Arzela's Theorem see [40, Ch.2, Section 7]). It means that the family of C^1-smooth functions $\{x_{x^0}\}_{x^0\in Per(f)}$ is precompact in the C^1-norm.

(3) Let $\Omega(f) \setminus Per(f) \neq \emptyset$. It is valid if there is a natural number $p \in \tau(f)$ such that $p/q \notin \{2^i\}_{i\geq 0}$ (see Proposition 2.1). Item 2 makes it possible to prove the existence of C^1-smooth fibres started from points of the set $(\Omega(f) \setminus Per(f)) \times \{a_2\}$.

In fact, let $x^* \in \Omega_p(f) \setminus Per(f)$. Then by item 2 there is a sequence of C^1-smooth on I_2 functions $\{x_{x_n^0}\}_{n\geq 1}$ satisfying the initial conditions $x_{x_n^0}(a_2) = x_n^0$, where $x_n^0 \in Per(f) \bigcap \Omega_p(f)$, and such that this functional sequence converges in the C^1-norm to a C^1-smooth function $x = x_{x^*}(y)$. In addition, the graph of this function passes through the point (x^*, a_2), the inequalities $|x_{x^*}(y) - x^*(y)| \leq \delta$, where $x^*(y) \equiv x^*$, and $|x'_{x^*}(y)| \leq \varepsilon/(M^* - 1)$ hold for every $y \in I_2$.

(4) The constructed above C^1-smooth function x_{x^*} is unique, and it does not depend on the choice of the convergent to x_{x^*} sequence of C^1-smooth functions $\{x_{x_n^0}\}_{n\geq 1}$ for $x_n^0 \in Per(f) \bigcap \Omega_p(f)$.

In fact, take another sequence of points $u_m^0 \in Per(f) \bigcap \Omega_p(f)$ convergent to x^*. By Proposition 3.2, items 1 and 2 of the proof this sequence of points defines the unique sequence of C^1-smooth functions $\{x_{u_m^0}(y)\}_{m\geq 1}$ satisfying the initial conditions $x_{u_m^0}(a_2) = u_m^0$ and the inequalities $|x_{u_m^0}(y) - u_m^0(y)| < \delta$, where $u_m^0(y) \equiv u_m^0$, and $|x'_{u_m^0}(y)| < \varepsilon/(M^* - 1)$ for any $y \in I_2$. Using the item 2 of the proof, we distinguish a convergent in the C^1-norm subsequence of functions $\{x_{u_{m(n)}^0}\}_{n\geq 1}$. Thus, the subsequence $\{x_{u_{m(n)}^0}\}_{n\geq 1}$ is fundamental in the C^1-norm. Form the new functional sequence $\{x_{x_n^0}(y)\}_{n\geq 1} \bigcup \{x_{u_{m(n)}^0}(y)\}_{n\geq 1}$. By the above this functional sequence is fundamental in the C^1-norm and converges to the C^1-smooth function $x = x_{x^*}(y)$. Thus, uniqueness of the function x_{x^*} is proved.

(5) The important conclusion follows from the above considerations: the set of C^1-smooth functions $\{x = x_{x^0}(y)\}_{x^0 \in \Omega(f)}$ that define C^1-smooth fibres of the invariant local lamination in the cylinder M is compact in the C^1-norm.

(6) Parameterize fibres $\{\gamma_{x^0}\}_{x^0 \in Per(f)}$ by the parameter $y \in I_2$. Then by (26) these fibres are given by the formula

$$\begin{cases} \varphi_{y,m}(x) - x = 0, \\ y = y; \end{cases} \tag{33}$$

moreover, the initial condition $x_{x^0}(a_2) = x^0$ is fulfilled for every $x^0 \in Per(f)$. By (33) for Jacobian J on the set of fibres over f-periodic points the formula holds:

$$J = \frac{\partial}{\partial x}\varphi_{y,m}(x) - 1. \tag{34}$$

Using item 1, formulas (28) and (34), we find a number $c > 0$ such that $|J| \geq c$ on every fibre γ_{x^0}. It implies the following property: every fibre γ_{x^0} for $x^0 \in Per(f)$ is locally C^1- diffeomorphic to an interval of the straight line, moreover, by items 1, 4 and 5 every fibre γ_{x^0} for $x^0 \in \Omega(f)$ is also locally C^1- diffeomorphic to an interval of the straight line so that Definition 2.7 of a C^1-smooth local lamination is satisfied. The support $\bigcup_{x^0 \in \Omega(f)} \gamma_{x^0}$ of the defined above local lamination does not coincide with M, when $f \in \mathbf{L}_1$; and $\bigcup_{x^0 \in \Omega(f)} \gamma_{x^0} = M$, when $f \in \mathbf{L}_2$ (see Proposition 2.1) Thus, in the first case we have a C^1-smooth lamination, in the second one we have a C^1-smooth foliation. Theorem 3.1 is proved. ■

Using Theorem 3.1, we use a curvilinear projection H defined on the support of the local lamination $L^1_{loc}(F)$. Let (x, y) be a point of the support of $L^1_{loc}(F)$. Then there is a fibre γ_{x^0} such that $(x, y) \in \gamma_{x^0}$, where $x^0 \in \Omega(f)$. We set

$$H(x, y) = x^0. \tag{35}$$

Corollary 3.3: *Let $\Phi \in T^1_L(M)$ be a map of the form (8). Let $\delta > 0$. Then there is an ε-neighbourhood $B^1_\varepsilon(\Phi)$ of the map Φ in the space $C^1_L(M)$ such that every map $F \in B^1_\varepsilon(\Phi)$ obtained from Φ by means of the C^1-smooth perturbation $\mu = \mu(x, y)$, where μ satisfies the condition $(ii_{\mu,\varepsilon})$, has the following property: the curvilinear projection H (defined by the equality (35)) of the support of the local lamination $L^1_{loc}(F)$ on $\Omega(f)$ is C^1-smooth injection on x for every $y \in I_2$, moreover,*

$$\frac{\partial}{\partial x} H(x, y) \neq 0 \quad \textit{on the support of } L^1_{loc}(F). \tag{36}$$

Proof: Let (x, y) be a point of the support of $L^1_{loc}(F)$. Then $(x, y) \in \gamma_{x^0}$ for some $x^0 \in \Omega(f)$. Since γ_{x^0} is locally C^1-diffeomorphic to an interval of a straight line, then there is a neighbourhood $U = U((x, y))$ such that the arc $\gamma_{x^0} \bigcap U((x, y))$ is C^1-diffeomorphic to an interval of the segment $\{x^0\} \times I_2$. Denote by $\chi_{x^0,U}$ this diffeomorphism. Then by formula (35) the value $H(x, y)$ can be presented in the following form:

$$H(x, y) = pr_1 \circ \chi_{x^0,U}(x, y). \tag{37}$$

Since pr_1 is C^1-smooth map, then formula (37) implies C^1-smoothness of the curvilinear projection H. For all points $x', x'' \in \Omega(f)$, where $x' \neq x''$, the equality holds:

$$\gamma_{x'} \bigcap \gamma_{x''} = \emptyset.$$

This equality jointly with Definition 2.7 of the local lamination $L^1_{loc}(F)$ means that the C^1-smooth map H is surjection of the support of the local lamination $L^1_{loc}(F)$ on $\Omega(f)$; moreover, H is injection on x for every $y \in I_2$. Correctness of the inequality (36) follows immediately from the presentation of $H(x, y)$ by formula (37). Corollary 3.3 is proved. ■

The following claim is the direct corollary of Definition 2.6, Theorem 2.8, Theorem 3.1 and Corollary 3.3.

Theorem 3.4: *Let $\Phi \in T^1_L(M)$ be a map of the form (8). Let $\delta > 0$. Then there is an ε-neighbourhood $B^1_\varepsilon(\Phi)$ of the map Φ in the space $C^1_L(M)$ such that every map $F \in B^1_\varepsilon(\Phi)$ obtained from Φ by means of the C^1-smooth perturbation $\mu = \mu(x, y)$, where μ satisfies the condition $(ii_{\mu,\varepsilon})$, is geometrically integrable on the support of the local lamination $L^1_{loc}(F)$ with the quotient $f_{|\Omega(f)}$ by means of a C^1-smooth curvilinear projection H such that for every $y \in I_2$ the map H is an injection on x satisfying the inequality (36).*

4. Example of maps with one-dimensional ramified attractors

In this part of the paper we consider C^1-smooth maps $F_k : M \to M$ ($k \in \mathbb{N}$; $k \geq 2$, $M = S^1 \times [0, 1]$) of type (11) satisfying (12).

We use here unperturbed C^1-smooth skew products $\Phi_k : M \to M$, where

$$\Phi_k(x, y) = (f_k(x), g_x(y)). \tag{38}$$

Since $f_k \in \mathbf{L}_2$ then the following claim holds (compare with Theorem 3.1).

Theorem 4.1: *Let $\Phi_k \in T_L^1(M)$ be a map of the form (38) satisfying (12). Let $\delta > 0$. Then there is an ε-neighbourhood $B_\varepsilon^1(\Phi_k)$ of the map Φ_k in the space $C_L^1(M)$ such that every map $F_k \in B_\varepsilon^1(\Phi_k)$ obtained from Φ_k by means of the C^1-smooth perturbation $\mu = \mu(x, y)$, where μ satisfies the condition $(ii_{\mu,\varepsilon})$, has an invariant C^1-smooth foliation $L_{loc}^1(F)$. Fibres $\gamma_{x'}$ of the foliation start from the points $(x', 0)$ of the set $S^1 \times \{0\}$ and are pairwise disjoint graphs of C^1-smooth functions $x = x_{x'}(y)$ on the unit interval. Moreover, every curvilinear fibre is ε'-close in the C^1-norm (for some $\varepsilon' > 0$, $\varepsilon' = \varepsilon'(\delta)$) to the vertical closed interval that starts from the same initial point of the set $S^1 \times \{0\}$ just as the curvilinear fibre.*

To construct maps $g_x(y)$ on $S^1 \times [0, 1]$ ($S^1 = [0, 1]^*$) for the cylinder map (11) satisfying (12), we use a C^1-smooth Urysohn function $y = h_k(x)$, where $x \in [0, 1]^*$ such that

$$h_k(0) = h_k(1) = \frac{3}{4}, \quad h_k\left(\frac{1}{k}\left[\frac{k}{2}\right]\right) = 0;$$

$$h_k'(0) = h_k'(1) = h_k'\left(\frac{1}{k}\left[\frac{k}{2}\right]\right) = 0.$$

Here $[\cdot]$ means integer part of a number. Moreover, h_k is strictly monotone on each interval $[0, 1/k[k/2])$ and $[1/k[k/2], 1]$; and each of these intervals contains the unique inflection point $x_*' \in (0, 1/k[k/2])$ and $x_*'' \in (1/k[k/2], 1)$. For example, we can use the function given by the following formula (for its graph see Figure 2):

$$h_k(x) = \begin{cases} \dfrac{3}{4}\cos\dfrac{\pi k}{2\left[\frac{k}{2}\right]}x, & \text{if } x \in \left[0, \dfrac{1}{k}\left[\dfrac{k}{2}\right]\right); \\[2ex] \dfrac{3}{4}\cos\dfrac{\pi k}{2\left(k - \left[\frac{k}{2}\right]\right)}(x-1), & \text{if } x \in \left[\dfrac{1}{k}\left[\dfrac{k}{2}\right], 1\right]. \end{cases} \tag{39}$$

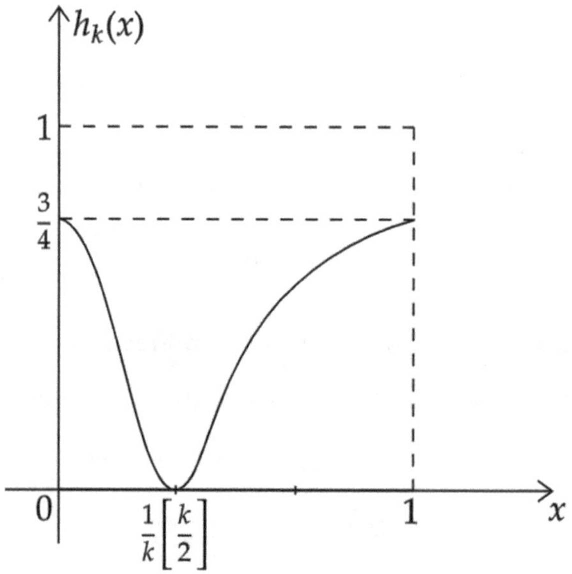

Figure 2. The graph of Urysohn function h_k for $k = 3$.

Let δ satisfy the inequalities $0 < \delta < \overline{\delta} = 2^{-1}\min\{|(h_k'(x_*'))^{-1}|, |(h_k'(x_*''))^{-1}|\}$. By the above $|(h_k'(x_*'))^{-1}|, |(h_k'(x_*''))^{-1}| < +\infty$. Using this δ, we choose $\varepsilon > 0$ so that $\varepsilon < \delta$. Then by item 1 of the proof of Theorem 3.1 the equality $\varepsilon' = \delta$ holds in the case under consideration. This implies the inequality $\varepsilon' < 2^{-1}\min\{|(h_k'(x_*'))^{-1}|, |(h_k'(x_*''))^{-1}|\}$.

Then by Theorem 4.1 the graph of the function $y = h_k(x)$ intersects every fibre $\gamma_{x'}$ in the unique point without tangency, and the equality $y = h_k(x_{x'}(y))$ is equivalent to $y = y(x')$, where $x' \in S^1$; moreover, the function $y = y(x')$ is C^1-smooth on S^1.

We use two connected sets

$$D' = \left\{ (x,y) \in \bigcup_{x' \in [0,1]^*} \gamma_{x'} : 0 \le y \le y(x') \right\};$$

$$D'' = \left\{ (x,y) \in \bigcup_{x' \in [0,1]^*} \gamma_{x'} : y(x') < y \le 1 \right\}.$$

Define C^1-smooth function $g(x,y)$ for maps (11) satisfying (12). For this goal we set

$$g_x(y) = \begin{cases} y, & \text{if } (x,y) \in \gamma_{x'} \bigcap D', \\ h_k(x) + \sin(y - h_k(x)), & \text{if } (x,y) \in \gamma_{x'} \bigcap D''. \end{cases} \tag{40}$$

Reduce maps (11) satisfying (12) and (40) to skew products on the cylinder M. Using Theorem 2.9 and Corollary 3.3, we obtain that the map $\widetilde{H}$ defined by the equality (23) is C^1-diffeomorphism. Therefore, every map (11) satisfying (12) and (40), is C^1-smoothly conjugate (under C^1-smooth diffeomorphism $\widetilde{H}$) to the skew product

$$\widehat{\Phi}_k(x',y) = (f_k(x'), g_x(y)), \tag{41}$$

where $g_x(y)$ is presented by formula (40) for x satisfying

$$x = pr_1 \circ \widetilde{H}^{-1}(x',y). \tag{42}$$

Finally, we denote by $\widehat{g}_{x'}$ fibre maps of the skew product $\widehat{\Phi}_k$. Graphs of fibre maps for different $x' \in [0,1]^*$ are presented in Figure 2(a,b).

As usually, denote by $\widehat{g}_{x',n}$ fibre maps of iterations $\widehat{\Phi}_k^n$ $(n \ge 2)$ of the map $\widehat{\Phi}_k$.

We need some properties of skew products $\widehat{\Phi}_k = (f_k, \widehat{g}_{x'})$ $(k \ge 2)$.

Lemma 4.2: *For each $x' \in [0,1]^*$ and $n \ge 2$ the equality $\widehat{g}_{x',n}(y) = y$ is fulfilled on the closed interval $[0, c_n^k(x')] \subseteq [0, 3/4]$, where*

$$c_n^k(x') = \min\{y(x'), y(f_k(x')), \dots, y(f_k^{n-1}(x'))\}, \tag{43}$$

and points $y(x'), y(f_k(x')), \dots, y(f_k^{n-1}(x'))$ lie on the graph of the C^1-smooth function $y = y(x')$. Moreover, for all $y \in (c_n^k(x'), 1]$ the inequality $\widehat{g}_{x',n}(y) < y$ holds.

Proof: Let $n = 2$. Then by formulas (40)–(42) the equality $\widehat{g}_{x',2}(y) = \widehat{g}_{f_k(x')} \circ \widehat{g}_{x'}(y)$ holds for every $x' \in [0,1]^*$, and $\widehat{g}_{x',2}(y) = y$ on the intersection of two intervals

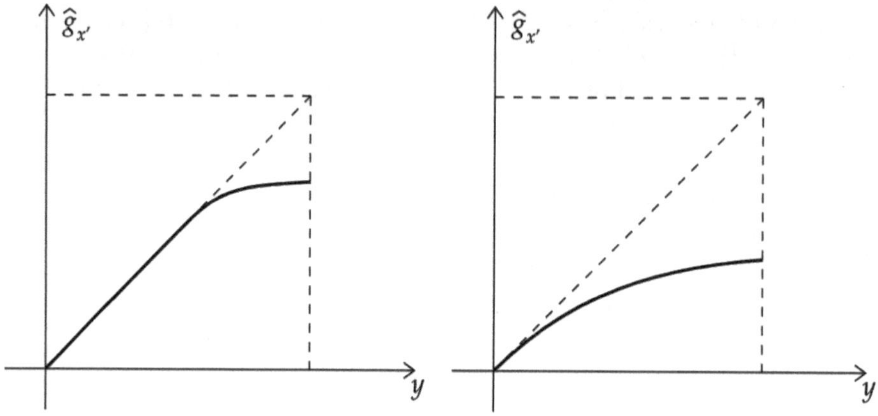

(a) (b)

Figure 3. (a) The graph of $\widehat{g}_{x'}$ for $x' \neq 1/k[k/2]$. (b) The graph of $\widehat{g}_{x'}$ for $x' = 1/k[k/2]$.

$[0, y(x')] \bigcap [0, y(f_k(x'))] = [0, c_2^k(x')]$ (see formula (43)), where $0 \leq c_2^k(x') \leq 3/4$; in addition, $\widehat{g}_{x',2}(y) < y$ if $y \in (c_2^k(x'), 1]$. Thus, Lemma 4.2 holds for $n = 2$. Suppose that Lemma 4.2 is valid for $n-1$ if $n \geq 3$.

Prove its correctness for n. In fact, using formula (9), for any $x' \in [0, 1]^*$ we obtain the equality $\widehat{g}_{x',n}(y) = \widehat{g}_{f_k^{n-1}(x')} \circ \widehat{g}_{x',n-1}(y)$. Then by the inductive assumption the equality $\widehat{g}_{x',n}(y) = y$ is correct on the intersection of closed intervals $[0, c_{n-1}^k(x')] \bigcap [0, y(f_k^{n-1}(x'))] = [0, c_n^k(x')]$ $(c_n^k(x') \in [0, 3/4])$, as it follows from formula (39)); in addition, $\widehat{g}_{x',n}(y) < y$ if $y \in (c_n^k(x'), 1]$. Lemma 4.2 is proved. ■

The following claim is the direct corollary of equality (43).

Corollary 4.3: *For any $x' \in [0, 1]^*$ the number sequence $\{c_n^k(x')\}_{n \geq 1}$ monotonically decreases in n so that the inequality is satisfied:*

$$c_n^k(x') \geq c_{n+1}^k(x') \quad \textit{for all } n \in \mathbb{N}.$$

Note that the point $x'_* = 1/k[k/2]$ is the homoclinic point for the fixed point $x' = 0 \pmod 1$ of f_k (for the definition of a homoclinic point see, e.g. [11]). Moreover, by (39) $h_k(1/k[k/2]) = 0$, there are no other zeros of h_k (see Figure 1), and by equalities (40)–(42) $\widehat{g}_{x'_*}(y) = \sin y$ for $y \in [0, 1]$ (see Figure 2 (b)). It means that $c_n^k(x') = 0$ if and only if for some i satisfying $0 \leq i \leq n-1$, the equality holds: $f_k^i(x') = 1/k[k/2]$.

Therefore, the trajectories of all points of the circle can be divided into two classes by separability (or, conversely, its absence) from the point $x'_* = 1/k[k/2]$. For example, the closure of every uniformly recurrent (in particular, periodic) trajectory is separated from this homoclinic point; and the closure of a transitive trajectory is not separated from x'_*. Set

$$c^k(x') = \inf_{n \geq 1} \{c_n^k(x')\} \quad \text{for all } x' \in [0, 1]^*.$$

By Corollary 4.3 the equality holds:

$$c^k(x') = \lim_{n\to+\infty} c^k_n(x'). \tag{44}$$

Thus, we obtain the following statement.

Lemma 4.4: *The equality $c^k(x') = 0$ holds if and only if the point $x'_* = 1/k[k/2]$ belongs to the closure of the trajectory $\overline{\{f^n_k(x')\}_{n\ge 0}}$ of a point $x' \in [0,1]^*$.*

Proof: By the equality (43) there is a nondecreasing sequence of natural numbers s_n satisfying $c^k_n = y(f^{s_n}_k(x'))$. Then by definitions of the function $y = y(x')$ or Uryson function h_k (see Figure 1), continuity of these functions and formula (44) the equality $c^k(x') = 0$ for a point $x' \in [0,1]^*$ holds if and only if

$$\lim_{n\to+\infty} f^{s_n}_k(x') = \frac{1}{k}\left[\frac{k}{2}\right].$$

Thus, Lemma 4.4 is proved. ■

The following statement is the direct corollary of Lemma 4.4.

Corollary 4.5: *If x' is uniformly recurrent, in particular, periodic point, then $c^k(x') > 0$.*

In fact, in conditions of Corollary 4.5 closure of the trajectory of x' does not contain $x'_* = 1/k[k/2]$. Therefore, by Lemma 4.4 the inequality $c^k(x') > 0$ holds.

Lemma 4.6: *For every $n \ge 1$ the auxiliary function $\Omega^{\widehat{\Phi}_k}_n : [0,1]^* \to 2^{[0,1]}$ for the Ω-function of the skew product $\widehat{\Phi}_k \in T^1_L(M)$ $(k \ge 2)$ is continuous.*

Proof: By Lemma 4.2 we have:

$$\begin{aligned} \Omega(\widehat{g}_{x'}) &= [0, y(x')] = \mathrm{Fix}(\widehat{g}_{x'}); \\ \Omega(\widehat{g}_{x',n}) &= [0, c^k_n(x')] = \mathrm{Fix}(\widehat{g}_{x',n}), \quad n \ge 2. \end{aligned} \tag{45}$$

Here $\mathrm{Fix}(\cdot)$ is the set of fixed points of a map.

Then by Definition 2.4, formulas (43) and (45) the graph of an auxiliary function $\Omega^{\widehat{\Phi}_k}_n$ $(n \ge 1)$ in the phase space M is the set, boundary of which is the union of the circle $[0,1]^* \times \{0\}$, the closed intervals $\{0 \pmod 1\} \times [0, 3/4]$ and some curve. This curve consists of points $\{(x', c^k_n(x'))\}_{x'\in[0,1]^*}$ and can be defined like this.

Consider the image of the graph C^* of the C^1-smooth function $y = y(x')$ with respect to the C^1-diffeomorphism $(\widehat{\Phi}_k)_n$, where $(\widehat{\Phi}_k)_n(x', y) = (id(x'), \widehat{g}_{x',n}(y))$ for $(x', y) \in M$ (see formulas (14), (39)– (42)). Then by Lemma 4.2 we obtain:

$$c^k_n(x') = \min\{((\widehat{\Phi}_k)_n(C^*))(x'), \ldots, ((\widehat{\Phi}_k)_n(C^*))(f^{n-1}_k(x'))\} \tag{46}$$

By the above and formula (46) the curve of points $\{(x', c^k_n(x'))\}_{x'\in[0,1]^*}$ is continuous. Continuity of this curve implies upper and lower semicontinuity of multifunctions $\Omega^{\widehat{\Phi}_k}_n$, i.e.

for every $n \geq 1$ the multifunction $\Omega_n^{\widehat{\Phi}_k}$ is continuous (for definitions and criteria see [41, Ch.4, Section 43, II]). Lemma 4.6 is proved. ■

Corollary 4.7: *Let x_1', x_2' be different points of the same periodic orbit $Orb_{x_0'}(f_k) = \{x_0', f_k(x_0'), \ldots, f_k^{m-1}(x_0')\}$ generated by a point $x_0' \in Per(f_k)$ with (least) period $m > 1$ of the quotient f_k (here $\tau(f_k) = \mathbb{N}$ [34]) of the skew product $\widehat{\Phi}_k \in T_L^1(M)$ ($k \geq 2$). Then $c^k(x_0') > 0$, and for every $n \geq m$ the equalities hold:*

$$\Omega(\widehat{g}_{x_1',n}) = \Omega(\widehat{g}_{x_2',n}) = [0, c_n^k(x_0')] = [0, c^k(x_0')].$$

Proof: Using formula (43), we obtain for $j = 1, 2$:

$$c_n^k(x_j') = \min\{x' \in (Orb_{x_0'}(f_k) \bigcup \{f_k(x_j'), \ldots, f_k^{n-1}(x_j')\})\}. \tag{47}$$

Since $\{f_k(x_j'), \ldots, f_k^{n-1}(x_j')\} \subset Orb_{x_0'}(f_k)$, then by (47) we obtain the equality

$$c_n^k(x_1') = c_n^k(x_2') = c_n^k(x_0') \quad \text{if } n \geq m.$$

For the periodic orbit $Orb_{x_0'}(f_k)$, as for a fiinite set, the equality

$$c_n^k(x_0') = c^k(x_0')$$

holds. Hence,

$$c_n^k(x_1') = c_n^k(x_2') = c^k(x_0').$$

By Corollary 4.5 the inequality $c^k(x_0') > 0$ is fulfilled, Corollary 4.7 is proved. ■

Using Theorem 2.5, Lemmas 4.2, 4.4, 4.6 and corollaries 4.5, 4.7, we obtain the following claim.

Proposition 4.8: *The set $\bigcup_{(x',y)\in M} \omega_{\widehat{\Phi}_k}((x',y))$ is closed, and the equality holds:*

$$\bigcup_{(x',y)\in M} \omega_{\widehat{\Phi}_k}((x',y)) = \bigcup_{x'\in[0,1]^*} \{x'\} \times [0, c^k(x')] = \overline{Per(\widehat{\Phi}_k)}. \tag{48}$$

Here a closed interval $[0, c^k(x')]$ is degenerate if and only if the closure of the f_k-trajectory of a point x' contains the point $x_*' = 1/k[k/2]$.

For any neighbourhood U of the set $\bigcup_{(x',y)\in M} \omega_{\widehat{\Phi}_k}((x',y))$ there exists an $m = m(U)$ such that for the trajectory of an arbitrary point $(x',y) \in M$ the inclusion holds:

$$\{\widehat{\Phi}^n(x',y)\}_{n\geq m(U)} \subset U. \tag{49}$$

Correctness of the second equality in formula (48) follows immediately from item (2.1.2) of Proposition (2.1) and Lemma 4.2, since $(x',y) \in Per(\widehat{\Phi}_k)$ if and only if $x' \in Per(f_k)$, and y is a fixed point of $\widehat{g}_{x',n(x')}$, where $n(x')$ is (the least) period of x' (see also [31]).

An additional property of the Birkhoff constant, described in the inclusion (49) (compare with the claim of Theorem 2.5), follows from the monotonicity of fibres maps $\widehat{g}_{x',n}(y)$ with respect to $y \in [0, 1]$ for every $x' \in [0, 1]^*$ and every $n \in \mathbb{N}$ (see formulas (40)–(42) and Lemma 4.2).

Lemma 4.6 also implies the closure of the graph of an auxiliary multifunction $\Omega_n^{\widehat{\Phi}_k}$ for all $n \in \mathbb{N}$.

Corollary 4.9: *For every $n \geq 1$ a suitable function $\overline{\Omega}_n^{\widehat{\Phi}_k}$, approximating functions $\Omega_{n,1}^{\widehat{\Phi}_k}$ and $\overline{\Omega}_{n,1}^{\widehat{\Phi}_k}$ for the Ω-function of the skew product $\widehat{\Phi}_k \in T_L^1(M)$ $(k \geq 2)$ are continuous; moreover, the equalities hold:*

$$\overline{\Omega}_n^{\widehat{\Phi}_k} = \Omega_n^{\widehat{\Phi}_k}; \ \Omega_{n,1}^{\widehat{\Phi}_k} = \overline{\Omega}_{n,1}^{\widehat{\Phi}_k}.$$

Remark 4.1: Different admissible combinations of continuity or discontinuity properties of auxiliary and suitable functions for Ω-functions of C^1-smooth skew products with Ω-stable (structurally stable) quotients define a natural decomposition of the space of these maps (see [22,25] for skew products of interval maps).

Our next goal is to describe the structure of the nonwandering set of skew products $\widehat{\Phi}_k$ on the cylinder. Point out, for the comparison, that papers [23–25] contain general results on the structure of the nonwandering set of skew products of interval maps with Ω-stable (in the C^1-norm) quotients.

The continuity property of auxiliary functions, proved in Lemma 4.6, helps to reduce the influence of the memory of skew products (see equalities (18)), allowing fibres to be used only over periodic points of quotients.

Recall the following property of maps f_k (see formula (12) and Figure 3): each interval $[i/k^n, (i+1)/k^n]$ (here $n \geq 1$, $0 \leq i \leq k^n - 1$) contains a periodic point with a least period, which divides n (every such point is a fixed point of f_k^n). Denote by $Per(f_k, n)$ the set of f_k-periodic points with least periods that divide n.

Further we use subsequences of graphs (in the phase space M) of auxiliary, suitable, approximating multifunctions $\{\Omega_{n_i}^{\widehat{\Phi}_k}\}_{i\geq 1}$, $\{\overline{\Omega}_{n_i}^{\widehat{\Phi}_k}\}_{i\geq 1}$, $\{\Omega_{n_i,1}^{\widehat{\Phi}_k}\}_{i\geq 1}$ and $\{\overline{\Omega}_{n_i,1}^{\widehat{\Phi}_k}\}_{i\geq 1}$ respectively for return times $n_i = i!$ and the definition of the topological limit of a set sequence (see [41, Ch. 1, Section 29, I–VI; Ch. 4, Section 43, II]).

Let X be a compact metric space (we consider here cases $X = M$ and $X = [0, 1]$), and $(2^X)_m$ be the space of all closed subsets of X with the Hausdorff metric dist_X (see [41, Ch. 2, Section 21, VII]).

For any two nonempty closed sets $A, B \subset X$ we set

$$\mathrm{dist}_X(A, B) = \max \left\{ \sup_{x\in A} \rho(x, B), \ \sup_{y\in B} \rho(y, A) \right\},$$

where ρ is a metric in X, $\rho(t, C)$ is the distance between a point $t = x$ or y and a set $C = B$ or A respectively.

Note also that by compactness of X the space 2^X with the exponential topology and the space $(2^X)_m$ with the topology generated by Hausdorff metric dist_X are isomorphic [41, Ch. 4, Section 42, II].

Theorem 4.10: *For a skew product* $\widehat{\Phi}_k \in T^1_L(M)$ $(k \geq 2)$ *there exist topological limits* $Lim_{i\to+\infty}\Omega_{n_i}^{\widehat{\Phi}_k}$ $Lim_{i\to+\infty}\overline{\Omega}_{n_i}^{\widehat{\Phi}_k}$, $Lim_{i\to+\infty}\Omega_{n_i,1}^{\widehat{\Phi}_k}$, $Lim_{i\to+\infty}\overline{\Omega}_{n_i,1}^{\widehat{\Phi}_k}$, $Lim_{i\to+\infty}\Omega_{n_i}^{\widehat{\Phi}_k|P_i}$, $Lim_{i\to+\infty}\Omega_{n_i,1}^{\widehat{\Phi}_k|P_i}$, *where* $\widehat{\Phi}_k|P_i$ *is the restriction of* $\widehat{\Phi}_k$ *on the set* $P_i = Per(f_k, n_i) \times [0,1]$, *and equalities hold:*

$$\Omega(\widehat{\Phi}_k) = \underset{i\to+\infty}{Lim}\ \Omega_{n_i}^{\widehat{\Phi}_k} = \underset{i\to+\infty}{Lim}\ \overline{\Omega}_{n_i}^{\widehat{\Phi}_k} = \underset{i\to+\infty}{Lim}\ \Omega_{n_i,1}^{\widehat{\Phi}_k} = \underset{i\to+\infty}{Lim}\ \overline{\Omega}_{n_i,1}^{\widehat{\Phi}_k} =$$

$$= \underset{i\to+\infty}{Lim}\ \Omega_{n_i}^{\widehat{\Phi}_k|P_i} = \underset{i\to+\infty}{Lim}\ \Omega_{n_i,1}^{\widehat{\Phi}_k|P_i} = \overline{\bigcup_{x'\in Per(f_k)} \{x'\} \times \Omega(\widehat{g}_{x',n(x')})} = \overline{Per(\widehat{\Phi}_k)}.$$

Here $n(x')$ *means the least period of* $x' \in Per(f_k)$.

Moreover, the nonwandering set $\Omega(\widehat{\Phi}_k)$ *is continuum, and the equality holds*

$$\Omega(\widehat{\Phi}_k) = \bigcup_{(x',y)\in M} \omega_{\widehat{\Phi}_k}((x',y)).$$

Proof: (1) We begin from the existence of $Lim_{i\to+\infty}\Omega_{n_i}^{\widehat{\Phi}_k|P_i}$.

In fact, since $n_i = i!$ then by Lemma 4.6 (see formula (45)) for $i \geq 1$ the inclusion holds

$$\Omega_{n_i}^{\widehat{\Phi}_k|P_i} \subset \Omega_{n_{i+1}}^{\widehat{\Phi}_k|P_{i+1}}.$$

It implies existence of the topological limit $Lim_{i\to+\infty}\Omega_{n_i}^{\widehat{\Phi}_k|P_i}$ and correctness of the equalities (see [41, Ch. 1, Section 29, II]):

$$\underset{i\to+\infty}{Lim}\ \Omega_{n_i}^{\widehat{\Phi}_k|P_i} = \overline{\bigcup_{x'\in Per(f_k)} \{x'\} \times \Omega(\widehat{g}_{x',n(x')})} = \overline{Per(\widehat{\Phi}_k)}. \tag{50}$$

Using equalities (18), we immediately obtain from here that there is $Lim_{i\to+\infty}\Omega_{n_i,1}^{\widehat{\Phi}_k|P_i}$ satisfying:

$$\underset{i\to+\infty}{Lim}\ \Omega_{n_i,1}^{\widehat{\Phi}_k|P_i} = \overline{Per(\widehat{\Phi}_k)}. \tag{51}$$

(2) Prove that there exists $Lim_{i\to+\infty}\Omega_{n_i}^{\widehat{\Phi}_k}$, and the equality is valid:

$$\underset{i\to+\infty}{Lim}\ \Omega_{n_i}^{\widehat{\Phi}_k|P_i} = \underset{i\to+\infty}{Lim}\ \Omega_{n_i}^{\widehat{\Phi}_k}. \tag{52}$$

In fact, let ε_* be a positive number. Since M is a compact, then the existence of the topological limit $Lim_{i\to+\infty}\Omega_{n_i}^{\widehat{\Phi}_k|P_i}$ is equivalent to the following property: for $\varepsilon_* > 0$ there is $i_* \in \mathbb{N}$ such that for every $i \geq i_*$ the inequality holds:

$$\mathrm{dist}_M\left(\Omega_{n_i}^{\widehat{\Phi}_k|P_i},\ \underset{i\to+\infty}{Lim}\ \Omega_{n_i}^{\widehat{\Phi}_k|P_i}\right) < \varepsilon_*. \tag{53}$$

Let the inequality $i \geq i_*$ be fulfilled. Use continuity of auxiliary multifunctions $\Omega_{n_i}^{\widehat{\Phi}_k}$ (see Lemma 4.6) and for ε_* find $\theta_i > 0$ $(\theta_i < \varepsilon_*)$ such that for all $x_1', x_2' \in [0,1]^*$

satisfying $|x'_1 - x'_2| < \theta_i$ the inequality holds:

$$\mathrm{dist}_{[0,1]}\left(\Omega_{n_i}^{\widehat{\Phi}_k}(x'_1),\ \Omega_{n_i}^{\widehat{\Phi}_k}(x'_2)\right) < \varepsilon_*. \tag{54}$$

Apply claim (2.1.5) of Proposition 2.1 and point out a periodic orbit $Orb_{x'_*}(f_k)$ (of some point $x'_* \in Per(f_k)$) that approximates $[0,1]^*$ up to θ_i. Note that (least) period of x'_* is not less then k^{n_i}. Let this period divide the number $n_{i(x'_*)}$, where $i(x'_*) > i \geq i_*$. Then the set $Per(f_k, n_{i(x'_*)})$ uniformly approximates $[0,1]^*$ up to θ_i. It means that for a point $x'_1 \in [0,1]^*$ there exists, at least, one point $x'_2 \in Per(f_k, n_{i(x'_*)})$ such that $|x'_1 - x'_2| < \theta_i$. Hence, the inequality (54) is valid for these points. Jointly with compactness of M and Lemma 4.6 it implies the inequality

$$\mathrm{dist}_M\left(\Omega_{n_i}^{\widehat{\Phi}_k},\ \Omega_{n_{i(x'*)}}^{\widehat{\Phi}_k|P_{i(x'_*)}}\right) < \varepsilon_* \quad \text{(see [41, Ch. 2, Section 21, X]).}$$

Using the inequality (53), we obtain from here that there exists $Lim_{i\to+\infty}\,\Omega_{n_i}^{\widehat{\Phi}_k}$, and the equality (52) is fulfilled.

(3) Applying equalities (18) and repeating arguments of the previous item 2, we verify that there exists the topological limit $Lim_{i\to+\infty}\,\Omega_{n_i,1}^{\widehat{\Phi}_k}$, and the equality holds

$$\underset{i\to+\infty}{Lim}\ \Omega_{n_i,1}^{\widehat{\Phi}_k} = \underset{i\to+\infty}{Lim}\ \Omega_{n_i,1}^{\widehat{\Phi}_k|P_i}. \tag{55}$$

(4) Using equalities (50)–(52), (55) and Corollary 4.9, we make sure that

$$\begin{aligned}\underset{i\to+\infty}{Lim}\ \Omega_{n_i}^{\widehat{\Phi}_k} &= \underset{i\to+\infty}{Lim}\ \overline{\Omega}_{n_i}^{\widehat{\Phi}_k} = \underset{i\to+\infty}{Lim}\ \Omega_{n_i,1}^{\widehat{\Phi}_k} = \underset{i\to+\infty}{Lim}\ \overline{\Omega}_{n_i,1}^{\widehat{\Phi}_k} = \underset{i\to+\infty}{Lim}\ \Omega_{n_i}^{\widehat{\Phi}_k|P_i} = \\ &= \underset{i\to+\infty}{Lim}\ \Omega_{n_i,1}^{\widehat{\Phi}_k|P_i} = \overline{\bigcup_{x'\in Per(f_k)} \{x'\}\times\Omega(\widehat{g}_{x',n(x')})} = \overline{Per(\widehat{\Phi}_k)}.\end{aligned} \tag{56}$$

By (56), compactness of M and connectedness (in M) of graphs of continuous auxiliary functions $\Omega_{n_i}^{\widehat{\Phi}_k}$ the set $Lim_{i\to+\infty}\,\Omega_{n_i}^{\widehat{\Phi}_k}$ is continuum (see [41, Ch. 5, Section 47, II]).

(5) Using equalities (48) and (56), we obtain the equalities

$$\underset{i\to+\infty}{Lim}\ \Omega_{n_i}^{\widehat{\Phi}_k} = \bigcup_{(x',y)\in M} \omega_{\widehat{\Phi}_k}((x',y)) = \bigcup_{x'\in[0,1]^*} \{x'\}\times[0, c^k(x')] = \overline{Per(\widehat{\Phi}_k)}. \tag{57}$$

Prove the equality

$$\Omega(\widehat{\Phi}_k) = \underset{i\to+\infty}{Lim}\ \Omega_{n_i}^{\widehat{\Phi}_k} = \overline{Per(\widehat{\Phi}_k)}. \tag{58}$$

In fact, let (x',y) be an arbitrary point of the open set $M \setminus \overline{Per(\widehat{\Phi}_k)}$. We separate the point (x',y) and the closed set $\overline{Per(\widehat{\Phi}_k)}$ by neighbourhoods $U((x',y))$ and $U = U(\overline{Per(\widehat{\Phi}_k)})$ respectively with disjoint closures. Then by Proposition 4.8 (see inclusion (49)) we find the Birkhoff constant $m = m(U)$ such that for every $n \geq m(U)$ the inclusion is valid:

$$\widehat{\Phi}_k^n(U((x',y))) \subset U.$$

Since the closures of neighbourhoods $U((x',y))$ and U do not intersect, then

$$U((x',y)) \bigcap \widehat{\Phi}_k^n(U((x',y))) = \emptyset$$

for all $n \geq m(U)$. Hence, (x',y) is a wandering point of $\widehat{\Phi}_k$, and equality (58) is valid. Theorem 4.10 is proved.

■

Use Definition 2.3 and reformulate Theorem 4.10 for the Ω-function of the skew product $\widehat{\Phi}_k$.

Corollary 4.11: *The Ω-function $\Omega^{\widehat{\Phi}_k} : [0,1]^* \to 2^{[0,1]}$ of a skew product $\widehat{\Phi}_k \in T_L^1(M)$ ($k \geq 2$) possesses the following properties:*

(4.11.1) *it is upper semicontinuous;*

(4.11.2) *in each point $x' \in [0,1]^*$ with f_k-trajectory, the closure of which does not contain $x'_* = 1/k[k/2]$, $\Omega^{\widehat{\Phi}_k}$ is discontinuous function, and $\Omega^{\widehat{\Phi}_k}(x') = [0, (c^k)(x')]$, where $c^k(x') > 0$;*

(4.11.3) *in each point $x' \in [0,1]^*$ with f_k-trajectory, the closure of which contains $x'_* = 1/k[k/2]$, $\Omega^{\widehat{\Phi}_k}$ is continuous function, and $\Omega^{\widehat{\Phi}_k}(x') = \{0\}$;*

(4.11.4) *the graph of the Ω-function $\Omega^{\widehat{\Phi}_k}$ in M is one-dimensional topological continuum; its ramification points form an everywhere dense subset of $[0,1]^* \times \{0\}$ of the cardinality equal to the continuum c, and each ramification point has the order 3 (see Figure 4);*

(4.11.5) *the graph of the Ω-function $\Omega^{\widehat{\Phi}_k}$ in M is the global chaotic attractor of $\widehat{\Phi}_k$ that coincides with the closure of the set $Per(\widehat{\Phi}_k)$ and with the closed set $\bigcup_{(x',y)\in M} \omega_{\widehat{\Phi}_k}((x',y))$; moreover, the second eigenvalue of each $\widehat{\Phi}_k$-periodic point equals 1.*

Proof: (1) Since $[0,1]$ is compact, $[0,1]^*$ is Hausdorff, and $\Omega(\widehat{\Phi}_k)$ is closed, it follows from Definition 2.3 that the Ω-function of $\widehat{\Phi}_k$ is upper semicontinuous (see also [25]). Thus, the claim (4.11.1) is correct.

(2) Correctness of claims (4.11.2) and (4.11.3) follows from Theorem 4.10 (see equalities (57)–(58)) and Lemma 4.4. Note also that a point $x' \in [0,1]^*$ with f_k-trajectory, the closure of which does not contain $x'_* = 1/k[k/2]$, is a discontinuity point of the Ω-function. In fact, every such point is not a point of lower semicontinuity, because for a sequence $\{x'_i\}_{i\geq 1}$ of f_k-transitive points, convergent to x', the equality holds:

$$\lim_{i\to+\infty} \Omega^{\widehat{\Phi}_k}(x'_i) = \{0\}, \quad \text{and} \quad \{0\} \neq \Omega^{\widehat{\Phi}_k}(x').$$

(3) Since the graph of the Ω-function $\Omega^{\widehat{\Phi}_k}$ in M coincides with the nonwandering set $\Omega(\widehat{\Phi}_k)$, then by Theorem 4.10 this graph is a continuum. By item (4.11.3) there is

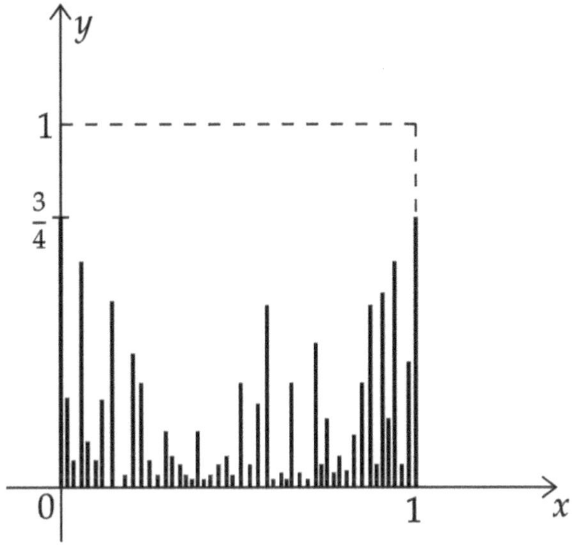

Figure 4. The graph of the Ω-function $\Omega^{\widehat{\Phi}_k}$ (it coincides with the global attractor).

everywhere dense set of points of $[0, 1]^*$ (for example, the set of f-transitive points) such that the value of the Ω-function in every point of this set consists of the unique point 0. This means that the graph of the Ω-function does not contain a homeomorphic image of a circle, and, hence, this graph is one-dimensional continuum. Moreover, by item (4.11.2) and Definition 1.2 each point $(x'; 0)$, where $x' \in (0, 1)$, and the closure of f_k-trajectory of x' does not contain $x'_* = 1/k[k/2]$, is a ramification point of the order 3. The set of ramification points contains also points $(x'; 0)$, where f_k-uniformly recurrent points x' form f_k-minimal sets other than periodic orbits. Therefore, the cardinality of the set of ramification points equals $\mathfrak{c}$. Thus, claim (4.11.4) is valid.

(4) Prove that the graph of the Ω-function (i.e. the nonwandering set $\Omega(\widehat{\Phi}_k)$) is the global attractor. In fact, by the definition of the map $\widehat{\Phi}_k$ (see formulas (12), (40)) the phase space M is the maximal absorbing domain so that strict inclusions are fulfilled:

$$M \supset \widehat{\Phi}_k(M) \supset \ldots \supset \widehat{\Phi}_k^n(M) \supset \ldots.$$

Since every set $\widehat{\Phi}_k^n(M)$ $(n \geq 1)$ is a compact, then there exists a topological limit

$$L^*(\widehat{\Phi}_k) = \underset{n \to +\infty}{Lim} \widehat{\Phi}_k^n(M) = \bigcap_{n=1}^{+\infty} \widehat{\Phi}_k^n(M).$$

Prove that $L^*(\widehat{\Phi}_k) = \Omega(\widehat{\Phi}_k)$. In fact, using the inclusion $\Omega(\widehat{\Phi}_k) \subset M$ and the invariance of $\Omega(\widehat{\Phi}_k)$, we obtain inclusions $\Omega(\widehat{\Phi}_k) \subset \widehat{\Phi}_k^n(M)$ $(n \geq 1)$. Hence, $\Omega(\widehat{\Phi}_k) \subset L^*(\widehat{\Phi}_k)$. Suppose that there is a point $(x'_*, y_*) \in L^*(\widehat{\Phi}_k) \setminus \Omega(\widehat{\Phi}_k)$. Let U be a neighbourhood of the nonwandering set $\Omega(\widehat{\Phi}_k)$, $U((x'_*, y_*))$ be a neighbourhood of the point (x'_*, y_*) with disjoint closures. Using equalities (57)–(58) and Proposition 4.8, for the neighbourhood U we find a natural number $m = m(U)$ such that for the trajectory of an arbitrary point $(x', y) \in M$ we have the inclusion $\{\widehat{\Phi}_k(x', y)\}_{n \geq m} \subset U$. Then by the choice of neighbourhoods U and $U((x'_*, y_*))$ we obtain

$$\{\widehat{\Phi}_k(x', y)\}_{n \geq m} \bigcap U((x'_*, y_*)) = \emptyset. \tag{59}$$

for an arbitrary point $(x', y) \in M$. The equality (59) contradicts the property $(x'_*, y_*) \in L^*(\widehat{\Phi}_k) \setminus \Omega(\widehat{\Phi}_k)$. Thus, by Definition 1.1 the graph of the Ω-function (i.e. the non-wandering set $\Omega(\widehat{\Phi}_k)$) is the global attractor. By Theorem 4.10 the attractor coincides with the set $\bigcup_{(x', y) \in M} \omega_{\widehat{\Phi}_k}((x', y))$.

Moreover, the attractor is chaotic (see Definition 1.1). In fact, the following inequalities hold for the topological entropy of maps $\widehat{\Phi}_{k|\Omega(\widehat{\Phi}_k)}$ and f_k (for details see [15]):

$$h(\widehat{\Phi}_{k|\Omega(\widehat{\Phi}_k)}) \geq h(f_k) > 0.$$

Let (x', y) be a periodic point of $\widehat{\Phi}_k$, $n = n(x')$ be its least period. Detailed calculations of elements of the Jacobi matrix $D\widehat{\Phi}_k^n$ of maps $\widehat{\Phi}_k^n$ are given in [20]. The matrix $D\widehat{\Phi}_k^n$ is lower triangular. Therefore, second eigenvalue $\lambda_2(x', y)$ of $D\widehat{\Phi}_k^n$ in the point (x', y) satisfies equalities

$$\lambda_2(x', y) = \frac{\partial}{\partial y} \widehat{g}_{x', n}(y) = 1$$

for $y \in [0, c^k(x')]$ (see Lemma 4.2 and Corollary 4.7). Corollary 4.11 is proved. ∎

Remark 4.2: Claim (4.11.2) of Corollary 4.11 shows that the set of discontinuity points of the Ω-function $\Omega^{\widehat{\Phi}_k}$ contains everywhere dense set of first category of f_k-periodic points. At the same time, claim (4.11.3) of Corollary 4.11 shows that the set of continuity points of the Ω-function $\Omega^{\widehat{\Phi}_k}$ contains everywhere dense residual set of f_k-transitive points. Therefore, the Ω-function $\Omega^{\widehat{\Phi}_k}$ is the dynamical analog of the Riemann function in classical analysis.

Consider intersections of sets of fibres over f_k-transitive points and f_k-periodic points with the set $M \setminus \Omega(\widehat{\Phi}_k)$. Applying Corollary 4.11, we obtain the following statement.

Corollary 4.12: *Trajectories of points of everywhere dense residual subset of the set $M \setminus \Omega(\widehat{\Phi}_k)$ are attracted by the circle $S^1 \times \{0\}$; and there is everywhere dense boundary set of first category in $M \setminus \Omega(\widehat{\Phi}_k)$ such that trajectories of its points are attracted by only periodic orbits of points $\{(x', c^k(x'))\}$, where $x' \in Per(f_k)$.*

Definition 4.13: We say that a map $F : M \to M$ with a global chaotic attractor A possesses the property of *dense intermittency* (in the complement to the attractor) of attraction sets of different ω-limit sets, the union of which coincides with the global attractor A, if every point $(\overline{x}', \overline{y}) \in M \setminus A$ has a neighbourhood $U((\overline{x}', \overline{y})) \subset M \setminus A$ satisfying the property:

in the intersection of $U((\overline{x}', \overline{y}))$ with an arbitrary circle $y = const$ (if it is not empty) between arbitrary two different points $(x'_1; y)$ and $(x'_2; y)$ of attraction sets of different ω-limit sets there is a point $(x'_3; y)$ that belongs to the attraction set of an ω-limit set different from each of previous two ω-limit sets.

As it follows from Corollary 4.12, maps $\widehat{\Phi}_k \in T^1_L(M)$ $(k \geq 2)$ possess the property of dense intermittency of attraction sets. This property for smooth endomorphisms with two-dimensional phase space is more strong than its analog discovered in [37] for diffeomorphisms with three-dimensional phase space.

Recall that the skew product $\widehat{\Phi}_k \in T^1_L(M)$ is C^1-smoothly congugate with some map $F_k \in C^1_L(M)$ of type (11) satisfying (12). This property makes it possible to reformulate obtained results for maps from the space $C^1_L(M)$.

Theorem 4.14: *Let $\Phi_k \in T^1_L(M)$ be a map of the form (38) satisfying (12) and (40). Let δ be a positive number, $\delta < \overline{\delta}$ (the choice of $\overline{\delta}$ is described above). Then there is an ε-neighbourhood $B^1_\varepsilon(\Phi_k)$ of the map Φ_k in the space $C^1_L(M)$ such that every map $F_k \in B^1_\varepsilon(\Phi_k)$ obtained from Φ_k by means of the C^1-smooth perturbation $\mu = \mu(x, y)$, where μ satisfies the condition $(ii_{\mu,\varepsilon})$, has one-dimensional global chaotic attractor $A(F_k)$, which is a ramified continuum with everywhere dense ramification points set of the cardinality c on the invariant circle $S^1 \times \{0\}$. Moreover, the following properties are fulfilled:*

(4.14.1) $A(F_k) = \overline{Per(F_k)} = \bigcup_{(x,y)\in M} \omega_{F_k}(x, y)$, *and all F_k-periodic points are not hyperbolic;*

(4.14.2) $A(F_k)$ *consists of two types of C^1-smooth arcs: on the circle $S^1 \times \{0\}$ the map F_k is mixing, and on each nondegenerate arc of the second type (the set of these arcs has continuum cardinality) the map F^n_k is not mixing for every $n \geq 1$;*

(4.14.3) F_k *possesses the property of dense intermittency (in the complement to the attractor) of attraction sets of different ω-limit sets, the union of which coincides with the global attractor $A(F_k)$.*

Theorem 4.14 shows that the set of maps with the above properties is open in $C^1_L(M)$.

Notes

1. A map $\varphi \in C^1(S^1)$ is said to be *expanding* if for every $x \in S^1$ the inequality holds $|\varphi'(x)| > 1$ [48, Ch. 1, Section 1.3].
2. A quasiminimal set is the closure of a recurrent (but not periodic) trajectory [47, Ch. 5, Section 5].
3. The maximal smoothness of a curve can be higher than r.
4. The inclusion (21) means the invariance of a local lamination.

Acknowledgements

The author thanks the reviewers for useful remarks and comments.

Disclosure statement

No potential conflict of interest was reported by the author(s).

Funding Acknowledgement

This work is supported by the grant of the Russian Science Foundation (RSF) (project No 24-21-00242, https://rscf.ru/en/project/24-21-00243/).

References

[1] G. Acosta, R. Hernández-Gutiérrez, I. Naghmouchi, and P. Oprocha, *Periodic points and transitivity on dendrites*, Ergod. Theory Dyn. Syst. 37(7) (2017), pp. 2017–2033.

[2] R.L. Adler, A.G. Konheim, and M.H. McAndrew, *Topological entropy*, Trans. Am. Math. Soc. 114 (2) (1965), pp. 309–319.

[3] V.S. Afraimovich, V.V. Bykov, and L.P. Shilnikov, *Attractive nonrough limit sets of Lorenz-attractor type*, Tr. Mosk. Mat. Obs. 44 (1982), pp. 150–212.

[4] V.S. Anishenko and T.E. Vadivasova, *Lectures on nonlinear dynamics*, Regular and Chaotic Dynamics, M.-Izhevsk, 2011, 516 p.

[5] D.V. Anosov, *Structurally stable systems*, Tr. Mat. Inst. Steklova, [*English Translation*], Proc. Steklov Inst. Math. 169 (1986), pp. 61–95. Original work published 1985.

[6] D.V. Anosov and E.V. Zhuzhoma, *Nonlocal asymptotic behavior of curves and leaves of laminations on universal coverings*, Tr. Mat. Inst. Steklova, [*English Translation*], Proc. Steklov Inst. Math. 2(249) (2005), pp. 1–219. Original work published 2005.

[7] M. Barge, *Homoclinic intersections and indecomposability*, Proc. Am. Math. Soc. 101 (3) (1987), pp. 541–544.

[8] M. Barge, *Nearly one-dimensional Hénon attractors and inverse limits*, Nonlinearity 8 (1) (1995), pp. 29–42.

[9] S.S. Bel'mesova and L.S. Efremova, *On the concept of integrability for discrete dynamical systems. Investigation of wandering points of some trace map*, in *Nonlinear Maps and their Appl., Springer Proc. Math. Statist.*, Vol. 112, Springer, Cham, 2015, pp. 127–158,

[10] V.N. Belykh, *Chaotic and strange attractors of a two-dimensional map*, Sb. Math. 186(3) (1995), pp. 311–326.

[11] L. Block, *Homoclinic points of mappings of the interval*, Proc. Am. Math. Soc. 72(3) (1978), pp. 576–580.

[12] A.M. Blokh, R.J. Fokkink, J.C. Mayer, L.G. Oversteegen, and E.D. Tymchatyn, Fixed Point Theorems for Plane Continua with Applications, Mem. Amer. Math. Soc., 224, 1053, Amer. Math. Soc., Providence, RI, 2013, 97 pp.

[13] J. Boroński and S. Štimac, *Densely branching trees as models for Hénon-like and Lozi-like attractors*, 2021, 26 p., arXiv:2104.14780v1.

[14] B.H. Bowditch, *Hausdoff dimension, and dendritic limit sets*, Math. Ann. 332 (3) (2005), pp. 667–676.

[15] R. Bowen, *Entropy for group endomorphisms and homogeneous spaces*, Trans. Am. Math. Soc. 153 (1971), pp. 401–414.

[16] J.J. Charatonik, *On ramification points in the classical sense*, Fundam. Math. 51 (3) (1962), pp. 229–252.

[17] J. Činč and P. Oprocha, *Pseudo-arc as the attractor in the disc – topological and measure-theoretic aspects*, 34 p., arXiv:2107.10347v1.

[18] W. de Melo and S. van Strien, One-Dimensional Dynamics, A series of modern surveys in mathematics, Springer-Verlag, 1993.

[19] L.S. Efremova, *On the concept of the Ω-function for the skew product of interval mappings*, Itogi Nauki Tekh. Ser. Sovrem. Mat. Prilozh. Temat. Obz., Vseross. Inst. Nauchn. I Tekhn. Inform. (VINITI), [*English Translation*], Journ. Math. Sci. (N.-Y.) 105 (2001), pp. 1779–1798. Original work published 1999.

[20] L.S. Efremova, *Differential properties and attracting sets of the simplest skew product of interval mappings*, Mat. Sb., [*English Translation*], Sb. Math. 201 (2010), pp. 873–907. Original work published 2010.

[21] L.S. Efremova, *Example of the smooth skew product in the plane with the one-dimensional ramified continuum as the global attractor*, ESAIM: Proc. 36 (2012), pp. 15–25.

[22] L.S. Efremova, *A decomposition theorem for the space of C^1-smooth skew products with complicated dynamics of the quotient map*, Mat.Sb., [*English Translation*], Sb. Math. 204 (2013), pp. 1598–1623. Original work published 2013.

[23] L.S. Efremova, *Multivalued functions and nonwandering set of skew products of maps of an interval with complicated dynamics of quotient map*, Izv. Vysshikh Uchebnykh Zaved. Mat., [*English Translation*], Russ. Math. 60 (2016), pp. 77–81. Original work published 2016.

[24] L.S. Efremova, *Nonwandering sets of C^1-smooth skew products of interval maps with complicated dynamics of quotient map*, Problemy Matem. Analiza, Novosibirsk, [*English Translation*], J. Math. Sci. (New York) 219 (2016), pp. 86–98. Original work published 2016.

[25] L.S. Efremova, *Dynamics of skew products of maps of an interval*, Uspekhi Mat. Nauk, [*English Translation*], Russ. Math. Surv. 72(1) (2017), pp. 101–178. Original work published 2017.

[26] L.S. Efremova, *The trace map and integrability of the multifunctions*, J. Phys. Conf. Ser. 990 (2018), Article ID 012003.

[27] L.S. Efremova, *Periodic behavior of maps obtained by small perturbations of smooth skew products*, Discontinuity Nonlinearity Complex 9(4) (2020), pp. 519–523.

[28] L.S. Efremova, *Small perturbations of smooth skew products and sharkovsky's theorem*, J. Differ. Equ. Appl. 26(8) (2020), pp. 1192–1211. Special issue on the occasion of the 82nd birthday of O.M. Sharkovsky.

[29] L.S. Efremova, *Small C^1-smooth perturbations of skew products and the partial integrability property*, Appl. Math. Nonlinear Sci. 5(2) (2020), pp. 317–328.

[30] L.S. Efremova, *Geometrically integrable maps in the plane and their periodic orbits*, Lobachevskii J. Math. 42(10) (2021), pp. 2315–2324.

[31] L.S. Efremova, *Simplest skew products on n-dimensional ($n \geq 2$) cells, cylinders and tori*, Lobachevskii J. Math. 43(7) (2022), pp. 1598–1618.

[32] L.S. Efremova, *Introduction to completely geometrically integrable maps in high dimensions*, Mathematics 11(4) (2023), pp. 926. https://doi.org/10.3390/math11040926.

[33] L.S. Efremova and E.N. Makhrova, *The dynamics of monotone maps of dendrites*, Mat. Sb., [*English Translation*], Sb. Math. 192(6) (2001), pp. 807–821. Original work published 2001.

[34] L.S. Efremova and E.N. Makhrova, *One-dimensional dynamical systems*, Uspekhi Mat. Nauk, [*English Translation*], Russ. Math. Surv. 76(1) (2021), pp. 821–881. Original work published 2021.

[35] M. Hénon, *A two-dimensional mapping with a strange attractor*, Commun. Math. Phys. 50(1) (1976), pp. 69–77.

[36] M.V. Jakobson, *Smooth mappings of the circle into itself*, Mat. Sb. (N.S.), [*English Translation*], Math. USSR – Sbornik 14(2) (1971), pp. 161–185. Original work published 1971.

[37] I. Kan, *Open sets of diffeomorphisms having two attractors, each with an everywhere dense basin*, Bul. Am. Math. Soc. 31(1) (1994), pp. 68–74.

[38] H. Kato, *Expansive homeomorphisms and indecomposability*, Fundam. Math. 139 (1991), pp. 49–57.

[39] H. Kato, *Concerning continuum-wise fully expansive homeomorphisms of continua*, Topol. Appl. 53(3) (1993), pp. 239–258.

[40] A.N. Kolmogorov and S.V. Fomin, Elements of the Theory of Functions and Functional Analysis, Vol. 1, Graylock Press, Rochester, N.Y., 1957. Original work published 1954.

[41] K. Kuratowski, Topology, Vol. 1, Academic Press, New York-London, 1966. 2, 1968.

[42] S.P. Kuznetsov, *Dynamical chaos and uniformly hyperbolic attractors: From mathematics to physics*, Physics-Uspekhi 54(2) (2011), pp. 119–144.

[43] R. Lozi, *Un attracteur étrange(?) du type attracteur de Hénon*, J. Phys. (Paris) 39(Call C5) (1978), pp. 9–10.

[44] E.N. Makhrova, *Monotone maps on dendrites*, Discontinuity Nonlinearity Complex 9(4) (2020), pp. 541–552.

[45] Y.N. Minsky, *On rigidity, limit sets and ends of hyperbolic 3-manifolds*, J. Am. Math. Soc. 7 (3) (1994), pp. 539–588.

[46] I. Naghmouchi, *Dynamical properties of monotone dendrite maps*, Topol. Appl. 159(1) (2012), pp. 144–149.

[47] V.V. Nemytskii and V.V. Stepanov, *Qualitative Theory of Differential Equations*, OGIZ, M.–L., [*English translation*], Princeton Univ. Press, Princeton, NJ, 1960, Original work published 1947.

[48] Z. Nitecki, Differentiable Dynamics. An Introduction to the Orbit Structure of Diffeomorphisms, The MIT Press, Cambridge, MA-London, 1971.

[49] R.V. Plykin, *Sources and sinks of a-diffeomorphisms of surfices*, Math. USSR-Sbornik 23(2) (1974), pp. 233–253.
[50] R.V. Plykin, *On the geometry of hyperbolic attractors of smooth cascades*, Rus. Math. Surv. 39(6) (1984), pp. 85–131.
[51] A.N. Sharkovsky, *On a theorem of G. D. Birkhoff*, Dopov. Akad. Nauk UkrRSR Ser. A 5 (1967), pp. 429–432.
[52] M.V. Shashkov and L.P. Shilnikov, *On the existence of a smooth invariant foliation in Lorenz-type mappings*, Differ. Equ. 30 (1994), pp. 536–544.
[53] M. Shub, *Endomorphisms of compact differentiable manifolds*, Am. J. Math. 91 (1969), pp. 175–199.

Dynamics of three-dimensional A-diffeomorphisms with two-dimensional attractors and repellers

Marina Barinova, Vyacheslav Grines and Olga Pochinka

ABSTRACT

In this paper, we consider a class of A-diffeomorphisms given on a 3-manifold, assuming that all the basic sets of the diffeomorphisms are two dimensional. It is known that such basic sets are either attractors or repellers and they are two types only, surface or expanding (contracting). One of the results of the paper is the proof that different types of two-dimensional basic sets do not coexist in the non-wandering set of the same 3-diffeomorphism. Also, the existence of an energy function is constructively proved for systems of the class under consideration. It is illustrated by examples that the two-dimensionality of the basic sets is essential in this matter and a decrease in the dimension can lead to the absence of the energy function for a diffeomorphism.

1. Introduction and formulation of results

Smale [19] in 1967 introduced so-called *A-diffeomorphisms* f on a compact manifold M^n, whose non-wandering set $NW(f)$ is hyperbolic and coincides with the closure of the periodic points of f. For such diffeomorphisms, he proves the Spectral Theorem: the non-wandering set $NW(f)$ of an A-diffeomorphism $f : M^n \to M^n$ can be represented uniquely as a disjoint union of a finite number of subsets

$$NW(f) = \Lambda_1 \cup \Lambda_2 \cup \cdots \cup \Lambda_k,$$

where every Λ_i is a compact, f-invariant and topologically transitive (contains an orbit of f which is dense in Λ_i). Such subsets are called *basic sets* of the diffeomorphism f. A basic set Λ_i, which is a periodic orbit, is called a *trivial* basic set. Otherwise, Λ_i is called *non-trivial*, in this case it contains infinitely many periodic points and $\dim\ W^s_x \cdot \dim\ W^u_x \neq 0$ for $x \in \Lambda_i$.

Recall that for diffeomorphism $f : M^n \to M^n$ a compact f-invariant subset $\mathcal{A}$ of M^n is called an *attractor* if there exists a compact neighbourhood $U_{\mathcal{A}}$ (*a trapping neighbourhood*) of $\mathcal{A}$ such that $f(U_{\mathcal{A}}) \subset \text{int}\ U_{\mathcal{A}}$ and $\mathcal{A} = \bigcap_{i=0}^{+\infty} f^i(U_{\mathcal{A}})$. A *repeller* is defined as an attractor for f^{-1}. An attractor $\mathcal{A}$ of a diffeomorphism f is called *expanding*, if $\dim \mathcal{A} = \dim\ W^u_x$, $x \in \mathcal{A}$. An expanding attractor for f^{-1} is called *a contracting repeller* for f.

Plykin [17, Theorem 1] proved that each hyperbolic attractor (repeller) consists of unstable (stable) manifolds of its points. Moreover, every basic set of an A-diffeomorphism $f: M^n \to M^n$ of topological dimension n or $n-1$ is either attractor or repeller. Herewith, if at least one basic set Λ_i has topological dimension n, then $M^n = \Lambda_i$ and f is an Anosov diffeomorphism.

In the present paper, we consider A-diffeomorphisms $f : M^3 \to M^3$, where M^3 is a connected closed manifold and $NW(f)$ consists of two-dimensional basic sets. It follows from [9, Theorems 1, 2] and [4, Corollary 1.2] that a basic set Λ of such a diffeomorphism has exactly one of the following properties:

- every connected component of Λ is homeomorphic to 2-torus (Λ is called *surfaced*);
- Λ is an expanding attractor (contracting repeller) (Λ is called *non-surfaced*), then it follows from [17, Theorem 2] Λ has a local structure of a direct product of a Cantor set and a 2-disc.

The first result of this paper shows that all basic sets of a diffeomorphism under considerations have the same type.

Theorem 1.1: *Let M^3 be a closed 3-manifold and $f : M^3 \to M^3$ be an A-diffeomorphism, whose non-wandering set consists of two-dimensional basic sets. Then either all basic sets are surfaced, or they are all non-surfaced.*

Ch. Conley [5] in 1978 proved an existence of a Lyapunov function[1] for Ω-stable diffeomorphisms[2] in sense of the following definition.

A Lyapunov function for an Ω-stable diffeomorphism $f : M^n \to M^n$ is a continuous function $\varphi : M^n \to \mathbb{R}$ with following properties:

- $\varphi(f(x)) < \varphi(x)$ if $x \notin NW(f)$;
- $\varphi(f(x)) = \varphi(x)$ if $x \in NW(f)$.

The central property of a Lyapunov function for a diffeomorphism $f : M^n \to M^n$ – the decreasing on the wandering set – naturally implies the expectation that a smooth Lyapunov function has no critical points (where the gradient vanishes) there. Such a function, being a Morse function out of non-trivial basic sets, was named *an energy function* for f. Indeed the condition of coinciding of the non-wandering set of the diffeomorphism with the set of the critical points of its Lyapunov function is quite strong and, starting from dimension $n = 2$, there are diffeomorphisms (both with and without non-trivial basic sets) that do not possess an energy function [1, 10, 14, 16]. In [3, 11, 12] were selected classes of Ω-stable 3-diffeomorphisms with non-trivial basic sets (one- and two-dimensional surfaced, unique expanding attractor or contracting repeller) which admit an energy function. The following theorem, which will be proven in Section 4, expands these classes.

Theorem 1.2: *Let M^3 be a closed 3-manifold and $f : M^3 \to M^3$ be an A-diffeomorphism, whose non-wandering set consists of two-dimensional basic sets. Then f possesses an energy function.*

The first example of a 3-diffeomorphism with a non-trivial basic set and without an energy function is constructed in Section 5.

2. On two-dimensional basic sets of axiom A 3-diffeomorphism

As was mentioned in the introduction, every two-dimensional basic set Λ of an A-diffeomorphism $f : M^3 \to M^3$ has exactly one of the following types:

- every connected component of Λ is homeomorphic to 2-torus (Λ is called *surfaced*);
- Λ is an expanding attractor (contracting repeller) (Λ is called *non-surfaced*).

Below we describe the topological structure of the basin of such basic sets.

2.1. Surfaced basic sets

Lemma 2.1: *If $\mathcal{A}$ is a two-dimensional surfaced attractor of an A-diffeomorphism $f : M^3 \to M^3$, then every connected component of the set $W^s_{\mathcal{A}} \setminus \mathcal{A}$ is homeomorphic to $\mathbb{T}^2 \times \mathbb{R}$.*

Proof: It was proved in [9, Theorem 1] that any surfaced attractor of an A-diffeomorphism $f : M^3 \to M^3$ is cylindrically embedded in M^3. So there exists a compact neighbourhood U of the attractor $\mathcal{A}$ and a homeomorphism $h : U \to \mathbb{T}^2 \times [-1, 1]$ such that $h(\mathcal{A}) = \mathbb{T} \times \{0\}$. Let us prove that there exists a number k such that $f^k(U) \subset \operatorname{int} U$.

Without loss of generality, we will assume that each connected component of the set $W^s_{\mathcal{A}} \setminus \mathcal{A}$ is f-invariant, in the opposite case we can consider a suitable degree of f. Let $\tilde{U}$ be a trapping neighbourhood for the attractor $\mathcal{A}$. As $\bigcap_{n\in\mathbb{N}} f^n(\tilde{U}) = \mathcal{A}$ then we can assume that $\tilde{U} \subset \operatorname{int} U$, in the opposite case we can consider instead $\tilde{U}$ a suitable iteration of $\tilde{U}$ by f. Let us consider an open cover of $W^s_{\mathcal{A}}$ by sets $P = \{f^n(\tilde{U}) : n \in \mathbb{Z}\}$. Since U is a compact set covered by P then there exists a finite subcover $\tilde{P}$ of P such that $U \subset \tilde{P}$. By the construction of P there exists a number $k > 0$ such that $U \subset f^{-k}(\tilde{U})$ so $f^k(U) \subset \operatorname{int} U$.

Let $V_{\mathcal{A}}$ be a connected component of $W^s_{\mathcal{A}} \setminus \mathcal{A}$. Without loss of generality, we will assume that $h(V_{\mathcal{A}} \cap U) = \mathbb{T}^2 \times (0, 1]$. Let $U_{\mathcal{A}} = h^{-1}(\mathbb{T}^2 \times (0, 1])$, $T_{\mathcal{A}} = h^{-1}(\mathbb{T}^2 \times \{1\})$, and $T^*_{\mathcal{A}} = f^k(T_{\mathcal{A}})$. Notice that $T^*_{\mathcal{A}} \subset U_{\mathcal{A}}$. Moreover, $\mathcal{A}$ and $T_{\mathcal{A}}$ belong to different connected components of the set $cl(U_{\mathcal{A}}) \setminus T^*_{\mathcal{A}}$. Indeed, if we assume the contrary, then, according to [8, Lemma 3.1], $T^*_{\mathcal{A}} = \partial D$ for some compact set $D \subset U_{\mathcal{A}}$. As $V_{\mathcal{A}} = U_{\mathcal{A}} \sqcup U^*_{\mathcal{A}}$ for $U^*_{\mathcal{A}} = V_{\mathcal{A}} \setminus U_{\mathcal{A}}$, $\partial U_{\mathcal{A}} = \mathcal{A} \sqcup T_{\mathcal{A}}$ and $f^k(\mathcal{A}) = \mathcal{A}$ then (see Figure 1) $f^k(U^*_{\mathcal{A}}) = D$ and $f^k(U_{\mathcal{A}}) = V_{\mathcal{A}} \setminus D$, that contradicts the fact that $f^k(U_{\mathcal{A}}) \subset U_{\mathcal{A}}$.

Thus the torus $T^*_{\mathcal{A}}$ separates $T_{\mathcal{A}}$ and $\mathcal{A}$ from each other in the set $cl(U_{\mathcal{A}})$. In such a case, by [8, Theorem 3.2], the closure of each connected component of $U_{\mathcal{A}} \setminus T^*_{\mathcal{A}}$ is homeomorphic to $\mathbb{T}^2 \times [0, 1]$. Let $p_1 : \mathbb{T}^2 \times [0, 1] \to \mathbb{T}^2$, $p_2 : \mathbb{T}^2 \times [0, 1] \to [0, 1]$ be canonical projections. Let $K_{\mathcal{A}} = U_{\mathcal{A}} \setminus \operatorname{int} f^k(U_{\mathcal{A}})$, then $V_{\mathcal{A}} = \bigcup_{n\in\mathbb{Z}} f^{kn}(K_{\mathcal{A}})$. Let $\psi_0 : K_{\mathcal{A}} \to \mathbb{T}^2 \times [0, 1]$ be a homeomorphism and $\phi_{01} : \mathbb{T}^2 \to \mathbb{T}^2$ be a diffeomorphism given by the formula $\phi_0(w) = p_1(\psi_0(f^k(\psi_0^{-1}(w, 0))))$, $w \in \mathbb{T}^2$. Let us extend ψ_0 to a homeomorphism $\psi_1 : f^k(K_{\mathcal{A}}) \to \mathbb{T}^2 \times [1, 2]$ by the formula

$$\psi_1(x) = (\phi_0(p_1(\psi_0(f^{-k}(x)))), \quad p_2(\psi_0(f^{-k}(x))) + 1), \; x \in f^k(K_{\mathcal{A}}).$$

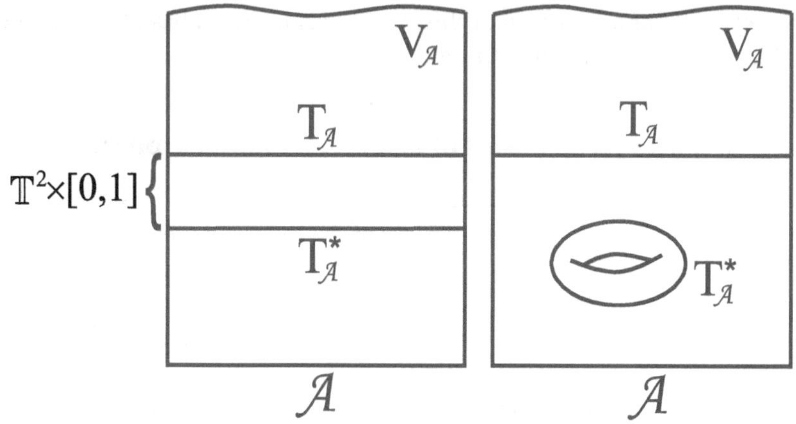

Figure 1. Mutual arrangement of tori $T_{\mathcal{A}}$, $T^*_{\mathcal{A}}$ and $\mathcal{A}$.

Similarly we can extend ψ_1 to $\psi_2 : f^{2k}(K_{\mathcal{A}}) \to \mathbb{T}^2 \times [2,3]$, etc. in the positive side. Also we can extend ψ_0 in the negative side. As a result, we get a homeomorphism $\psi : V_{\mathcal{A}} \to \mathbb{T}^2 \times \mathbb{R}$. ■

2.2. Non-surfaced basic set

Now let $\mathcal{A}$ be a co-dimension 1 expanding attractor of an A-diffeomorphism $f : M^n \to M^n$. For any point $x \in \mathcal{A}$, dim $W^s_x = 1$ allows one to introduce the notation $(y,z)^s$ $([y,z]^s)$ for an open (closed) arc of the stable manifold W^s_x bounded by points $y, z \in W^s_x$. The set $W^s_x \setminus \{x\}$ consists of two connected components, and by virtue of [6, Lemma 2.1], [7, Lemma 1.5] at least one of them has a non-empty intersection with the set $\mathcal{A}$. A point $p \in \mathcal{A}$ is called *boundary* if one of the connected components of the set $W^s_p \setminus \{p\}$ does not intersect $\mathcal{A}$. Let's denote this component $W^{s\varnothing}_p$. The set of boundary points of the basic set is finite. The union of the unstable manifolds of all boundary points $p_1, \ldots, p_r$ of the attractor $\mathcal{A}$ whose components $W^{s\varnothing}_{p_1}, \ldots, W^{s\varnothing}_{p_r}$ belong to the same connected component W_B of the set $W^s_{\mathcal{A}} \setminus \mathcal{A}$ is called a *bunch* B of the attractor $\mathcal{A}$, corresponded to W_B, the number r is called a *degree of the bunch*. It follows from [18, Theorem 2.1] that every expanding attractor $\mathcal{A}$ of an A-diffeomorphism $f : M^3 \to M^3$ has bunches of degrees 1 or 2 only.

Lemma 2.2: *Let $\mathcal{A}$ be a two-dimensional non-surfaced attractor of an A-diffeomorphism $f : M^3 \to M^3$ and W_B be a connected component of the set $W^s_{\mathcal{A}} \setminus \mathcal{A}$ corresponded to a bunch B. Then there are two possible cases for the structure of W_B:*

- *if B has degree 2, then W_B is homeomorphic to $\mathbb{S}^2 \times \mathbb{R}$;*
- *if B has degree 1, then W_B is homeomorphic to $\mathbb{R}P^2 \times \mathbb{R}$, where $\mathbb{R}P^2$ is a real projective plane.*

Proof: Let p, q be boundary periodic points of the bunch B ($p = q$ in the case when B has the degree 1) and W^u_p, W^u_q be unstable manifolds of the points p and q respectively. Without loss of generality we can assume that p and q are fixed points of the diffeomorphism f. Then

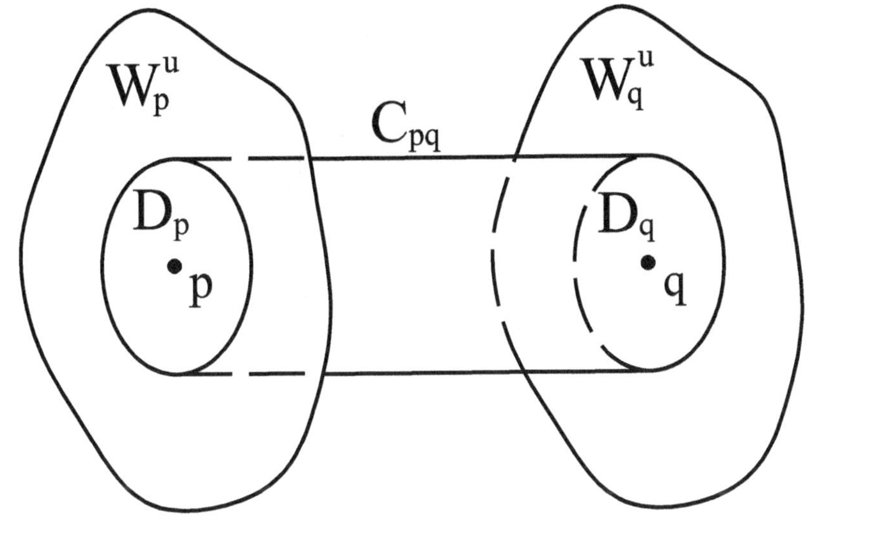

Figure 2. Connecting cylinder.

by [7, Section 2] and [20, Lemmas 2.1, 2.3] there exists a homeomorphism $\varphi : W^u_p \setminus \{p\} \to W^u_q \setminus \{q\}$ which assigns to a point $x \in W^u_p \setminus \{p\}$ the point $\varphi(x) \in W^u_q \cap W^s_x$ such that the arc $(x, \varphi(x))^s \subset W^s(x)$ does not intersect $\mathcal{A}$. It is directly verified that $\varphi \circ f|_{W^u_p \setminus \{p\}} = f \circ \varphi|_{W^u_p \setminus \{p\}}$ and, hence, φ is continuously extended up to $\varphi : W^u_p \to W^u_q$ by $\varphi(p) = q$.

Further let us consider cases (a) $p \neq q$ and (b) $p = q$ separately.

(a) Let B be the 2-bunch of the attractor $\mathcal{A}$. The restriction $f|_{W^u_p}$ has exactly one hyperbolic repelling fixed point p, hence there exists a smooth 2-disc $D_p \subset W^u_p$ such that $p \in f(D_p) \subset \text{int } D_p$. Then the set $C_{pq} = \bigcup_{x \in \partial D_p} [x, \varphi(x)]^s$ is diffeomorphic to a closed cylinder $\mathbb{S}^1 \times [0, 1]$. The set C_{pq} is called the *connecting cylinder*. The circle $\varphi(\partial D_p)$ bounds in W^u_q a 2-disc D_q such that $q \in D_q \subset \text{int } (f(D_q))$. The set $S_{pq} = D_p \cup C_{pq} \cup D_q$ is homeomorphic to a 2-dimensional sphere and it is called the *characteristic sphere* corresponding to the bunch B (see Figure 2).

Let us construct a foliation of W_B on 2-spheres. For this aim let us consider a point $x_\alpha \in W^{s\varnothing}_\alpha$, $\alpha \in \{p, q\}$ and the fundamental segment $I_\alpha \subset W^{s\varnothing}_\alpha$ with the endpoints x_α and $f(x_\alpha)$. Choose a tubular neighbourhood $V(I_\alpha)$ diffeomorphic to $\mathbb{D}^2 \times [0, 1]$ by means a diffeomorphism $h_\alpha : \mathbb{D}^2 \times [0, 1] \to V(I_\alpha)$ and such that $f(h_\alpha(\mathbb{D}^2 \times \{0\})) \subset h_\alpha(\mathbb{D}^2 \times \{1\})$. Due to λ-lemma one can assume that every disc $h_\alpha(\mathbb{D}^2 \times \{0\})$ intersects $W^s(x)$ for every $x \in (D_p \setminus p)$ exactly once (in the opposite case one can choose an appropriate iteration of $V(I_\alpha)$).

The set $K_p = C_{pq} \setminus \text{int} f^{-1}(C_{pq})$ is diffeomorphic to $\mathbb{S}^1 \times [0, 1] \times [0, 1]$ by means a diffeomorphism $h : \mathbb{S}^1 \times [0, 1] \times [0, 1] \to K_p$. Denote by $S_t, t \in [0, 1]$ the two-dimensional sphere bounded by the discs $h_p(\mathbb{D}^2 \times \{t\})$, $h_q(\mathbb{D}^2 \times \{t\})$ and the cylinder $h(\mathbb{S}^1 \times [0, 1] \times \{t\})$. Each S_t can be considered smooth (see, e.g. [13, Statement 10.58]). Iterations of these spheres by $f^k, k \in \mathbb{Z}$ give the desired foliation on W_B.

(b) Let B be a bunch of the degree 1. Then $p = q$ and $\varphi : W^u_p \to W^u_p$ is an orientation-preserving periodic homeomorphism of period 2 with a unique fixed point p. As $f|_{W^u_p \setminus \{p\}}$

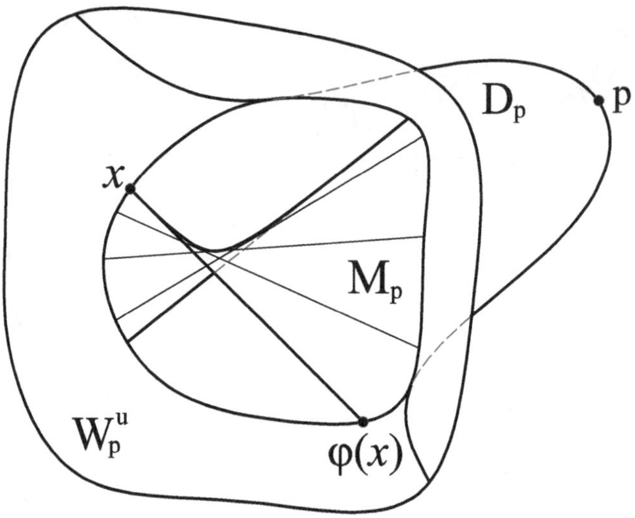

Figure 3. Connecting mobius band.

is smoothly conjugated with the linear expansion then the orbit space $(W^u_p \setminus \{p\})/f$ is diffeomorphic to the 2-torus $\mathbb{T}^2$. Denote by $\rho : W^u_p \setminus \{p\} \to \mathbb{T}^2$ the natural projection. Since the homeomorphism φ commutes with f then it can be projected to $\mathbb{T}^2$ as $\hat{\varphi} = \rho\varphi\rho^{-1} : \mathbb{T}^2 \to \mathbb{T}^2$. Thus $\hat{\varphi}$ is a periodic diffeomorphism whose all points have the period 2. Then it is topologically conjugate to the shift on the torus $\Psi(e^{i2x\pi}, e^{i2y\pi}) = (e^{i2\pi(x+\frac{1}{2})}, e^{i2y\pi})$ (see, e.g. [2, Theorem 1.2]). Hence there exists a foliation $\hat{F}$ by circles $\hat{F} = \{\hat{l}\}$ on $\mathbb{T}^2$ such that $\hat{\varphi}|_{\hat{l}}$ is conjugated with the rotation on 180 degrees. As φ is a periodic map, then $F_p = \{\rho^{-1}(\hat{l})\}$ is a foliation by circles on $W^u_p \setminus \{p\}$ with the following properties:

- each leaf $l \in F_p$ is homeomorphic to $\mathbb{S}^1$ and bounded a disc D_p containing p such that $f^{-1}(D_p) \subset \text{int}\, D_p$;
- $\varphi|_l$ is conjugated with the rotation on 180 degrees.

Let $l \in F_p, x \in l$ and $c \subset l$ be a closed segment of l bounded by the points x and $\varphi(x)$. Let $M_p = \bigcup_{x \in c}[x, \varphi(x)]^s$ (see Figure 3). Then M_p is diffeomorphic to a mobius band, which we will called *connecting mobius band*. Thus $M_p \cup D_p$ is homeomorphic to the real projective plane $\mathbb{R}P^2$. Then a foliation of W_B on $\mathbb{R}P^2$ can be obtained analogically to the case a) so W_B is diffeomorphic to $\mathbb{R}P_2 \times \mathbb{R}$.

■

3. On *A*-diffeomorphisms with two-dimensional basic sets

In this section, we prove Theorem 1.1.

Proof: First notice that due to [15, Proposition 3.11], the non-wandering set $NW(f)$ of considered diffeomorphism f contains at least one attractor and at least one repeller. Let $NW(f) = \Lambda_1 \cup \cdots \cup \Lambda_k \cup \Lambda_{k+1} \cup \cdots \cup \Lambda_{k+m}$, where $\Lambda_i,\ i = 1, \ldots, k$ is two-dimensional attractor and Λ_{k+j}, $j = 1, \dots, m$ is two-dimensional repeller. Introduce the following denotations: $V_i^s = W^s_{\Lambda_i} \setminus \Lambda_i$, $i = 1, \dots, k$, and $V^u_{k+j} = W^u_{\Lambda_{k+j}} \setminus \Lambda_{k+j}$, $j = 1, \dots, m$. Then V_i^s is a basin of the attractor Λ_i without the attractor and V^u_{k+j} is a basin of the repeller Λ_{k+j} without the repeller. Denote unions of such sets by $V^s = \bigcup_{i=1}^k V_i^s$ and $V^u = \bigcup_{j=1}^m V^u_{k+j}$. Due to S. Smale [19], $M^3 = \bigcup_{i=1}^m W^s_{\Lambda_i} = \bigcup_{j=1}^m W^u_{\Lambda_{k+j}}$ and, hence, $V^s = V^u = M^3 \setminus NW(f)$. Then the number of connected components of V^s is equal to the number of connected components of V^u and they are pairwise coincide.

It follows from Lemmas 2.1 and 2.2 that every connected component of the set V_i^s (V^u_{k+j}) is homeomorphic to $\mathbb{T}^2 \times \mathbb{R}$ if and only if Λ_i (Λ_{k+j}) is a surface attractor (repeller). Then an attractor Λ_i and a repeller Λ_{k+j}, such that $W^s_{\Lambda_i} \cap W^u_{\Lambda_{k+j}} \neq 0$, have the same type.

Let $\Lambda_1 \cup \dots \cup \Lambda_p \cup \Lambda_{k+1} \cup \dots \cup \Lambda_{k+q}$, $p \leqslant k$, $q \leqslant m$ be a union of all surface basic set of diffeomorphism f. It follows from explanation above that $\bigcup_{i=1}^p (W^s_{\Lambda_i} \setminus \Lambda_i) = \bigcup_{j=1}^q (W^u_{\Lambda_{k+j}} \setminus \Lambda_{k+j})$ and, hence,

$$\bigcup_{i=1}^p W^s_{\Lambda_i} \cup \bigcup_{j=1}^q \Lambda_{k+j} = \bigcup_{j=1}^q W^u_{\Lambda_{k+j}} \cup \bigcup_{i=1}^p \Lambda_i = \bigcup_{i=1}^p W^s_{\Lambda_i} \cup \bigcup_{j=1}^q W^u_{\Lambda_{k+j}}.$$

Let $\tilde{M}^3 = \bigcup_{i=1}^p W^s_{\Lambda_i} \cup \bigcup_{j=1}^q \Lambda_{k+j}$. Due to [19] $cl(W^s_{\Lambda_i}) \setminus W^s_{\Lambda_i} = \bigcup_{\Lambda_{k+j}: W^u_{\Lambda_{k+j}} \cap W^s_{\Lambda_i} \neq \emptyset} W^s_{\Lambda_{k+j}}$ that implies that $\tilde{M}^3$ is closed. On the other side $W^s_{\Lambda_i}$ and $W^u_{\Lambda_{k+j}}$ are open as they are basins of an attractor and repeller, accordingly. Thus $\tilde{M}^3$ is closed and open subset of M^3 simultaneously and, hence, coincides with M^3 as it is connected.

Thus the presence of surface basic sets in $NW(f)$ leads to the fact that all basic sets of f are surface, which completes the proof of the theorem. ■

3.1. *Example of a 3-diffeomorphism with two-dimensional basic sets of the both types*

The homogeneous property proved in the previous section can be broken if the non-wandering set contains a basic set of dimensions 0 or 1. Notice that all such examples are not structurally stable due to results of Grines and Zhuzhoma [7, Theorem 7.1]: if a non-wandering set of a structurally stable diffeomorphism of closed 3-manifold contains an expanding attractor or a contracting repeller then all other basic sets are trivial.

Let us construct an example of an Ω-stable diffeomorphism $f : \mathbb{T}^3 \# \mathbb{T}^3$ with both types of two-dimensional basic sets. We will start with an Anosov diffeomorphism $f_C : \mathbb{T}^2 \to \mathbb{T}^2$ induced by a hyperbolic matrix $C \in GL(2, \mathbb{Z})$ and a sink-source system $f_{NS} : \mathbb{S}^1 \to \mathbb{S}^1$ on a circle $\mathbb{S}^1$. Then the direct product $F = f_C \times f_{NS} : \mathbb{T}^3 \to \mathbb{T}^3$ of the diffeomorphisms f_C and f_{NS} has the non-wandering set consisting of one attracting torus $\mathbb{T}_A$ and one repelling torus $\mathbb{T}_R$ (see Figure 4).

Due to [3, Lemma 1] there exists a diffeotopic to the identity diffeomorphism $b : \mathbb{T}^2 \to \mathbb{T}^2$ such that $b \circ f_C$ is a DA-diffeomorphism with one-dimensional repeller and a sink. Let $h_t : \mathbb{T}^2 \to \mathbb{T}^2$, $t \in [0, 1]$ be a diffeotopy from b to the identity map, $\mathbb{S}^1 = \{e^{i\pi\varphi}, \varphi \in [-1, 1]\}$ and $F_b : \mathbb{T}^2 \times \mathbb{S}^1 \to \mathbb{T}^2 \times \mathbb{S}^1$ be a diffeomorphism given by the formula

$$F_b(w, \varphi) = (h_{|\varphi|}(w), \varphi).$$

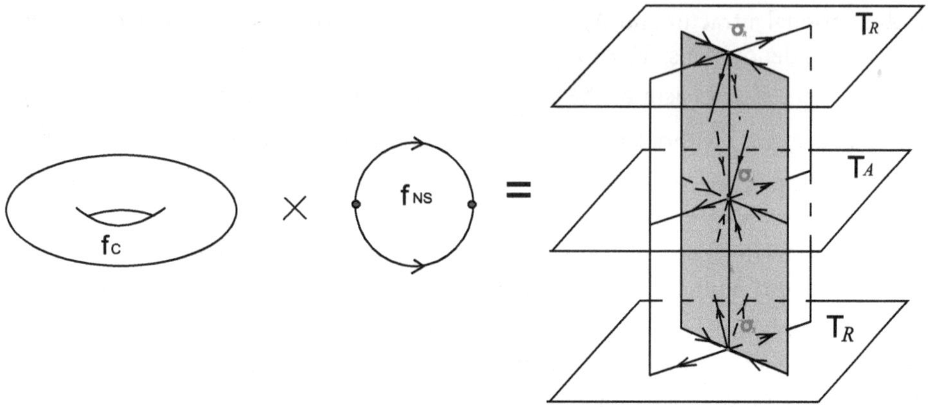

Figure 4. Diffeomorphism $F = f_C \times f_{NS}$.

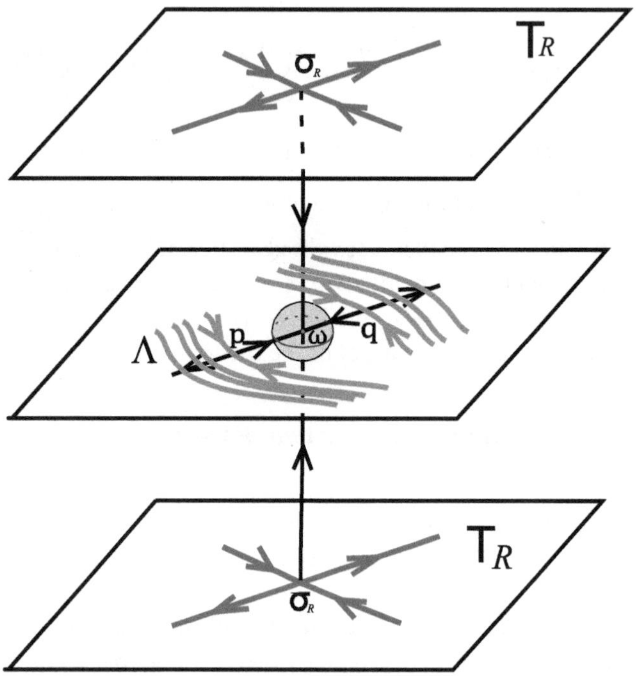

Figure 5. Diffeomorphism F_1.

Then the diffeomorphism $F_1 = F_b \circ F : \mathbb{T}^3 \to \mathbb{T}^3$ has three basic sets (see Figure 5): a two-dimensional repelling torus $\mathbb{T}_R$, a fixed sink ω and one-dimensional saddle basic set Λ.

Let $f_D : \mathbb{T}^3 \to \mathbb{T}^3$ be an Anosov diffeomorphism induced by a hyperbolic matrix $D \in SL(3, \mathbb{Z})$. Similar to [3, Lemma 2], there exists a diffeotopic to the identity diffeomorphism $B : \mathbb{T}^3 \to \mathbb{T}^3$ such that the diffeomorphism $F_2 = B \circ f_D$ is a DA-diffeomorphism with two-dimensional attractor and a source α (see Figure 6).

As ω and α are hyperbolic points then there are 3-balls B_ω, B_α around their such that $F_1(B_\omega) \subset B_\omega$, $F_2^{-1}(B_\alpha) \subset B_\alpha$. By the operation of the connected sum of two copies of $\mathbb{T}^3$

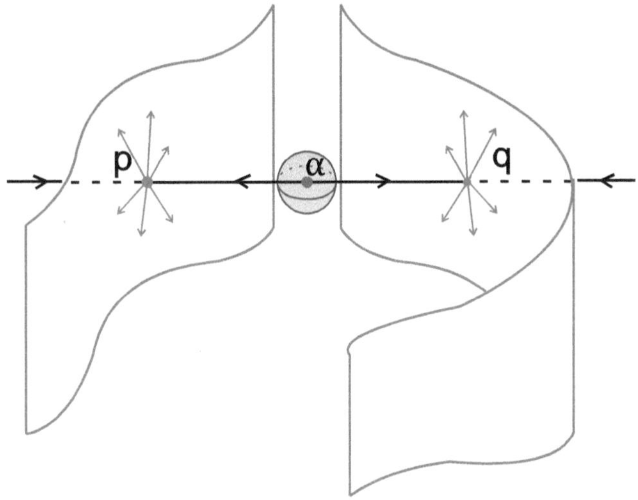

Figure 6. Diffeomorphism F_2.

along B_ω, B_α we get $f : \mathbb{T}^3\#\mathbb{T}^3 \to \mathbb{T}^3\#\mathbb{T}^3$ with the dynamic F_1 on $\mathbb{T}^3 \setminus B_\omega$ and with the dynamic F_2 on $\mathbb{T}^3 \setminus B_\alpha$.

4. Energy function for 3-diffeomorphisms with expanding attractors and contracting repellers

In this section we will prove Theorem 1.2. Namely we consider a closed 3-manifold M^3, Ω-stable diffeomorphism $f : M^3 \to M^3$, all of whose basic sets are two-dimensional and construct an energy function for f.

Proof: Let $NW(f) = \mathcal{A}_1 \cup \cdots \cup \mathcal{A}_l \cup \mathcal{R}_1 \ldots \cup \mathcal{R}_m$, where $\mathcal{A}_i$, $i \in \{1, \ldots, l\}$, is a two-dimensional attractor and $\mathcal{R}_j, j \in \{1, \ldots, m\}$ is a two-dimensional repeller. It follows from Theorem 1.1 that $\mathcal{A}_i$ and $\mathcal{R}_j$ simultaneously are either surfaces or non-surfaces basic sets. The first case was considered in [11] and an energy function was constructed there for such diffeomorphisms. Let us prove the existence of an energy function for the second type of diffeomorphisms.

Let $\mathcal{A} = \mathcal{A}_1 \cup \cdots \cup \mathcal{A}_l$, $\mathcal{R} = \mathcal{R}_1 \ldots \cup \mathcal{R}_m$ and $V = M^3 \setminus (\mathcal{A} \cup \mathcal{R})$. We will construct an energy function $\varphi : M^3 \to [0, 1]$ for f such that $\varphi(\mathcal{A}) = 0$ and $\varphi(\mathcal{R}) = 1$. Let us explain how to construct it on the set V. Let v be a connected component of V, denoted by k_v its period.

Lemma 2.2 implies that v is diffeomorphic to $S \times \mathbb{R}$, where S is either 2-sphere or real projective plane. Moreover, there exists a diffeomorphism $\chi : S \times (0, 1) \to V$ such that a foliation on 2-spheres or real projective planes $\{S_t = \chi(\mathbb{S}^2 \times \{t\}), t \in (0, 1)\}$ is f^{k_v}-invariant. Let us define a function φ on v by the formula $\varphi(S_t) = t$. After that let us extend it to the set $\bigcup_{i=0}^{k_v-1} f^i(v)$ by the formula

$$\varphi(f^i(S_t)) = t + \frac{i(\varphi(f^{k_v}(S_t)) - t)}{k_v}.$$

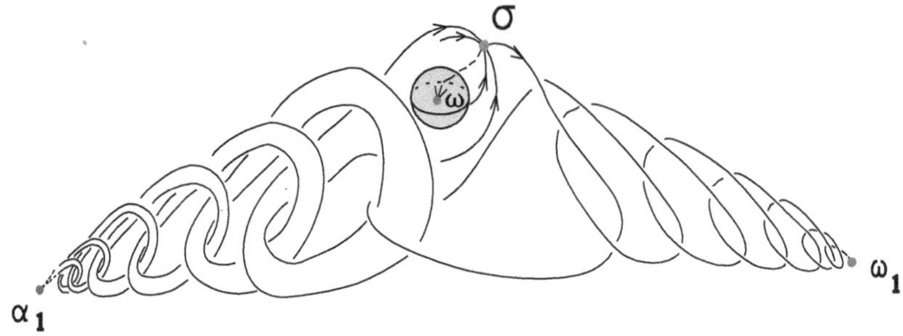

Figure 7. Diffeomorphism F_3.

Carrying out similar reasoning, we can extend the function φ to the whole manifold M^3. The obtained function $\varphi : M^3 \to [0,1]$ is a Lyapunov function for f, smooth on $M^3 \setminus NW(f))$, continuous on M^3 and has no wandering critical points.

Due to the lemma [3, Lemma 5], there exist functions $g_A : [0,1] \to [0,1]$ and $g_R : [0,1] \to [0,1]$ such that the function $\psi = (1 - g_R) \circ g_A \circ \varphi$ is an smooth energy function for the diffeomorphism f. ■

5. Example of an Ω-stable 3-diffeomorphism with an expanding two-dimensional attractor without an energy function

As it was mentioned in the introduction the first example of a diffeomorphism without an energy function is a Pixton example (see Figure 7). We will show how to make an example of a 3-diffeomorphism with non-trivial basic set which does not possess an energy function from the Pixton example and a DA-diffeomorphism.

For this aim it is enough to consider the Pixton example – a Morse–Smale diffeomorphism $F_3 : \mathbb{S}^3 \to \mathbb{S}^3$ with 4 fixed points: two sinks ω and ω_1, a source α_1, and a saddle σ such that the saddle separatrices are wild, except one 1-dimensional. The system does not possess an energy function. The second diffeomorphism is F_2 from Section 3.1. Analogically to 3.1 the hyperbolic points ω and α have 3-balls B_ω, B_α around them such that $F_3(B_\omega) \subset B_\omega$, $F_2^{-1}(B_\alpha) \subset B_\alpha$. By the operation of the connected sum of $\mathbb{S}^3$ and $\mathbb{T}^3$ along B_ω, B_α we get $f : \mathbb{T}^3 \to \mathbb{T}^3$ with the dynamic F_3 on $\mathbb{S}^3 \setminus B_\omega$ and with the dynamic F_2 on $\mathbb{T}^3 \setminus B_\alpha$.

Then the diffeomorphism f does not possess an energy function which is a Morse function outside the expanding attractor.

Notes

1. A fact of the existence of a Lyapunov function for a wider class of dynamical systems is called the Fundamental Theorem of Dynamical Systems.
2. Let $Diff(M^n)$ is a space of diffeomorphisms given on a manifold M^n. A diffeomorphism $f \in Diff(M^n)$ is called Ω-*stable* if there exists $\varepsilon > 0$ such that every diffeomorphism $g \in Diff(M^n)$ such that $||g - f||_{C^1} < \varepsilon$ possesses the property $g|_{NW(g)}$ and $f|_{NW(g)}$ are topologically conjugated.

Disclosure statement

No potential conflict of interest was reported by the author(s).

Funding

The paper is supported by Russian Science Foundation (Grant No. 21-11-00010), except Sections 3 and 4 devoted to the existence of an energy function for diffeomorphisms under consideration, which was supported by the Laboratory of Dynamical Systems and Applications NRU HSE , grant of the Ministry of Science and Higher Education of the Russian Federation (ag. 075-15-2022-1101).

ORCID

Marina Barinova http://orcid.org/0000-0002-4406-583X
Vyacheslav Grines http://orcid.org/0000-0003-4709-6858
Olga Pochinka http://orcid.org/0000-0002-6587-5305

References

[1] M. Barinova, On existence of an energy function for ω-stable surface diffeomorphisms, Lobachevskii J. Math. 42 (2021), pp. 3317–3323.
[2] D. Baranov, V. Grines, O. Pochinka, and E. Chilina, *On classification of periodic maps on the 2-torus*. arXiv preprint arXiv:2112.01256. 2021.
[3] M. Barinova, V. Grines, O. Pochinka, and B. Yu, Existence of an energy function for three-dimensional chaotic "sink-source" cascades, Chaos: Interdiscip. J. Nonlinear Sci. 31 (2021), Article ID 063112.
[4] A.W. Brown, Nonexpanding attractors: Conjugacy to algebraic models and classification in 3-manifolds, J. Mod. Dyn. 4 (07 2010), pp. 517–548.
[5] C. Conley, Isolated Invariant Sets and the Morse Index, Vol. 38, American Mathematical Society, CBMS, Providence, RI, 01 1978.
[6] V. Grines, On topological conjugacy of diffeomorphisms of a two-dimensional manifold onto one-dimensional orientable basic sets i, Trans. Mosc. Math. Soc. 32 (1975), pp. 31–56.
[7] V. Grines and E. Zhuzhoma, On structurally stable diffeomorphisms with codimension one expanding attractors, Trans. Am. Math. Soc. 357(2) (2004), pp. 617–667.
[8] V.Z. Grines, V.S. Medvedev, and E.V. Zhuzhoma, New relations for Morse–Smale systems with trivially embedded one-dimensional separatrices, Sb. Math. 194 (2003), pp. 979–1007.
[9] V.Z. Grines, V.S. Medvedev, and E.V. Zhuzhoma, On surface attractors and repellers in 3-manifolds, Math. Notes 78(5–6) (2005), pp. 757–767.
[10] V. Grines, F. Laudenbach, and O. Pochinka, Dynamically ordered energy function for Morse–Smale diffeomorphisms on 3-manifolds, Proc. Stekl. Inst. Math. 278 (01 2012), pp. 27–40.
[11] V. Grines, M. Noskova, and O. Pochinka, Construction of an energy function for *a*-diffeomorphisms of two-dimensional non-wandering sets on 3-manifolds, Middle Volga Math. Soc. J. 17 (2015), pp. 12–17.
[12] V. Grines, M. Noksova, and O. Pochinka, The construction of an energy function for three-dimensional cascades with a two-dimensional expanding attractor, Trans. Mosc. Math. Soc. 76 (2015), pp. 237–249.
[13] V. Grines, T. Medvedev, and O. Pochinka, Dynamical Systems on 2- and 3-Manifolds, Vol. 46, Springer, Switzerland, 01 2016.
[14] T.V. Medvedev and O.V. Pochinka, The wild Fox–Artin arc in invariant sets of dynamical systems, Dyn. Syst. 33(4) (2018), pp. 660–666.
[15] S.E. Nowhouse, Hyperbolic limit sets, Trans. Am. Math. Soc. 167 (1972), pp. 125–150.
[16] D. Pixton, Wild unstable manifolds, Topology 16 (12 1977), pp. 167–172.

[17] R. Plykin, The topology of basis sets for smale diffeomorphisms, Math. USSR-Sb. 13 (1971), pp. 297–307.
[18] R.V. Plykin, On the geometry of hyperbolic attractors of smooth cascades, Russian Math. Surv. 39(6) (1984), pp. 85–131.
[19] S. Smale, Differentiable dynamical systems, Bull. Am. Math. Soc. 73 (11 1967), pp. 747–817.
[20] E.V. Zhuzhoma and V.S. Medvedev, On non-orientable two-dimensional basic sets on 3-manifolds, Sb. Math. 193(6) (2002), pp. 869–888.

Chaotic behaviour of countable products of homeomorphism groups

Nina Zhukova and Alexander Korotkov

ABSTRACT
Relationships between a chaotic behaviour and closely related properties of topological transitivity, sensitivity to initial conditions, density of closed orbits of homeomorphism groups and their countable products are investigated. We provide a large number of new examples of chaotic groups of homeomorphisms of countable products of various metrizable topological spaces, including infinite-dimensional topological manifolds, whose factors can be as noncompact surfaces, so triangulable closed manifolds of an arbitrary dimension.

1. Introduction

In theory of chaos, the chaotic behaviour of infinite products of transformation groups has not been investigated. The reason is that, by definition, the chaotic behaviour of a group of homeomorphisms assumes the density of the set of finite orbits (see, e.g. [8, 9, 19]). This requirement may not be fulfilled when moving to the infinite product of spaces. Following [6], we give a more general definition of the chaotic behaviour of groups of homeomorphisms by weakening the above condition by requiring the density of the set of closed orbits (Definition 1.3). Here by a closed orbit we mean an orbit which is a closed subset of the respective topological space. This allows us to investigate the chaotic behaviour of arbitrary infinite products of homeomorphism groups.

1.1. Devaney chaos

Let $T : X \to X$ be a continuous map of metric space. The family $\{T^n\}_{n\in\mathbb{N}}$ denoted by the pair (X, T) is called a dynamical system. Devaney [13] proposed the following notion of chaos, which is usually called Devaney chaos.

Definition 1.1: A dynamical system (X, T) is called *Devaney chaotic* if it satisfies the following three properties:

(1) (X, T) is transitive;
(2) the set of periodic points of (X, T) is dense in X;
(3) (X, T) has sensitive dependence on initial conditions.

Sensitive dependence on initial conditions is widely understood as being the central idea of chaos. In [4], it was shown that in Devaney's definition of chaos, the sensitive dependence follows from transitivity and density of periodic orbits. It was found in [3] that neither transitivity nor density of periodic trajectories are deducible from the remaining two conditions.

In [8], Cairns et al. introduced the following notion of a chaotic group action as a generalization of chaotic dynamical systems (Definition 1.2). They showed that, if a group G acts chaotically on a compact Hausdorff space, then G is residually finite. Moreover, the reverse is also true, i.e. for every residually finite group G there exists a Hausdorff space on which G acts chaotically. As in [8], we don't assume any topology on the group G, but we assume that each element of G acts on a topological space X as a homeomorphism of X, and the set X is infinite. All group actions are assumed to be faithful, i.e. the only element of a group G which acts as identity homeomorphism is the neutral element in G.

Definition 1.2: A group of homeomorphisms G of a Hausdorff topological space X is called chaotic if the following two conditions are met:

(1) *Topological transitivity*: For every pair of nonempty open subsets U and V in X, there exists an element $g \in G$ such that $g(U) \cap V \neq \varnothing$.
(2) *Finite orbits dense*: The union of finite orbits is dense in X.

Following [6], we give and use in this work a different definition of the chaotic behaviour of an arbitrary group G.

Definition 1.3: A group of homeomorphisms G of a topological space X is called *chaotic* (or G has a chaotic behaviour) on X if the following two conditions are met:

(1) there exists a dense non-closed orbit of the group G in X (*the existence of a dense orbit*);
(2) the union of closed orbits is dense in X (*the density of closed orbits*).

Recall that a separable space homeomorphic to a complete metric space is referred to as a Polish space. Note that Definition 1.3 is more general than Definition 1.2 in the wide class of Polish spaces. Emphasize that in the case when G is a countable homeomorphism group of a metrizable compact space X, Definitions 1.2 and 1.3 are equivalent (Proposition 3.4 and Lemma 7.2).

If (X, d) is a metric space, then we define the notion of a sensitive dependence of a homeomorphism group G on initial conditions (in Section 6).

1.2. The organization of this work. Main results

For the convenience of the reader, we provide the basic notions in Sections 2 and 5.

Let A be any set and let X_α, $\alpha \in A$, be any topological spaces. We prove that the direct product of groups $G = \prod_{\alpha \in A} G_\alpha$ is topologically transitive on the Tychonoff product of topological spaces $X = \prod_{\alpha \in A} X_\alpha$ if and only if every homeomorphism group G_α, $\alpha \in A$, is topologically transitive on the respective factor X_α (Theorem 3.2). The analogous statement is proved for the existence of dense orbits (Theorem 3.3). We get also an analog of the Birkhoff theorem (Proposition 3.4).

We investigate density of closed orbits in Section 4 and show that the direct product of groups $G = \prod_{\alpha \in A} G_\alpha$ has a dense subset of closed orbits in $X = \prod_{\alpha \in A} X_\alpha$ if and only if for every $\alpha \in A$, the group G_α has a dense subset of closed orbits in X_α (Theorem 4.1).

In Section 5, we recall the definition of the product of a countable family of metric spaces.

Section 6 is devoted to the sensitive dependence of group actions on initial conditions. Recall that a topological space X is a Baire space if every countable intersection of open dense subsets of X is dense in X [18, Def. 8.2]. A topological space X is referred to a completely metrizable space, if it admits an agreed complete metric [18, Def. 3.1]. According to the Baire category theorem, every completely metrizable space is a Baire space. Emphasize that every Polish space, and, in particular, every compact metric space are completely metrizable spaces. We prove the following theorem which is a generalization of the main result of [4] in the case of invertible dynamical systems.

Theorem 1.4: *Let (X, d) be a metric Baire space. Assume that every closed orbit is a nowhere dense set in X. If a homeomorphism group G acts chaotically on X, then the group G is sensitive to initial conditions.*

Let G_i, $i \in J \subset \mathbb{N}$, be a homeomorphism group of a metric space X_i, and on the Tychonoff product $X = \prod_{i \in J} X_i$ the canonical action of the direct product of groups $G = \prod_{i \in J} G_i$ is given. We prove that, in contrast to the transitivity and density of closed orbits, in order for the canonical action of the group G on the product X to be sensitive to initial conditions, it is sufficient to have one group G_n, $n \in J$, which is sensitive to initial conditions on X_n (Theorem 6.7). In the case when the index set J is finite, this condition is also necessary (Theorem 6.8).

In Section 7, we prove the following theorem.

Theorem 1.5: *For every set A of indexes, let G_α, $\alpha \in A$, be a homeomorphism group of a topological space X_α, and on the Tychonoff product $X = \prod_{\alpha \in A} X_\alpha$ the canonical action of the product of groups $G = \prod_{\alpha \in A} G_\alpha$ is given. Then the group G acts chaotically on X if and only if every group G_α, $\alpha \in A$, acts chaotically on X_α.*

The application to Polish spaces is considered (Section 7.2). In particular, we prove the following theorem.

Theorem 1.6: *Let G_i, $i \in \mathbb{N}$, be a countable group of homeomorphisms of a metrizable compact space X_i. Assume that every G_i acts chaotically on X_i. Then:*

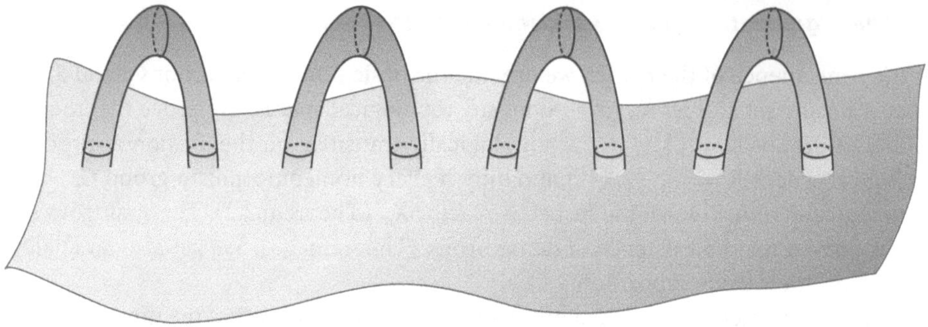

Figure 1. The Loch Ness monster.

(1) *the canonical action of the product of groups* $G = \prod_{i\in\mathbb{N}} G_i$ *is chaotic on the Tychonoff product* $X = \prod_{i\in\mathbb{N}} X_i$;
(2) *exists a dense subset* $F \subset X$ *which is the union of continual compact orbits, and every such orbit is a perfect subset of* X;
(3) *exists a dense continuum orbit of the group* G *in* X;
(4) *all groups* G_i, $i \in \mathbb{N}$, *and* G *are residually finite;*
(5) *all groups* G_i, $i \in \mathbb{N}$, *and* G *are sensitive to initial conditions;*
(6) *if each group* G_i *has a fixed point, then the union of the finite orbits of group* G *is dense in* X, *and* G *has a fixed point.*

Since every topological manifold is a Polish space, then all results of our work for Polish spaces are applicable to topological manifolds.

Sections 8 and 9 contain the construction of families of homeomorphism groups of various topological spaces.

In Section 8, we check the chaoticity of the group generated by the full N-shift of the space of bi-infinite sequences Σ^N of N symbols. This allows us to get series of new chaotic groups of homeomorphisms of different finite and infinite products of spaces Σ^{N_i}, $N_i \in \mathbb{N}$. Emphasize, that the space Σ^N is homeomorphic to the $(2N-1)$-ary Cantor set.

In Section 9, we construct numerous examples of chaotic homeomorphism groups of topological manifolds including noncompact manifolds. Using the method from [8], we construct a countable series of examples of chaotic groups of homeomorphisms, isomorphic to the group $\mathbb{Z}$, on every closed surface as well as on various noncompact surfaces, examples of which are shown in Figures 1 and 2.

Emphasize that all examples of chaotic group of homeomorphisms on noncompact topological manifolds are new and they are represented for the first time.

We use the obtained chaotic actions as building blocks for constructions of chaotic actions of homeomorphism groups on Tychonoff products of topological manifolds. Due to the result of G. Cairns and A. Kolganova [9] and Theorem 1.5, every triangulable closed manifold of arbitrary dimension can be taken as a factor on which arbitrary countably generated free groups act chaotically. In particular, we get a continuum set of examples of chaotic actions of homeomorphism groups on infinite dimensional topological manifolds.

Notations. If a group G acts on a set X, we denote by $g.x$ the action of an element $g \in G$ on a point $x \in X$. By $G.x$ we denote the orbit of x respectively G.

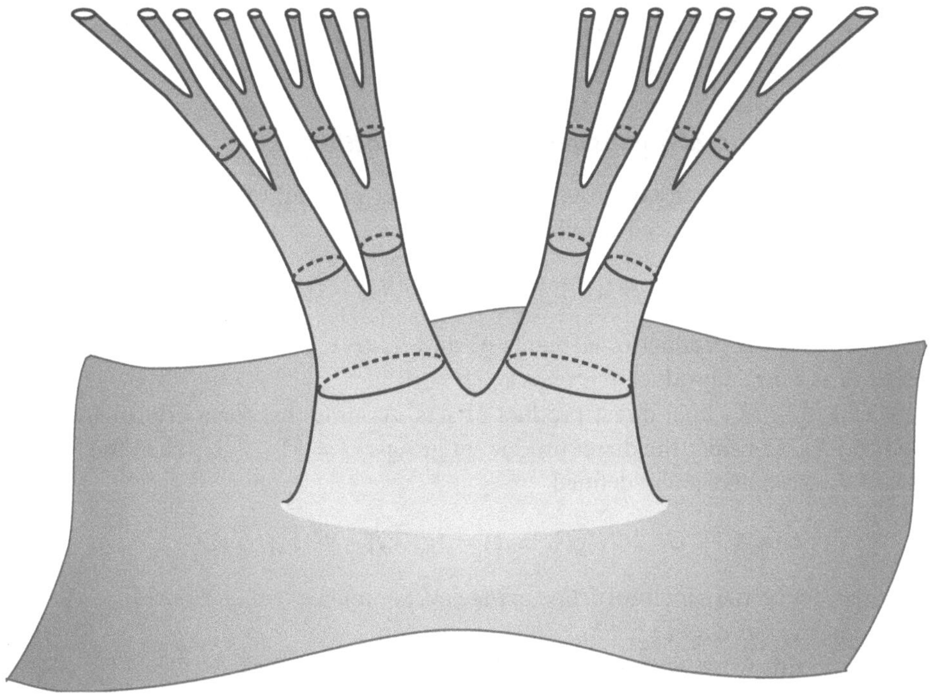

Figure 2. The surface homeomorphic to the plane without the Cantor set.

We use the notations $D_r(x) = \{y \in X \mid d(x,y) < r\}$ for the open ball of a radius r with the centre at x and $D_\varepsilon(B) = \{y \in X \mid d(y,B) < \varepsilon\} = \bigcup_{b \in B} D_\varepsilon(b)$ for the ε-neighbourhood of a subset B in a metric space (X,d).

Assumptions. Inclusions do not exclude equality. All neighbourhoods are assumed to be open. By a countable set we mean an infinite countable set as well as a finite set.

2. The canonical action of the direct product of groups

2.1. The Tychonoff product of topological spaces

Let A be an arbitrary set, let $\{X_\alpha \mid \alpha \in A\}$ be a family of any sets. The direct (Cartesian) product $X = \prod_{\alpha \in A} X_\alpha$ is the set of all maps of $x : A \to \bigcup_{\alpha \in A} X_\alpha$ such that $x(\alpha) \in X_\alpha$ for any $\alpha \in A$. If $x \in X$, then the point $x(\alpha) \in X_\alpha$ we will denote by the symbol x_α and call the α-coordinate of the element x. The symbol $\{x_\alpha\}$ will denote the point of the product X, α-coordinate of which is the point $x_\alpha \in X_\alpha$.

Let $\{(X_\alpha, \tau_\alpha) \mid \alpha \in A\}$ be a family of topological spaces. Assume that $X = \prod_{\alpha \in A} X_\alpha$ is provided by the weakest topology τ such that all projections $\pi_\alpha : \prod_{\alpha \in A} X_\alpha \to X_\alpha$, $\pi_\alpha(\{x_\alpha\}) := x_\alpha$, are continuous. This topology τ is called by the *Tychonoff* topology. Note that

$$\zeta = \{\pi_\beta^{-1}(U) \subset X \mid U \in \tau_\beta, \beta \in A\}$$

is a subbase of τ. The base of the Tychonoff topology formed by all possible finite intersections of subsets of ζ, is called the *canonical* base.

The topological space (X, τ) is called the *Tychonoff product of topological spaces* (X_α, τ_α) and is denoted by $(X, \tau) = \prod_{\alpha \in A}(X_\alpha, \tau_\alpha)$.

2.2. A direct product of groups and its canonical action

Let G_α, $\alpha \in A$, be a family of groups. On the direct product of sets $G = \prod_{\alpha \in A} G_\alpha$, the group operation is introduced as follows:

$$\Psi : G \times G \to G, \quad \Psi(\{g_\alpha\}, \{h_\alpha\}) = \{g_\alpha \cdot h_\alpha\} \quad \forall(\{g_\alpha\}, \{h_\alpha\}) \in G \times G,$$

where $g_\alpha \cdot h_\alpha$ is the product of elements g_α and h_α in G_α. The group $G = \prod_{\alpha \in A} G_\alpha$ is referred to as a direct product of groups G_α, $\alpha \in A$.

Let $X = \prod_{\alpha \in A} X_\alpha$ be a direct product of sets. Assume that for every $\alpha \in A$ a group G_α acts on X_α. Consider the direct product of groups $G = \prod_{\alpha \in A} G_\alpha$. Then the following action of the group G on X is defined

$$\Phi : G \times X \to G, \quad \Phi(\{g_\alpha\}, \{x_\alpha\}) = \{g_\alpha . x_\alpha\} \quad \forall(\{g_\alpha\}, \{x_\alpha\}) \in G \times X.$$

We call this action the *canonical action of the direct product of groups $G = \prod_{\alpha \in A} G_\alpha$ on the direct product of sets $X = \prod_{\alpha \in A} X_\alpha$*.

Further in this work, we assume that the direct product of groups $G = \prod_{\alpha \in A} G_\alpha$ acts on $X = \prod_{\alpha \in A} X_\alpha$ canonically.

3. Transitivity of the canonical product of homeomorphism groups

Definition 3.1: A homeomorphism group G of a topological space X is called topological transitive on X if for every nonempty open subsets U and V in X there exists such an element $g \in G$ that

$$g(U) \cap V \neq \varnothing.$$

Theorem 3.2: *Let A be an arbitrary set of indexes. For every $\alpha \in A$, a homeomorphism group G_α of the topological space X_α is topologically transitive on X_α if and only if the direct product of groups $G = \prod_{\alpha \in A} G_\alpha$ topologically transitive on the Tychonoff product of topological spaces $X = \prod_{\alpha \in A} X_\alpha$.*

Proof: Suppose that for every $\alpha \in A$, the group of homeomorphisms G_α of the topological space X_α is topologically transitive on X_α. Let $X = \prod_{\alpha \in A} X_\alpha$ be the Tychonoff product of topological spaces X_α. Show that for every nonempty open subsets U and V in X there exists an element $g = \{g_\alpha\} \in G$ such that $g(U) \cap V \neq \varnothing$. It is enough to prove this fact for every sets U and V from the canonical base of the Tychonoff topology on X. Let $U = \prod_{\alpha \in A} U_\alpha$, $V = \prod_{\alpha \in A} V_\alpha$ where U_α and V_α are nonempty open subsets in X_α, and for some finite subsets $A_1, A_2 \subset A$ the equalities are fulfilled

$$U_\alpha = X_\alpha \quad \forall \alpha \in A \setminus A_1, \quad V_\beta = X_\beta \quad \forall \beta \in A \setminus A_2.$$

The topological transitivity of G_α on X_α implies the existence of elements $g_\alpha \in G_\alpha$ satisfying $g_\alpha(U_\alpha) \cap V_\alpha \neq \varnothing$ for all $\alpha \in A$. Put $g = \{g_\alpha\} \in G$, then $g(U) \cap V = \prod_{\alpha \in A}(g_\alpha(U_\alpha) \cap V_\alpha) \neq \varnothing$. Thus the group G is topologically transitive on X.

The opposite. Suppose that the homeomorphism group $G = \prod_{\alpha \in A} G_\alpha$ is topologically transitive on the Tychonoff product $X = \prod_{\alpha \in A} X_\alpha$. Fix an arbitrary element $\delta \in A$. Let $\widetilde{U}_\delta$ and $\widetilde{V}_\delta$ be any nonempty open subsets in X_δ. Let $U = \prod_{\alpha \in A} U_\alpha$, $V = \prod_{\alpha \in A} V_\alpha$ where $U_\delta = \widetilde{U}_\delta$, $V_\delta = \widetilde{V}_\delta$, $U_\alpha = V_\alpha = X_\alpha$ for every $\alpha \in A \setminus \{\delta\}$. Note that U and V are nonempty open subsets in X. Since the group G acts topologically transitive on X, then there is an element $g = \{g_\alpha\} \in G$ such that $\varnothing \neq g(U) \cap V = \prod_{\alpha \in A}(g_\alpha(U_\alpha) \cap V_\alpha)$, therefore $g_\delta(U_\delta) \cap V_\delta \neq \varnothing$. This implies topological transitivity of the group G_δ on X_δ for every $\delta \in A$. ■

Theorem 3.3: *Let A be an arbitrary index set and let $X = \prod_{\alpha \in A} X_\alpha$ be the Tychonoff product of topological spaces X_α. Assume that G_α is a homeomorphism group of X_α. Then the direct product of groups $G = \prod_{\alpha \in A} G_\alpha$ has a dense orbit in X if and only if the group G_α has a dense orbit in X_α for every $\alpha \in A$.*

Proof: As known [15, Prop. 2.3.3], for every family of subsets $B_\alpha \subset X_\alpha$ in the product $X = \prod_{\alpha \in A} X_\alpha$ the closures satisfy the following relation:

$$\overline{\prod_{\alpha \in A} B_\alpha} = \prod_{\alpha \in A} \overline{B_\alpha}. \tag{1}$$

Since the orbit $G.x$ of $x = \{x_\alpha\} \in X$ is equal to the product of orbits $G_\alpha.x_\alpha$, i.e. $G.x = \prod_{\alpha \in A} G_\alpha.x_\alpha$, then taking into account (1) we get a chain of equalities $\overline{G.x} = \overline{\prod_{\alpha \in A} G_\alpha.x_\alpha} = \prod_{\alpha \in A} \overline{G_\alpha.x_\alpha}$. Therefore

$$\overline{G.x} = \prod_{\alpha \in A} \overline{G_\alpha.x_\alpha} \quad \forall x = \{x_\alpha\} \in X = \prod_{\alpha \in A} X_\alpha. \tag{2}$$

Using (2), it is easy to obtain a statement of the theorem being proved. ■

An analog of the Birkhoff transitivity theorem

If the action of the group G on the space X has a dense orbit, then it is topologically transitive. Indeed, let U and V are any open nonempty subsets of X and let $\overline{G.x} = X$ for some $x \in X$. Then there are elements $g_1, g_2 \in G$ such that $g_1.x \in U$ and $g_2.x \in V$. It follows that $g.U \cap V \neq \varnothing$ where $g = g_2 g_1^{-1} \in G$.

According to [10, Prop. 1], if the group G is topologically transitive on a Baire space X with a countable base, then there exists a point $x \in X$ with a dense orbit. Therefore we get the following analog of the Birkhoff theorem for homeomorphism groups of Baire spaces with a countable base.

Proposition 3.4: *If G is a homeomorphism group of a topological space X, then the existence of a dense orbit of G implies topological transitivity of G.*

When X is a Baire space with a countable base, the converse is also true.

In particular, for Polish spaces X, the existence of a dense orbit of G on X is equivalent to topological transitivity of G.

Example 3.5: Let $\{e_k\}_{k=\overline{1,n}}$ be a basis of the vector space $\mathbb{R}^n$ for $n \geq 1$. Define homeomorphisms of the Euclidean space $\mathbb{R}^n$ by the following equalities: $g_k(x) = x + e_k$, for $k = \overline{1,n}$, and $g_{n+1}(x) = \lambda x$ where $\lambda > 1$ for all $x \in \mathbb{R}^n$. Consider the homeomorphism group $G = \langle g_k \mid k = \overline{1, n+1} \rangle$. According to [24, Prop. 16], the homeomorphism group G is topologically transitive, and every its orbit is dense in $\mathbb{R}^n$. In other words, $\mathbb{R}^n$ is a minimal set of the group G. Note that there is no transitive subgroup of G with the number of generators less than $n+1$.

4. Density of closed orbits

Theorem 4.1: *Let A be an arbitrary index set and let $X = \prod_{\alpha \in A} X_\alpha$ be the Tychonoff product of topological spaces X_α. Assume that G_α is a homeomorphism group of X_α. Then the direct product of groups $G = \prod_{\alpha \in A} G_\alpha$ has a dense union of closed orbits in X if and only if for every $\alpha \in A$, the group G_α has a dense union of closed orbits in the topological space X_α.*

Proof: Let $x = \{x_\alpha\} \in X$, then $G.x = \prod_{\alpha \in A} G_\alpha.x_\alpha$. According to (2), we get

$$\overline{G.x} = G.x \iff \overline{G_\alpha.x_\alpha} = G_\alpha.x_\alpha \quad \forall \alpha \in A. \tag{3}$$

This means that an orbit $G.x$ is closed in X if and only if the orbit $G_\alpha.x_\alpha$ is closed in X_α for every $\alpha \in A$. Let B be the union of all closed orbits of G in X. Denote by B_α the union of all closed orbits of G_α in X_α. Therefore $B = \prod_{\alpha \in A} B_\alpha$. Suppose that B is dense in X, hence applying the equality (1), we get the following chain of equalities $X = \overline{B} = \overline{\prod_{\alpha \in A} B_\alpha} = \prod_{\alpha \in A} \overline{B_\alpha}$. Consequently $X_\alpha = \overline{B_\alpha}$, i.e. B_α is dense in X_α.

Conversely, let for every $\alpha \in A$ the subset of B_α be dense in X_α. As $B = \prod_{\alpha \in A} B_\alpha$, applying the equality (1), we have $\overline{B} = \prod_{\alpha \in A} \overline{B_\alpha} = \prod_{\alpha \in A} X_\alpha = X$. This means that the union of all closed orbits of G is dense in X. ■

5. Countable products of metric spaces

5.1. The direct product of a countable family of metric spaces

The direct product of two metric spaces

Let (X_1, d_1) and (X_2, d_2) be metric spaces. A metric d on the product of two metric spaces $X_1 \times X_2 = \{(x_1, x_2) \mid x_1 \in X_1, x_2 \in X_2\}$ may be introduced in the following ways ([14, Sec. 4.2]):

(1) $d(x,y) = \sqrt[p]{d_1^p(x_1,y_1) + d_2^p(x_2,y_2)}$ where $p \geq 1$.
(2) $d(x,y) = \max\{d_1(x_1,y_1), d_2(x_2,y_2)\}$.
(3) $d(x,y) = \sum_{i=1}^{2} \frac{1}{2^i} \frac{d_i(x_i,y_i)}{1+d_i(x_i,y_i)}$.

All these methods can be easily extended to the case of the product of any finite number of factors.

The direct product of a countable family of metric spaces

Based on any metric d on the set X, you can get a metric bounded by the number 1, by the formula

$$\widetilde{d}(x,y) = \frac{d(x,y)}{1 + d(x,y)} \quad \forall x, y \in X, \tag{4}$$

and metric topologies defined by the metrics d and $\widetilde{d}$ are coincided.

Let (X_i, d_i), $i \in \mathbb{N}$, be a countable family of metric spaces. According to the above, without loss of generality, we will assume that all metrics d_i are limited by the number 1. Metric on the product $X = \prod_{i\in\mathbb{N}} X_i$ can be given by the equality ([15, Th. 4.2.2]):

$$d(x,y) = \sum_{i=1}^{\infty} \frac{1}{2^i}\widetilde{d}_i(x_i, y_i), \tag{5}$$

where $\widetilde{d}_i$ is defined by the formula (4) if the metric d_i not limited by a number 1, otherwise $\widetilde{d}_i = d_i$.

Emphasize that the topology on X generated by the metric d, defined by (5), coincides with the Tychonoff topology of the product $X = \prod_{i\in\mathbb{N}} X_i$ of topological spaces X_i.

Definition 5.1: The metric d on $X = \prod_{i\in\mathbb{N}} X_i$ given by the formula (5) is called by the *direct product* of metrics $\widetilde{d}_i$. The metric space (X, d) is called the *direct product* of countable family of metric spaces $(X_i, \widetilde{d}_i)$ and it is denoted by $(X, d) = \prod_{i\in\mathbb{N}}(X_i, \widetilde{d}_i)$.

Remark 5.1: As it is known [14], the metric (5) is a special case of the metric $d(x,y) = \sum_{i=1}^{\infty} A_i\widetilde{d}_i(x_i, y_i)$, where the series $\sum_{i=1}^{\infty} A_i$ converges and all of its members are positive.

5.2. Nonmetrizability of the product of an uncountable family of topological spaces

Let (M, d) be a metric space and $D_r(x) = \{z \in M \mid d(x,z) < r\}$ be the ball of radius $r > 0$ centred at x. Recall that a topological space satisfies the first axiom of countability if it has a countable base of topology at each point. Every metric space (M, d) has a countable base $\Sigma_x = \{D_{1/n}(x) \mid n \in \mathbb{N}\}$ at each point $x \in M$. According to [15, Cor. 4.2.4], an uncountable product of metrizable spaces each of which contains at least two points does not satisfy the first axiom of countability. Thus an uncountable product of such spaces is not metrizable.

Therefore, we will consider further only products of a countable family of metric spaces in investigations of sensitivity of homeomorphism groups.

6. Sensitivity to initial conditions

6.1. Sufficient conditions for sensitivity to initial conditions

Definition 6.1: A homeomorphism group G of a metric space (X, d) is called *sensitive to initial conditions at a point* (or, for short, *sensitive at a point*) $x \in X$, if there exists a number

$\delta = \delta(x) > 0$ such that for every $\varepsilon > 0$ there exists an element $g \in G$ such that:

$$\mathrm{diam}(g.D_\varepsilon(x)) > \delta. \tag{6}$$

A group G is called *pointwise sensitive to initial conditions* (or, for short, *pointwise sensitive*), if it is sensitive at every point $x \in X$.

Usually, group actions are investigated on compact topological spaces, and the following definition of sensitivity to initial conditions is used.

Definition 6.2: A homeomorphism group G of a metric space (X, d) is called *sensitive to initial conditions* (or, for short, *sensitive*) if there exists a number $\delta > 0$ such that for each $x \in X$ and for every $\varepsilon > 0$ there exists an element $g \in G$ such that:

$$\mathrm{diam}(g.D_\varepsilon(x)) > \delta. \tag{7}$$

The number δ is referred to as *the sensitivity constant* for G.

Emphasize that every group G satisfying Definition 6.2, satisfies also Definition 6.1.

Example 6.3: Recall that a homeomorphism group G of a metric space (X, d) is said to be *expansive* on X, if there exists a constant $c > 0$ such that for every $x \neq y$ in X, there is $g \in G$ satisfying $d(g.x, g.y) > c$. Such a constant c is called an *expansivity constant* of this group [5]. Every expansive group G of homeomorphisms of a metric space (X, d) is sensitive to initial conditions, and the role of δ in Definition 6.2 plays the expansivity constant c.

Proposition 6.4: *Let G be a homeomorphism group of a metric Baire space X having a dense non-closed orbit. Then G is pointwise sensitive if and only if it has sensitive dependence on initial conditions.*

Proof: As the sensitivity implies pointwise sensitivity, proves the inverse. Let X be a metric Baire space. Assume that a group G is pointwise sensitive and has a dense non-closed orbit. For each $n \in \mathbb{N}$ consider the following subset:

$$V_n = \{x \in X \mid \exists\, \varepsilon > 0 : \mathrm{diam}(g.D_\varepsilon(x)) \leq 1/n \quad \forall g \in G\}. \tag{8}$$

Note that V_n is an open G-invariant subset in X, and $V_n \supset V_{n+1} \supset V_{n+2} \supset \cdots$ Since every open set contains a point with dense orbit, then if $V_n \neq \varnothing$ for some n, due to G-invariance of V_n, the set V_n is dense in X. Emphasize that each $x \in \bigcap_{n\in\mathbb{N}} V_n$ is not a sensitive point respectively G, that contradicts the assumption. Therefore $\bigcap_{n\in\mathbb{N}} V_n = \varnothing$, and there exists $m \in \mathbb{N}$ for which $V_m = \varnothing$, because otherwise $\bigcap_{n\in\mathbb{N}} V_n$ is dense in the Baire space X. Therefore, $V_n = \varnothing$ for every $n \geq m$. According to (8), this means that there exists $\delta = 1/m$ such that for every $\varepsilon > 0$ there exists an element $g \in G$ satisfying $\mathrm{diam}(g.D_\varepsilon(x)) > \delta$. By Definition 6.2, the group G is sensitive to initial conditions on X. This completes the proof. ■

Remark 6.1: For a continuous action of a topological group G on a compact metric space (X, d), as indicated by F. Polo [20, Prop. 1.3] one can prove Proposition 6.4 using ideas from [1]. In fact, we have implemented such a possibility under a more general assumption, replacing the compactness condition of the metric space X with the assumption that X is a Baire space.

Remark 6.2: As it is known, the sensitivity is not invariant with respect to topological conjugacy. However, if dynamical systems are defined on a compact topological space, in [4] it is shown that the sensitivity is invariant respectively topological conjugacy.

The following statement contains conditions that guarantee that sensitivity to initial conditions of the group of homeomorphisms of the topological space X does not depend on the choice of the metric that metrizes X.

Proposition 6.5: *Let X be a Baire space. Let (X, d) and (X, ρ) be metric spaces such that the metric topologies given by the metrics d and ρ coincide with the topology of the space X. Then the homeomorphism group G of X, having a dense non-closed orbit, is sensitive to initial conditions in (X, d) if and only if G is sensitive to initial conditions in (X, ρ).*

Proof: According to Proposition 6.4, it is sufficient to prove that pointwise sensitivity of G in the spaces (X, d) and (X, ρ) are equivalent.

Let τ_d and τ_ρ be metric topologies defined by metrics d and ρ on X respectively, and $\tau_d = \tau_\rho$. Recall that τ_d is the topology with a base $\sigma = \{D_r(x) \mid x \in X, r > 0\}$. Analogously, $\widetilde{\sigma} = \{\widetilde{D}_r(x) \mid x \in X, r > 0\}$ is a base of τ_ρ.

Assume that a homeomorphism group G of X is pointwise sensitive to initial conditions in (X, d). According to Definition 6.1, for every $x \in X$ there is a number $\delta = \delta(x) > 0$ such that for each $\varepsilon > 0$ there exists an element $g \in G$ satisfying the inequality $\mathrm{diam}_d(g.D_\varepsilon(x)) > \delta$, where $\mathrm{diam}_d(g.D_\varepsilon(x))$ denotes the diameter of the set $g.D_\varepsilon(x)$ in (X, d). Consider an arbitrary $\widetilde{\varepsilon} > 0$, then there will be $\varepsilon' > 0$ such that $D_{\varepsilon'}(x) \subset \widetilde{D}_{\widetilde{\varepsilon}}(x)$. Hence there is $\widetilde{g} \in G$ for which $\mathrm{diam}_d(\widetilde{g}.D_{\varepsilon'}(x)) > \delta$. Therefore there exists $y \in D_{\varepsilon'}(x)$ such that $d(\widetilde{g}.y, \widetilde{g}.x) > \delta/2$. This means that

$$\widetilde{g}.y \notin D_{\delta/2}(\widetilde{g}.x). \tag{9}$$

According to the equality $\tau_d = \tau_\rho$, there will be a number $\widetilde{\delta} = \widetilde{\delta}(x) > 0$ satisfying the following inclusion:

$$\widetilde{D}_{\widetilde{\delta}}(\widetilde{g}.x) \subset D_{\delta/2}(\widetilde{g}.x). \tag{10}$$

The relations (9)–(10) imply that $\widetilde{g}.y \notin \widetilde{D}_{\widetilde{\delta}}(\widetilde{g}.x)$, i.e. $\rho(\widetilde{g}.x, \widetilde{g}.y) > \widetilde{\delta}$. This means that $\mathrm{diam}_\rho(\widetilde{g}.D_{\widetilde{\varepsilon}}(x)) > \widetilde{\delta}$.

Thus, for every $x \in X$ there is a number $\widetilde{\delta} = \widetilde{\delta}(x) > 0$ such that for every $\widetilde{\varepsilon} > 0$ there exists an element $\widetilde{g} \in G$ satisfying the inequality $\mathrm{diam}_\rho(\widetilde{g}.D_{\widetilde{\varepsilon}}(x)) > \widetilde{\delta}$. This means that the group G is pointwise sensitive to initial conditions in the metric space (X, ρ).

Similarly, the sensitivity of the group G to initial conditions in the metric space (X, ρ) implies the sensitivity of G to initial conditions in (X, d). ■

6.2. *Proof of Theorem 1.4*

Suppose the opposite, let the group of homeomorphisms G of the metric space (X, d) acts chaotically, but G is not sensitive to initial conditions. According to Proposition 6.4, there exists a point $x \in X$ which is not sensitive to initial conditions.

Case 1: The orbit $G.x$ is not dense in X, i.e. the $\overline{G.x} \neq X$. Then $U = X \setminus \overline{G.x}$ is nonempty open subset in X. Therefore there exists $a \in U$, and $d(a, \overline{G.x}) > 0$. Put $\delta = \frac{1}{2}d(a, \overline{G.x})$. According to the condition (1) of Definition 1.3, there exists a dense orbit $G.z$ in X. Hence for every $\varepsilon > 0$, there is $y \in G.z \cap D_\varepsilon(x)$ and there exists $y' \in G.z \cap D_\delta(a)$. Hence there exists an element $g \in G$ for which $y' = g.y$. Show that $\text{diam}(g.D_\varepsilon(x)) > \delta$. Suppose the opposite, i.e. $\text{diam}(g.D_\varepsilon(x)) \leq \delta$. Using the triangle inequality in the metric space (X, d), we get

$$d(a, g.x) \leq d(a, y') + d(g.y, g.x) < \delta + \delta = 2\delta = d(a, \overline{G.x})$$

which contradicts the definition of distance $d(a, \overline{G.x})$. The contradiction proves inequality $\text{diam}(g.D_\varepsilon(x)) > \delta$. Therefore, x is sensitive to initial conditions.

Case 2: The orbit $G.x$ is dense in X, i.e. $\overline{G.x} = X$, and at point $x \in X$ the group G is not sensitive to initial conditions. Hence for every $\delta > 0$ there exists $\varepsilon_0 > 0$ such that $\text{diam}(g.D_{\varepsilon_0}(x)) \leq \delta$ for each $g \in G$. Therefore for every $y \in D_{\varepsilon_0}(x)$ and each $x' \in G.x$ there exists $y' \in G.y$ satisfying the inequality

$$d(x', y') \leq \delta. \tag{11}$$

As the union of closed orbits of G is dense in X, there is a point $y \in D_{\varepsilon_0}(x)$, through which the closed orbit $G.y$ passes. Consider $z \in U = X \setminus G.y$. Since the orbit $G.x$ is dense in X, for any $\varepsilon > 0$ there exists $x' \in G.x \cap D_\varepsilon(z)$. Due to the arbitrariness of $\varepsilon > 0$, given (11) and applying the triangle inequality, we get

$$d(z, y') \leq d(z, x') + d(x', y') < \varepsilon + \delta \Rightarrow z \in \overline{D_\delta(G.y)}. \tag{12}$$

Consequently,

$$X \subset \overline{D_\delta(G.y)} \Rightarrow X = \overline{D_\delta(G.y)}. \tag{13}$$

Take a sequence of positive numbers $\delta_n \to 0$. Hence for every δ_n there exists $\varepsilon_n > 0$ and $y_n \in D_{\varepsilon_n}(x)$ such that the orbit $G.y_n$ is closed, and $d(g.x, g.y_n) \leq \delta_n$ for each $g \in G$. Hence

$$X = \overline{D_{\delta_n}(G.y_n)} \quad \forall n \in \mathbb{N} \Rightarrow X = \bigcap_{n \in \mathbb{N}} \overline{D_{\delta_n}(G.y_n)}. \tag{14}$$

Consequently, for every point $v \in X$ we get $v \in \overline{D_{\delta_n}(G.y_n)}$ for every $n \in \mathbb{N}$. This means that for every $\varepsilon > 0$ there exists $z_n \in D_{\delta_n}(G.y_n)$ such that $d(v, z_n) < \varepsilon$. Therefore for every $n \in \mathbb{N}$, there exists $\widetilde{y}_n \in G.y_n$, satisfying the inequality $d(z_n, \widetilde{y}_n) < \delta_n$. Applying the triangle inequality, we get

$$d(v, \widetilde{y}_n) \leq d(v, z_n) + d(z_n, \widetilde{y}_n) < \varepsilon + \delta_n \quad \forall n \in \mathbb{N}.$$

As ε and δ_n are arbitrarily small quantities, it is necessary that

$$v \in \overline{\bigcup_{n \in \mathbb{N}} G.y_n} \Rightarrow X = \overline{\bigcup_{n \in \mathbb{N}} G.y_n}. \tag{15}$$

By the condition of theorem being proved, every closed orbit $G.y_n$ is a nowhere dense subset in X, hence a countable union $\bigcup_{n \in \mathbb{N}} G.y_n$ is also nowhere dense in the Baire space X. Therefore the equality in (15) is not possible. The contraction completes the proof. □

As in each Hausdorff space a finite orbit is closed, Theorem 1.4 implies the following statement.

Corollary 6.6: *Let X be a metric Baire space with a countable base. If a homeomorphism group G of X is chaotic in the sense of Definition 1.2, then the group G is sensitive to initial conditions.*

6.3. Sensitivity to initial conditions of the direct product of homeomorphism groups

Emphasize that all metric spaces are considered with the metric topology, unless otherwise specified. We use notations from Section 5. In the case when J is a finite subset of $\mathbb{N}$, the metric d of the product $\prod_{i\in J}(X_i, \widetilde{d}_i)$ is defined by the formula (5), in which $i \in J$. The sensitivity of the group $G = \prod_{i\in J} G_i$ is not changed if the metric d will be defined by the formula

$$d(x, y) = \sum_{i=1}^{n} d_i(x_i, y_i) \quad \forall x = \{x_i\}, y = \{y_i\},$$

where boundedness of metrics d_i is not assumed.

Theorem 6.7: *Let $(X, d) = \prod_{i\in J}(X_i, \widetilde{d}_i)$ where $(X_i, \widetilde{d}_i)$ is a metric space and let $J \subset \mathbb{N}$. Let G_i, $i \in J$, be a homeomorphism group of X_i. If there exists $n \in J$ such that G_n is sensitive to initial conditions on $(X_n, \widetilde{d}_n)$, then the direct product of groups $G = \prod_{i\in J} G_i$ is also sensitive to initial conditions on (X, d).*

Proof: Let $x = \{x_i\}$ be any point in X. Assume that a group G_n is sensitive to initial conditions on $(X_n, \widetilde{d}_n)$. This means that there exists a number $\sigma > 0$ such that for the point $x_n \in X_n$ and for any $\eta > 0$ there is an element $\widetilde{g} \in G_n$ such that $\mathrm{diam}(\widetilde{g}.D_\eta(x_n)) > \sigma$. Therefore there exists a point $\widetilde{y} \in D_\eta(x_n)$ satisfying the inequality $\widetilde{d}_n(\widetilde{g}.x_n, \widetilde{g}.\widetilde{y}) > \frac{\sigma}{2}$.

Consider $y = \{y_i\} \in X$ where $y_n = \widetilde{y}$ and $y_i = x_i$ for $i \in J \setminus \{n\}$. Pick an element $g = \{g_i\} \in G$ where g_i is an arbitrary element of G_i where $i \in J \setminus \{n\}$ and $g_n = \widetilde{g}$. Then

$$d(x, y) = \frac{\widetilde{d}_n(x_n, \widetilde{y})}{2^n} < \frac{\eta}{2^n} = \varepsilon \quad \textit{and} \quad d(g.x, g.y) = \frac{\widetilde{d}_n(g_n.x_n, g_n.\widetilde{y})}{2^n} > \frac{\sigma}{2^{n+1}} = \delta.$$

Hence $\mathrm{diam}(g.D_\epsilon(x)) > \delta$. Since η is any arbitrarily small positive number, then $\varepsilon = \frac{\eta}{2^n}$ is also any arbitrarily small positive number. As x is any point from X, the proven means the sensitivity of the group $G = \prod_{i\in J} G_i$ to initial conditions. ■

Theorem 6.8: *Let G_i, $i = \overline{1, m}$, be a homeomorphism group of a metric space (X_i, d_i). Assume that $(X, d) = \prod_{i=1}^{m}(X_i, d_i)$. Then the direct product of groups $G = \prod_{i=1}^{m} G_i$ is sensitive to initial conditions on (X, d) if and only if there exists n, $1 \leq n \leq m$, such that the group G_n is sensitive to initial conditions.*

Proof: Sufficiency is proved similarly to Theorem 6.7. Let's prove the necessity. To do this, assume that the product of groups $G = \prod_{i=1}^{m} G_i$ is sensitive to initial conditions and at the same time each group G_i is not sensitive to initial conditions. Therefore for every $\delta_i > 0$

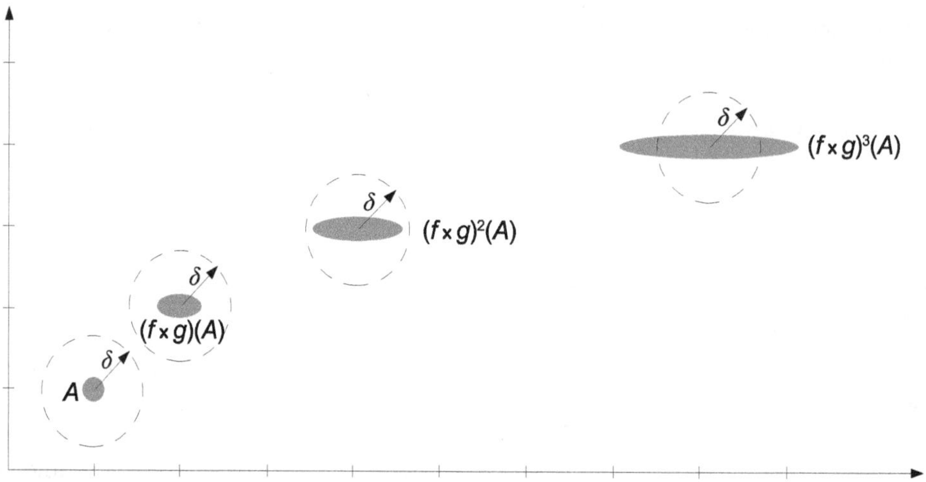

Figure 3. Example of insensitive group of homeomorphisms.

there exist point $x_i \in X_i$ and $\varepsilon_i > 0$ such that for every $g_i \in G_i$ the following inequality $\operatorname{diam}(g_i.D_{\varepsilon_i}(x_i)) \leq \delta_i$ is satisfied. Put $\varepsilon = \min\{\varepsilon_i \mid i = \overline{1,m}\} \Rightarrow \varepsilon > 0$, $\delta_i = \frac{\delta}{2m}$, where δ is the sensitivity constant of the action of the group G. Let $x = \{x_i\}$. Then for every $y = \{y_i\} \in D_\varepsilon(x)$ we get

$$d(g.x, g.y) = \sum_{i=1}^{m} d_i(g_i.x_i, g_i.y_i) \leq \sum_{i=1}^{m} \frac{\delta}{2m} = \frac{\delta}{2},$$

consequently $\operatorname{diam}(g.D_\varepsilon(x)) \leq \delta$ that contradicts the sensitivity of the group G to initial conditions. Thus there exists n, $1 \leq n \leq m$, such that the group G_n is sensitive to initial conditions. ■

Corollary 6.9: *Let $(X, d) = \prod_{i=1}^{m}(X_i, d_i)$, where (X_i, d_i) are metric spaces. Let G_i be a homeomorphism group of X_i. Then the expansiveness of one of the groups G_i entails the sensitivity of the product group $G = \prod_{i=1}^{m} G_i$ to initial conditions.*

Example 6.10: If the group of homeomorphisms G_1 is expansive on (X_1, d_1), and for $i = \overline{2,m}$ the group G_i is an isometry group of a metric space (X_i, d_i), then, according to Corollary 6.9, the group $G = \prod_{i=1}^{m} G_i$ is sensitive to initial conditions on the product $(X, d) = \prod_{i=1}^{m}(X_i, d_i)$.

Here is a concrete example. The straight line $\mathbb{R}^1$ and the plane $\mathbb{R}^2$ are considered as metric spaces with standard metrics d_1 and d, respectively, and $(\mathbb{R}^2, d) = (\mathbb{R}^1, d_1) \times (\mathbb{R}^1, d_1)$. Consider two groups of homeomorphisms $G_1 = \langle f \rangle$, where $f(x) = 2x$, $x \in \mathbb{R}^1$, and $G_2 = \langle g \rangle$, where $g(y) = y + 1$, $y \in \mathbb{R}^1$. Since the group G_2 is expansive, then, according to Corollary 6.9, the group $G = G_1 \times G_2$ of plane homeomorphisms is sensitive to initial conditions (see Figure 3). Note that the group G is not expansive.

7. Chaotic actions of groups

7.1. Proof of Theorem 1.5

Let A be an arbitrary set of indexes. Suppose that the group $G = \prod_{\alpha \in A} G_\alpha$ acts chaotically on the Tychonoff product $X = \prod_{\alpha \in A} X_\alpha$ of topological spaces X_α. By Definition 1.3, G has a dense non-closed orbit and a dense union of closed orbits. According to Theorem 3.3, the group G has a dense non-closed orbit in X if and only if for every $\alpha \in A$ the group G_α has a dense orbit in X_α. By Theorem 4.1, G has a dense union of closed orbits if and only if the union of closed orbits of the group G_α is dense on X_α for every $\alpha \in A$. Thus a chaotic behaviour of the group G on X is equivalent to a chaotic behaviour of each group G_α on X_α for every $\alpha \in A$.

Further we use notations introduced in Section 5.

Theorem 7.1: *For every subset $J \subset \mathbb{N}$, let G_i, $i \in J$, be a homeomorphism group of metric space $(X_i, \widetilde{d}_i)$ and on the product of metric spaces $(X, d) = \prod_{i \in J}(X_i, \widetilde{d}_i)$ the direct product of groups $G = \prod_{i \in J} G_i$ is given. Assume that (X, d) is a Baire space. If G acts on X chaotically, then:*

(1) *the group G is sensitive to initial conditions;*
(2) *for every $i \in J$ the group G_i is chaotic and sensitive.*

Proof: First note that by Theorem 1.5, every group G_i, $i \in J$, is chaotic. Since (X, d) is a Baire space, then every factor $(X_i, \widetilde{d}_i)$, $i \in J$, is also a Baire space. Hence, according to Theorem 1.4, chaoticity of G implies its sensitivity to initial conditions on X and chaoticity of G_i implies the sensitivity of G_i on $(X_i, \widetilde{d}_i)$. ∎

7.2. Chaotic products of countable homeomorphism groups of Polish spaces

Lemma 7.2: *Let H be a countable group of homeomorphisms of a Polish space X. Then:*

(1) *every closed orbit of H is discrete;*
(2) *if X is compact, then every closed orbit of H is finite.*

Proof: Assume that $H.x$ is a closed orbit of $x \in X$. Then $H.x$ is a Polish space as a closed subset of a Polish space X. It is shown that the induced topology on the orbit $H.x$ is discrete. Assume that it is not true, hence $H.x$ has a non-isolated point x_0. Since H is countable, then H is represented in the form $H = \{h_i \mid i \in \mathbb{N}\}$. As x_0 is non-isolated, there exists a sequence $z_n = g_n.x$, $g_n \in H$, such that $z_n \to x_0$ as $n \to +\infty$. Now check that every point of the orbit $H.x$ is non-isolated. Pick $y \in H.x$, then there is an element $g \in H$ for which $y = g.x_0$. Hence $y_n = g.z_n \to y$ as $n \to +\infty$, i.e. y is non-isolated.

As $h_i.x$ is not isolated in $H.x$, the set $U_i = H.x \setminus h_i.x$ is dense and open in $H.x$. Since $H.x$ is a Polish space, then $H.x$ is also a Baire space. According to the definition of Baire space, the intersection $\bigcap_{i \in \mathbb{N}} U_i$ is dense in $H.x$. As $\bigcap_{i \in \mathbb{N}} U_i = \varnothing$, we get a contradiction. Hence every point of $H.x$ is isolated. Thus the statement (1) is proved.

Suppose now that a Polish space X is compact. In this case, every closed orbit $H.x$, $x \in X$, of H is finite as a discrete compact subspace of the compact X. Therefore, the statement (2) is also true. ∎

Recall that a group G is referred to as *residually finite* or *finitely approximable*, if for every non-neutral element $g \in G$ there exists a normal subgroup, not containing g, of finite index in G.

Theorem 7.3: *If a countable group of homeomorphisms G is chaotic (in the sense of Definition 1.3) on a metrizable compact space X, then G is residually finite and sensitive to initial conditions.*

Proof: Lemma 7.2 and Proposition 3.4 imply that definitions of chaos 1.2 and 1.3 are equivalent. Hence the homeomorphism group G is chaotic in the sense of Definition 1.2. It follows from [8, Th. 1] that every chaotic group of homeomorphisms is residually finite. According to Theorem 1.4, chaoticity of the group G in the sense of Definition 1.3 implies sensitivity of G to initial conditions. ■

As it is known, the following groups are residually finite:

(1) matrix groups $SL(n, \mathbb{Z})$ for all $n \geq 2$;
(2) finitely generated linear groups;
(3) (finite or infinite) direct products of residually finite groups;
(4) countable generated free groups;
(5) finitely generated nilpotent groups;
(6) quotients of residually finite groups by finite normal subgroups;
(7) fundamental groups of compact 3-manifolds

and some others.

Moreover, groups having infinite simple subgroups and, in particular, simple groups are not residually finite [8].

Remark 7.1: In [8], the sensitivity to initial conditions is not investigated.

Proof of Theorem 1.6

Note that every compact metrizable space is Polish. As it is well known, unlike the Baire space, any Polish space is either countable or has the continuum cardinality.

Assume that every group G_i, $i \in \mathbb{N}$, is chaotic. According to Theorem 1.5, the canonical action of the direct product of groups $G = \prod_{i \in \mathbb{N}} G_i$ is chaotic, i.e. the statement (1) is true.

As the group G_i is chaotic, the union of closed orbits of G_i is dense in X_i. Due to countability of G_i and compactness of X_i, by Lemma 7.2, every closed orbit of G_i is finite. The union $F_i \subset X_i$ of closed orbits of the group G_i, containing greater than one point, is dense in X_i. Otherwise there exists an open subset $U \subset X_i$ such that $U \cap F_i = \varnothing$. Since G_i is chaotic, then the union of its closed orbits is dense in X_i. Therefore the union of one-point orbits of G_i is dense in U. Continuation of G_i implies $G|_U = \mathrm{id}_U$ that contradicts the chaoticity of G_i. Thus F_i is dense in X_i. Analogously to the proof Theorem 4.1, we get that the subset $F = \prod_{i \in \mathbb{N}} F_i$ of $X = \prod_{i \in \mathbb{N}} X_i$ is a union of closed orbits of the group $G = \prod_{i \in \mathbb{N}} G_i$, and F is dense in X. Every orbit $G.x$ where $x = \{x_i\} \in X$, $x_i \in F_i$, has continuum cardinality as a product of countable set of finite orbits $G_i.x_i$, $i \in \mathbb{N}$. By the Tychonoff theorem, the orbit $G.x$ is compact as the product of compacts orbits $G_i.x_i$.

Assume that there exists a point $x_0 \in G.x$ which is isolated in $G.x$. Now check that every point of the orbit $G.x$ is isolated. Otherwise there is non-isolated $y = g.x_0$ where $g \in G$. Hence there exists a sequence $y_n \to y$ as $n \to +\infty$. Then $z_n = g^{-1}.y_n \to x_0$ as $n \to +\infty$, that contradicts to our assumption. Therefore orbit $G.x$ has no isolated points, i.e. $G.x$ is perfect subset of X. Consequently, the statement (2) is proved.

According Definition 1.3, every chaotic group G_i has a dense orbit $G_i.v_i$, $v_i \in X_i$. Note that $G_i.v_i$ is a countable subset of X_i, hence the orbit $G.v = \prod_{i\in\mathbb{N}} G_i.v_i$ has continuum cardinality. Since $\overline{G.v} = \prod_{i\in\mathbb{N}} \overline{G_i.v_i} = \prod_{i\in\mathbb{N}} X_i = X$, the orbit $G.v$ is dense in X. Hence the statement (3) is proved.

The statements (4)–(5) are corollaries of Theorem 7.3.

Let for every $i \in \mathbb{N}$, the group G_i has a fixed point x_i^0. Let Y be the union of all finite orbits of the group $G = \prod_{i\in\mathbb{N}} G_i$. Note that $x_0 = \{x_i^0\}$ is a fixed point of G, hence $x_0 \in Y$. Observe that $z = \{z_i\} \in Y$ if and only if there exists a finite subset $A \subset \mathbb{N}$ such that $z_i = x_i^0$ for $i \in \mathbb{N} \setminus A$ and z_i has a finite orbit $G_i.z_i$ for $i \in A$. As the intersection of Y with every set from the canonical base of the Tychonoff topology of X is a nonempty set, Y is dense in X. Thus the statement (6) is proved. □

8. Products of groups generated by generalized horseshoe maps

8.1. Bi-infinite sequence of symbols

Let $S = \{1, 2, \ldots, N\}$ with $N \geq 2$. Equip the set S with discrete metric d. Note that the metric topology coincides with the discrete topology on S. Let $\{S_i \mid i \in \mathbb{Z}\}$ be a family of topological spaces, and $S_i = S$, $d_i = d$ for every $i \in \mathbb{Z}$. Let Σ^N be a Tychonoff product of this family: $\Sigma^N = \prod_{i\in\mathbb{Z}} S_i$. Every point of Σ^N can be represented as a bi-infinite sequence of N symbols:

$$\sigma \in \Sigma^N \Leftrightarrow \sigma = (\cdots, \sigma_{-k}, \ldots, \sigma_{-1}, \sigma_0, \sigma_1, \ldots, \sigma_k, \ldots),$$

where $\sigma_i \in \{1, 2, \ldots, N\}$ for every $i \in \mathbb{Z}$. The topological space Σ^N is compact (as product of compact spaces), totally disconnected (as product of totally disconnected spaces), perfect and uncountable. Also Σ^N is a metrizable space. The metric on Σ^N can be defined by the following way:

$$\rho(\sigma, \tau) = \sum_{i\in\mathbb{Z}} \frac{1}{2^{|i|}} \frac{d_i(\sigma_i, \tau_i)}{1 + d_i(\sigma_i, \tau_i)},$$

where d_i is the metric on the space S_i. The topology induced by the metric ρ on Σ^N is the same as the Tychonoff topology on the product $\prod_{i\in\mathbb{Z}} S_i$.

As is known, the space Σ^2 is homeomorphic to the standard ternary Cantor set [15, Example 3.1.28].

Now we define a map $g : \Sigma^N \to \Sigma^N$ as follows: $(g(\sigma))_i = \sigma_{i+1}$. The map g is referred to as *the full N-shift*. Consider a homeomorphism group $G = \langle g \rangle$ generated by g.

Proposition 8.1: *The homeomorphism group $G = \langle g \rangle$ is chaotic on the space Σ^N, and the union of finite orbits is infinite countable and dense in Σ^N. Besides, the group G is sensitive to initial conditions.*

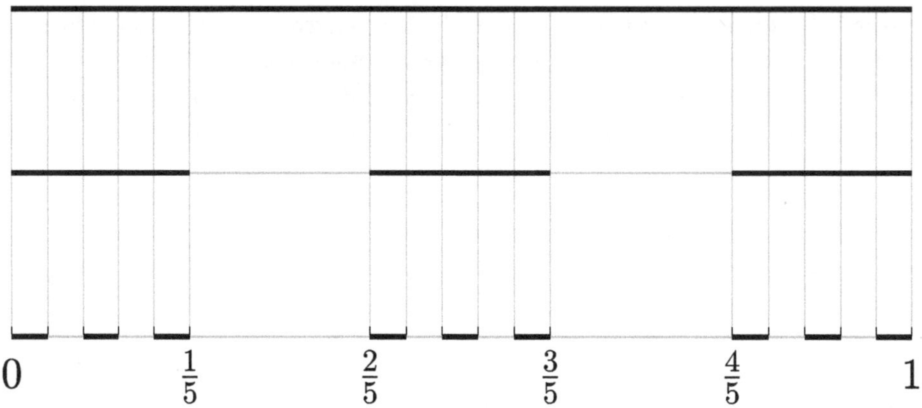

Figure 4. Three steps of construction of the 5-ary Cantor set.

Proof: According to [22, Prop. 3.9.4], the group G has a dense orbit.

Let $\sigma = (\sigma_i) \in \Sigma^N$. It is sufficiently to show that for every $\varepsilon > 0$ there is periodic point $\tau \in D_\varepsilon(\sigma)$ of the group G. Let $m > \log_2 \frac{1}{\varepsilon}$ and $m \in \mathbb{N}$. The point $\tau = \{\tau_i\}$, where $\tau_{k(2m+1)+j} = \sigma_j \, \forall k \in \mathbb{Z}$, $\forall j \in [-m, m] \cap \mathbb{Z}$, is periodic with the minimal period $2m+1$. Besides, $\tau \in D_\varepsilon(\sigma)$ because $\rho(\sigma, \tau) \leq \sum_{i=m+1}^{\infty} \frac{1}{2^i} = \frac{1}{2^m} < \varepsilon$. Thus the set of periodic orbits is dense in Σ^N.

By analogy with [17, Corollary 2.5.1], one can prove that the set of periodic orbits is infinite countable.

Moreover, the topological space Σ^N has a countable base, then it is a Polish space, hence Theorem 1.4 implies that the chaotic group of homeomorphisms G of Σ^N is sensitive to initial conditions. ■

Remark 8.1: Let J be any subset of $\mathbb{N}$ and $\Sigma_i^N = \Sigma^N$ for all $i \in J$. Applying [15, Prop. 2.3.7], we get that the Tychonoff product $\Sigma = \prod_{i \in J} \Sigma_i^N$ is homeomorphic to the space Σ^N.

8.2. *N-ary Cantor sets*

Here we recall the known generalization of the standard ternary Cantor set and the relation between this generalization and the space Σ^N.

Let $N \in \mathbb{N}$, $N > 2$ and N is odd. The N-ary Cantor set can be constructed in the similar way as the standard ternary Cantor set. Let $C_0 = [0, 1]$, $C_1 = [0, \frac{1}{N}] \cup [\frac{2}{N}, \frac{3}{N}] \cup \cdots \cup [\frac{N-1}{N}, \frac{N}{N}]$. Further we subdivide every closed interval of the set C_1 into N equal parts and delete open intervals with even numbers. The remaining set is denoted C_2. By repeating this process, we get sets C_i for every $i \geq 2$. The set $C = \bigcap_{i \in \mathbb{N}} C_i$ is the N-ary Cantor set [11, section 2.3]. In Figure 4, this definition for case $N = 5$ is illustrated.

Another equivalent approach to the definition of the N-ary Cantor set is known. Let $N \in \mathbb{N}$, $N > 2$ and N is odd. The set $\{\sum_{i \in \mathbb{N}} \frac{a_i}{N^i} \mid a_i \in \mathbb{Z}, 0 \leq a_i < N, a_i \, \text{iseven}\}$ is called the N-ary Cantor set [2, section 1.9.1].

The N-ary Cantor set is homeomorphic to the space $\Sigma^{\frac{N+1}{2}}$. The proof of this fact is fully similar the proof that the standard ternary Cantor set is homeomorphic to the space Σ^2 [15, Example 3.1.28].

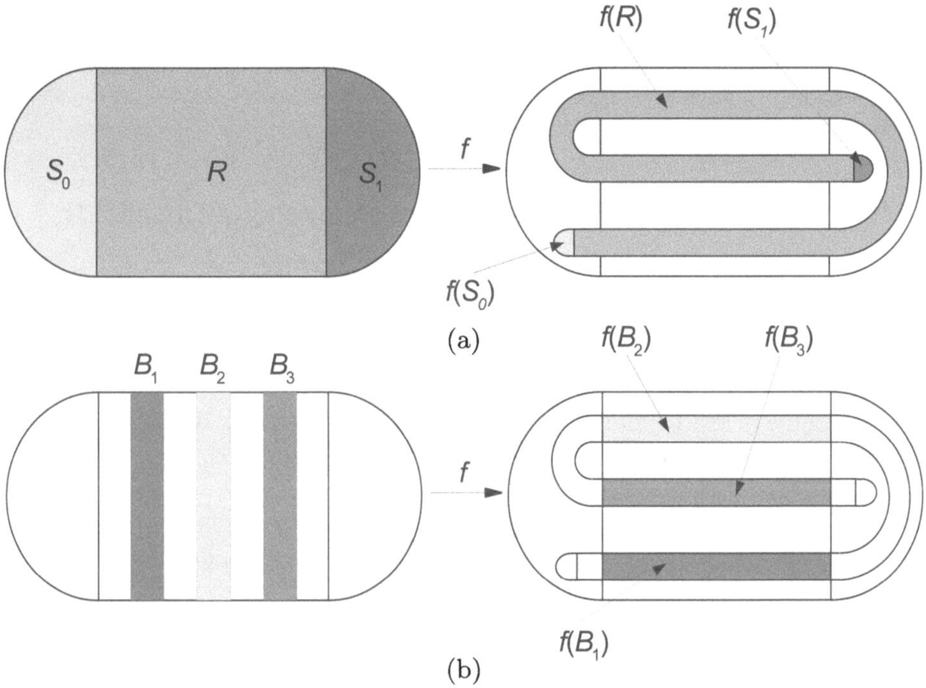

Figure 5. The generalized horseshoe map of length 3.

8.3. Generalized horseshoe maps

Original notion of the horseshoe map belongs to S. Smale. As is known, the invariant set Λ of the horseshoe map is homeomorphic to the standard ternary Cantor set and the restriction of the horseshoe map to the set Λ is conjugate to the 2-shift map on the space Σ^2 [17, Section 2.5].

The definition of generalized horseshoe map of length N can be found in [16, Definition 2.7]. In Figure 5, the definition for case $N = 3$ is illustrated.

Let $D = S_0 \cup R \cup S_1$. The generalized horseshoe map f of length 3 has the set D as its domain. The image of the set D respectively f is shown in Figure 5(a).

The invariant set Λ of the generalized horseshoe map of length N is homeomorphic to the $(2N - 1)$-ary Cantor set and, consequently, to the space Σ^N. The full N-shift is conjugate to the restriction of the generalized horseshoe map of length N to the invariant set Λ [22, Section 4.2.3], [16, Remark 2.9].

8.4. Chaotic groups of homeomorphisms on countable products of Cantor sets

Let A be any index set. Consider a family of chaotic groups $\{G_\alpha = \langle g_\alpha \rangle\}_{\alpha \in A}$, where g_α is the full N_α-shift of Σ^{N_α}. According to Theorem 1.5, the product of groups $G = \prod_{\alpha \in A} G_\alpha$ of homeomorphisms of the space $\Sigma = \prod_{\alpha \in A} \Sigma_\alpha^{N_\alpha}$ is chaotic.

Let the set A be countable. As we have seen above, every Σ^{N_α} is a compact metric space, hence Σ^{N_α} is a metric Baire space. Consequently, Σ is a metric Baire space, so according to Theorem 1.4, G is sensitive to initial conditions on Σ.

In the case when $N_\alpha = N$ for every $\alpha \in A \subset \mathbb{N}$, we get $\Sigma_\alpha^N := \Sigma^{N_\alpha} = \Sigma^N$ and $\Sigma = \prod_{\alpha \in A} \Sigma_\alpha^N$. According to Remark 8.1, Σ is homeomorphic to Σ^N. Thus we get new chaotic actions of free Abelian groups G with countable set of generators on the space Σ homeomorphic to $(2N-1)$-ary Cantor set.

9. Construction of chaotic actions of groups on topological manifolds

9.1. Chaotic actions of the group $\mathbb{Z}$ on every closed surface

By a closed surface we mean a connected compact topological two-dimensional manifold M without boundary.

When proving Theorem 9.1, we use the concept of a two-dimensional orbifold. A simple exposition of the theory of compact two-dimensional orbifolds can be found in [23]. The necessary information about orbifolds used by us is contained in [7].

Theorem 9.1: *For every closed surface M there exists a countable family of chaotic groups of homeomorphisms, isomorphic to the group $\mathbb{Z}$, such that the union of finite orbits of every such group is dense in M.*

Proof: Show that for each $k, m \in \mathbb{N}$, $k, m \geq 3$, the following matrix

$$A = A(k,m) = \begin{pmatrix} 1 & k \\ m & 1+km \end{pmatrix} \tag{16}$$

induces a chaotic homeomorphism g_A of the closed two-dimensional disk $\mathbb{B}^2$. Let f_A be the Anosov torus automorphism defined by A.

We represent the torus $\mathbb{T}^2$ as the square $[-1/2, 1/2] \times [-1/2, 1/2]$ in a Cartesian coordinate system Oxy on a plane $\mathbb{R}^2$ with identified opposite sides. In other words $\mathbb{T}^2$ has coordinates (x, y) where x and y are periodic of period one. For each fixed pair of numbers $k, m \in \mathbb{N}$, $k, m \geq 3$, consider subsets $P = [-1/2, 1/2] \times [-1/k, 1/k]$ and $Q = [-1/m, 1/m] \times [-1/2, 1/2]$ of $\mathbb{T}^2$. Note that P and Q are two overlapping annuli on the torus $\mathbb{T}^2$. Denote the union of the annuli by $R = P \cup Q$ and the intersection by $S = P \cap Q$, see Figure 6. Define maps $f : R \to R$ and $h : R \to R$ by the following equations:

$$f(x,y,k) = \begin{cases} (x+ky, y) & \text{if } (x,y) \in P, \\ (x,y) & \text{if } (x,y) \in R \setminus P, \end{cases} \tag{17}$$

$$h(x,y,m) = \begin{cases} (x, y+mx) & \text{if } (x,y) \in Q, \\ (x,y) & \text{if } (x,y) \in R \setminus Q. \end{cases} \tag{18}$$

The map $g = h \circ f : R \to R$ is called the *toral linked twist map* defined by composing f and h. According to [12, Theorem A], the constructed linked twist map $g : R \to R$ is topologically mixing and the periodic points of g are dense in R. Therefore, the homeomorphism group $G = \langle g \rangle$ generated by g is topologically transitive and chaotic on the topological space R. Note that $g = g(k,m)$ and $A = A(k,m)$ are related by the equality $g(x,y) = A\binom{x}{y}$ in some neighbourhood U of a point $(0,0)$, $U \subset S \subset \mathbb{T}^2$.

Let E_2 be the unit two-dimensional matrix. Let us identify the quotient space $\mathcal{N} = \mathbb{T}^2 / \pm E_2$ with the rectangle $[-1/2, 1/2] \times [0, 1/2]$ the sides of which are identified in the

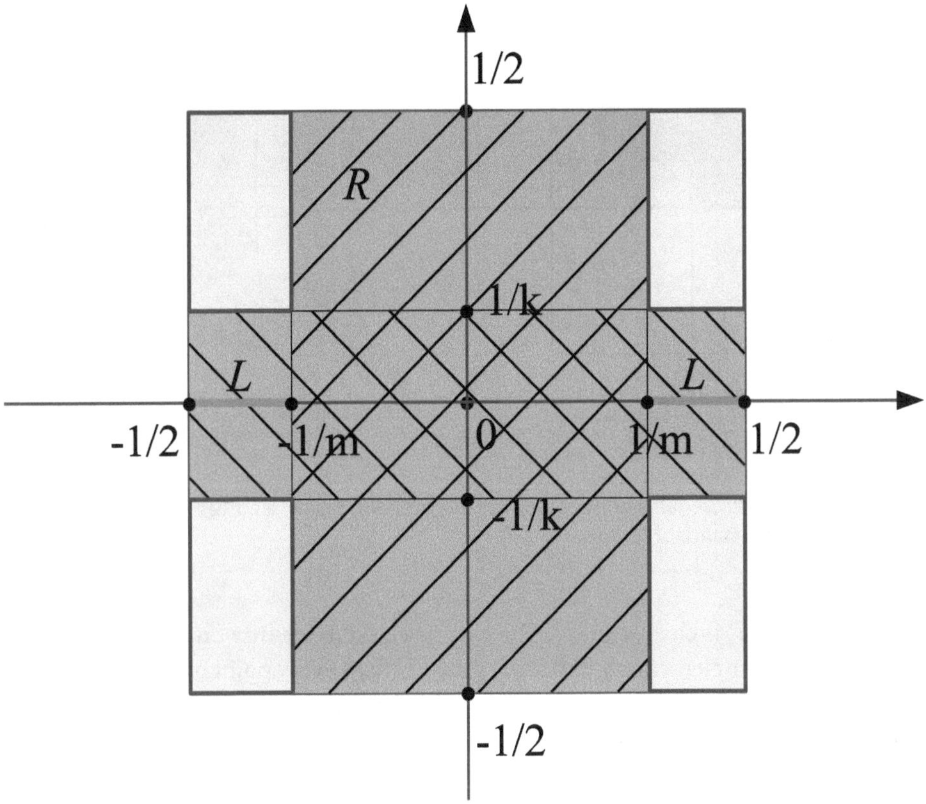

Figure 6. Representation of the torus on the unit square.

way indicated by the arrows in Figure 7. Denote by $p : \mathbb{T}^2 \to \mathcal{N}$ the quotient map. Note that $\mathcal{N}$ is the orbifold 'Pillow', which is homeomorphic to the standard sphere $\mathbb{S}^2$, and both subsets $p(R)$, coloured green, and $p(\mathbb{T}^2 \setminus R)$, coloured yellow, are homeomorphic to a closed disk $\mathbb{B}^2$ (see Figure 7 b). Identify the topological space of $\mathcal{N}$ with the sphere $\mathbb{S}^2$.

It is easy to check that $g|_{\partial R} = \mathrm{id}|_{\partial R}$. Therefore we can continue g for the entire torus $\mathbb{T}^2$ such that $g|_{\mathbb{T}^2 \setminus R} = \mathrm{id}|_{\mathbb{T}^2 \setminus R}$. The resulting torus homeomorphism will continue to denote by g. Emphasize that the points coloured yellow in Figure 6 are fixed relative to the homeomorphism g. Since the map g satisfies the equality $g(-x, -y) = -g(x, y)$ for every $(x, y) \in \mathbb{T}^2$, it induces a homeomorphism $g_{\mathbb{S}^2} : \mathbb{S}^2 \to \mathbb{S}^2$ satisfying the following equality $p \circ g = g_{\mathbb{S}^2} \circ p$. Therefore the restriction $g_{\mathbb{B}^2}$ of $g_{\mathbb{S}^2}$ to the closed disk $\mathbb{B}^2 = p(R)$ (coloured green in Figure 7b) satisfies the commutative diagram

$$\begin{array}{ccc} R & \xrightarrow{\ g\ } & R \\ {\scriptstyle p}\downarrow & & \downarrow{\scriptstyle p} \\ \mathbb{B}^2 & \xrightarrow{\ g_{\mathbb{B}^2}\ } & \mathbb{B}^2, \end{array} \tag{19}$$

and the projection $p : R \to \mathbb{B}^2$ is a surjective continuous open map. As a finite orbit of the group G maps onto a finite orbit of the group $\Gamma = \langle g_{\mathbb{B}^2} \rangle$ with respect to $p : R \to \mathbb{B}^2$, the union of finite orbits of Γ is dense in $\mathbb{B}^2$. A dense orbit of G maps onto a dense orbit of

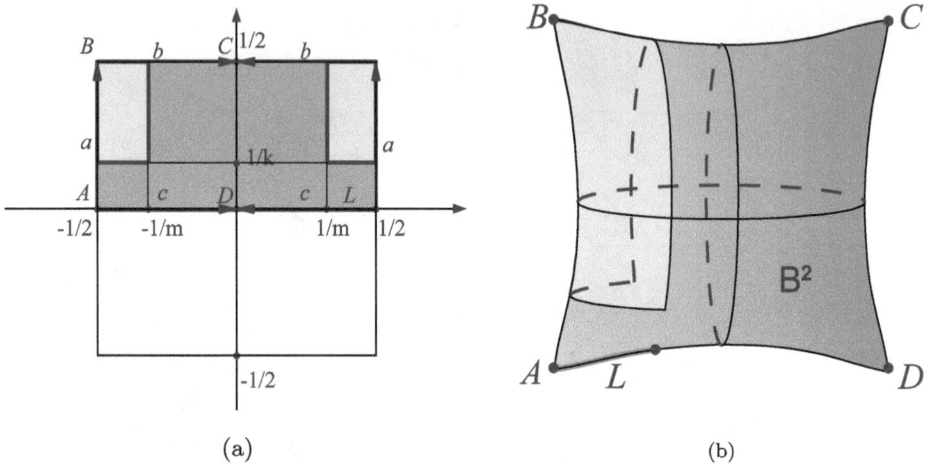

Figure 7. Orbifold 'Pillow': (a) representation of the orbifold 'Pillow' and (b) the orbifold 'Pillow' is homeomorphic to the standard sphere.

Γ. Thus the group Γ is chaotic in $\mathbb{B}^2$. Since $g_{\mathbb{S}^2}$ is equal to identity on another closed disc coloured yellow, complementary to $\mathbb{B}^2$, the group Γ fixes every point of the boundary $\partial\mathbb{B}^2$. Therefore it is possible to glue the boundary $\partial\mathbb{B}^2$ in an arbitrary way and to obtain a new surface M, and as a result, we can get every closed surface M. Denote the corresponding quotient map by $k : \mathbb{B}^2 \to M$. Since $\Gamma|_{\partial\mathbb{B}^2} = \mathrm{id}_{\partial\mathbb{B}^2}$, then Γ induces an isomorphic group of homeomorphisms $\widetilde{\Gamma}$ of the surface M, and the mapping $k : \mathbb{B}^2 \to M$ is a topological semi-conjugation of the groups Γ and $\widetilde{\Gamma}$. Emphasize that the interior $Int(\mathbb{B}^2)$ is invariant respectively Γ, and the restriction $k|_{Int(\mathbb{B}^2)}$ is a homeomorphism conjugating $\Gamma|_{Int(\mathbb{B}^2)}$ with $\widetilde{\Gamma}|_{k(Int(\mathbb{B}^2))}$. Therefore, the group $\widetilde{\Gamma}$ is chaotic on M. Further we write $\widetilde{\Gamma} = \widetilde{\Gamma}(k, m)$, since this group is defined by the matrix $A = A(k, m)$ of the form (16).

Thus we get a countable family of chaotic groups $\{\widetilde{\Gamma}(k, m) \mid k, m \in \mathbb{N}, k, m \geq 3\}$ on every closed surface M. For short, further we denote such families by $\{\widetilde{\Gamma}(k, m)\}$.

More specifically, we consider the following ways of gluing the disk boundary $\partial\mathbb{B}^2$.

Case I. Consider the boundary $\partial\mathbb{B}^2 \cong \mathbb{S}^1$ as a $4q$-polygon whose sides are glued together according to the scheme $a_1 b_1 a_1^{-1} b_1^{-1} \cdots a_q b_q a_q^{-1} b_q^{-1}$, $q \in \mathbb{N}$. The scheme is obtained as follows: the direction of movement along the sides of the polygon is selected, the sides are written out in a row, the glued sides are marked with one letter, the degree -1 means the direction of gluing opposite to the direction of movement along the side. As a result we get a closed surface homeomorphic to the sphere with q handles. Denote this surface by $\mathbb{S}^2_q$. Thus we get a countable family of chaotic groups $\{\widetilde{\Gamma}(k, m)\}$ on $\mathbb{S}^2_q$ for every $q \in \mathbb{N}$.

Case II. Consider the boundary $\partial\mathbb{B}^2 \cong \mathbb{S}^1$ as a $2q$-polygon whose sides are glued together according to the scheme $a_1 a_1 \cdots a_q a_q$, $q \in \mathbb{N}$. In this case, we get a closed non-orientable surface homeomorphic to the sphere with q Mobius bands which will be denoted by $N\mathbb{S}^2_q$. Thus we get an infinite series of chaotic groups $\{\widetilde{\Gamma}(k, m)\}$ on each surface $N\mathbb{S}^2_q$, $q \in \mathbb{N}$.

Case III. Let us glue the boundary $\partial\mathbb{B}^2 \cong \mathbb{S}^1$ into an arbitrary segment denoted by L_0 such that as the result we get a topological sphere $\mathbb{S}^2$. In the same way as above, we get an infinite series of chaotic groups $\{\widetilde{\Gamma}(k, m)\}$ on $\mathbb{S}^2$.

Since every closed surface M is homeomorphic to one of the canonical surfaces $\mathbb{S}^2$, $\mathbb{S}_q^2$ or $N\mathbb{S}_q^2$ where $q \geq 1$, then in the above way we obtain a countable family of chaotic homeomorphism groups $\{\widetilde{\Gamma}(k,m) \cong \mathbb{Z}\}$ on M.

It is well known that finite orbits of every Anosov automorphism of the torus $\mathbb{T}^2$ form a countable dense subset in $\mathbb{T}^2$. However, this is not true for the constructed homeomorphism g. In the next section, we will show that g has a continuum of fixed points. Hence, the set of all finite orbits of both groups Γ and $\widetilde{\Gamma}$ has the cardinality of the continuum. ■

Remark 9.1: In contrast to [8], we have constructed an infinite countable family of chaotic actions of the group $\mathbb{Z}$ on each closed surface.

9.2. Chaotic actions of the group $\mathbb{Z}$ on noncompact two-dimensional manifolds

Consider the toral linked twist map $g = g(k,m) : R \to R$ constructed in the previous section. Recall that R is the union of two annuli $P \cup Q$ on the torus $\mathbb{T}^2$. Let $p : R \to \mathbb{B}^2$ be the projection satisfying the commutative diagram (19), hence $p(\partial R) = \partial\mathbb{B}^2 \cong \mathbb{S}^1$.

At first we will pick out some point $z \in \partial\mathbb{B}^2$. Identify $\partial\mathbb{B}^2 \setminus \{z\}$ with the real line $\mathbb{R}^1 = \bigcup_{n=-\infty}^{+\infty}[n, n+1]$. Represent $\mathbb{B}^2 \setminus \{z\}$ as an polygon without a point z, the boundary of which is divided into a countable set of pairwise glued sides. Consider the following gluing rules.

Case IV. $[n, n+1] \sim a_n b_n a_n^{-1} b_n^{-1}\ \forall n \in \mathbb{Z}$. As a result of gluing, we will get a noncompact two-dimensional manifold M without boundary, homeomorphic to a plane with a countable family of handles which we denote by $\mathbb{R}^2_\infty$. Emphasize that M is homeomorphic to the Loch Ness monster (see Figure 1).

Case V. $[n, n+1] \sim a_n a_n\ \forall n \in \mathbb{Z}$. As a result of gluing, we will get a noncompact two-dimensional manifold M without boundary, homeomorphic to a plane with a countable family of Mobius bands which we denote by $N\mathbb{R}^2_\infty$.

Case VI. Let us glue $p(\partial R) = \partial\mathbb{B}^2 \cong \mathbb{S}^1$ to a segment L_0 as in Case III. As a result, we get a topological sphere $\mathbb{S}^2$. As above, let $k : \mathbb{B}^2 \to \mathbb{S}^2$ be the respective quotient map. The image $k(\partial\mathbb{B}^2)$ of $\partial\mathbb{B}^2$ we denote by $L_0 \subset \mathbb{S}^2$. Consider an arbitrary nonempty compact subset K_0 of L_0 and a point $z_0 \in L_0 \setminus K_0$. We get a noncompact two-dimensional manifold denoted by $\mathbb{R}^2(K_0) := \mathbb{S}^2 \setminus (K_0 \cup \{z_0\})$ homeomorphic to a plane without K_0. In Figure 2, you can see the noncompact surface $\mathbb{R}^2(K_0)$ where K_0 is the standard Cantor set.

Now we consider the restriction of $g : R \to R$, constructed in the previous section, to a circle $C = \{(x,0) \mid -1/2 \leq x \leq 1/2\}$. The image $g(C)$ is a closed curve on the torus $\mathbb{T}^2$. Observe that

$$g(x,0) = (x,0) \quad \forall x \in [-1/2, -1/m] \cup [1/m, 1/2].$$

Emphasize that $[-1/2, -1/m] \cup [1/m, 1/2]$ is a connected segment in $R \subset \mathbb{T}^2$. The set $L = p([-1/2, -1/m] \cup [1/m, 1/2])$ is a connected topological segment in the inside of the disk $\mathbb{B}^2$ highlighted in red in Figure 7(b), and $g|_L = \mathrm{id}_L$. Assume that the boundary $\partial\mathbb{B}^2$ of $\mathbb{B}^2$ is glued in one of the ways specified above in Cases I– VI. The result is a manifold M where M is either one of arbitrary closed surfaces $\mathbb{S}^2$, $\mathbb{S}_q^2$, $N\mathbb{S}_q^2$ where $q \geq 1$, or one of noncompact surfaces $\mathbb{R}^2_\infty$, $N\mathbb{R}^2_\infty$, $\mathbb{R}^2(K_0)$. Let $k : \mathbb{B}^2 \to M$ be the corresponding quotient map. As no points from L were not glued together, we identify the image $k(L)$ of L in M with L and consider L as subset of M. This fact allows us to remove an arbitrary closed subset K of L from L and to obtain a noncompact two-dimensional topological manifold $\widehat{M} = \widehat{M}(K) := M \setminus K$. Emphasize that we do not exclude the possibility $K = \varnothing$. The homeomorphism $\widetilde{g} = \widetilde{g}(k, m)$ induces a chaotic homeomorphism $\widehat{g} = \widehat{g}(k, m)$ of $\widehat{M}$, and the group $\widehat{\Gamma} = \langle \widehat{g} \rangle$ has a dense set periodic orbits in $\widehat{M}$.

Indicate some important classes of obtained manifolds $\widehat{M} = \widehat{M}(K)$.

If K is a finite subset of L, then $\widehat{M}$ is a noncompact manifold with a finite number of ends.

If K is an infinite subset of L with one limit point z', then $z' \in K$ and we get a noncompact manifold $\widehat{M} = M \setminus K$ with a discrete countable set of ends.

When K is the standard Cantor set on L, we get a noncompact manifold $\widehat{M}$ containing the Cantor set of ends. It is possible that $\widehat{M}$ has other ends obtained by removing a compact subset $K_0 \subset L_0$ where $L_0 \subset k(\partial \mathbb{B}^2)$ similarly to Case VI.

Emphasize that for every closed subset K, the set of handles or Mobius bands on $\widehat{M}$ may be arbitrary finite as well as infinite.

In Case VI, for $M = \mathbb{R}^2(K_0)$ we get $\widehat{M} = \widehat{M}(K_0, K)$. When $K_0 = \varnothing$ and K is the standard Cantor set, the surface $\widehat{M}(K_0, K)$ is homeomorphic to the surface in Figure 2.

On every such constructing noncompact topological two-dimensional manifold $\widehat{M}$ the homeomorphism $\widetilde{g} = \widetilde{g}(k, m)$ induces a chaotic homeomorphism $\widehat{g} = \widehat{g}(k, m)$, and the group $\widehat{\Gamma} = \langle \widehat{g} \rangle$ has a dense set periodic orbits in $\widehat{M}$.

Thus we get an infinite countable family of chaotic groups $\{\widehat{\Gamma}(k, m) = \langle \widehat{g}(k, m) \rangle\}$ on each surface $\widehat{M}$ constructed above.

9.3. Chaotic groups acting on closed n-dimensional manifold

Let G be any countably generated free group. In [9], Cairns et al. showed that every compact triangulable manifold M of an arbitrary dimension n greater than 1 admits a faithful chaotic action of the group G.

Emphasize that the free group Γ of two generators may be implemented as a chaotic group of homeomorphisms of every such manifold M [9].

9.4. Chaotic groups of homeomorphisms on products of manifolds

Since every open subset of a Polish space is also Polish, then all topological manifolds defined above in Sections 9 satisfies conditions of Theorems 1.4–1.5. For an arbitrary index set A, we use the constructed above chaotic actions of homeomorphism groups G_α, $\alpha \in A$, on topological manifolds M_α as building blocks for constructions of chaotic canonical action of the product of groups $G = \prod_{\alpha \in A} G_\alpha$ on $M = \prod_{\alpha \in A} M_\alpha$.

According to Theorem 1.6, if G_i, $i \in J \subset \mathbb{N}$, is a countable chaotic group of homeomorphisms of n_i-dimensional closed triangulable manifold M_i, $n_i \geq 2$, then the canonical action of the direct product of groups $G = \prod_{i \in J} G_i$ is chaotic on the product $M = \prod_{i \in J} M_i$. Here M is a infinite-dimensional topological manifold, if the set J is infinite countable, otherwise M is a finite-dimensional topological manifold. In both cases, by Theorem 1.4, every groups G and G_i have sensitivity to initial conditions. Emphasize that the dimension of M_i may be an arbitrary greater than 1.

Moreover, in the case $J = \mathbb{N}$, according to Theorem 1.6, there exists a dense subset $F \subset M$ which is the union of compact continuum orbits of the group G, and every such orbit is a perfect subset of M. Besides there exists a dense continuum orbit of G in M.

If, moreover, every M_i is a two-dimensional manifold with a chaotic action of the group $G_i \cong \mathbb{Z}$ constructed in Sections 9.1 and 9.2, then every group G_i has finite orbits and, in particular, a fixed point. Therefore, in this case, the group $G = \prod_{i\in\mathbb{N}} G_i$ on the infinite-dimensional topological manifold $M = \prod_{i\in J} M_i$ has a continuum set of finite orbits, and the union of finite orbits is dense in M.

Example 9.2: Consider the standard two-dimensional torus $\mathbb{T}^2 = \mathbb{R}^2/\mathbb{Z}^2$. Anosov torus automorphism $\mathbb{T}^2$, given by a matrix $A \in SL(2,\mathbb{Z})$, is denoted by g_A. As is well known [21], every matrix

$$A = \begin{pmatrix} a & b \\ c & d \end{pmatrix}, \tag{20}$$

where $ad-bc = 1$ and $a+d > 2$ define an Anosov automorphism g_A of the torus $\mathbb{T}^2$ preserving its orientation. The group $G = \langle g_A \rangle \cong \mathbb{Z}$ generated by g_A acts chaotically on $\mathbb{T}^2$. It is well known that there exists a countable set of finite orbits of G, and this set is dense in $\mathbb{T}^2$. There exists a countable family $\mathcal{A} = \{A_k \mid k \in \mathbb{N}\}$ such matrices defining Anosov automorphisms.

Consider the infinite-dimensional torus $\mathbb{T}^\infty = \prod_{i\in\mathbb{N}} \mathbb{T}_i^2$ where $\mathbb{T}_i^2 = \mathbb{T}^2$ for each $i \in \mathbb{N}$. Let g_{A_i}, $A_i \in \mathcal{A}$, be an Anosov automorphism on $\mathbb{T}_i^2$. Let $G_i = \langle g_{A_i} \rangle$ and $G = \prod_{i\in\mathbb{N}} G_i$. According to Theorem 1.6, we get a chaotic group of homeomorphisms G of the torus $\mathbb{T}^\infty$. Taking into account that A_i may be an arbitrary matrix belonging to $\mathcal{A}$, we see that different groups G form a continuum set. Emphasize that G also has continuum cardinality. As every group G_i has a fixed point, by Theorem 1.6, the union of finite orbits of group G is dense in $\mathbb{T}^\infty$, and G has a fixed point. Moreover, the union of its compact orbits of continuum cardinality is also dense in $\mathbb{T}^\infty$, and every such orbit is a perfect subset of $\mathbb{T}^\infty$. Note that every dense orbit has continuum cardinality. Besides, by Theorem 1.6, the group G is sensitive to initial conditions.

Disclosure statement

No potential conflict of interest was reported by the author(s).

Funding

This work is supported by Russian Science Foundation, project No 22-21-00304.

References

[1] E. Akin, J. Auslander, and K. Berg, *When is a transitive map chaotic?*, in *Convergence in Ergodic Theory and Probability*, V. Bergelson, P. March and J. Rosenblatt, eds., Gruyter, Berlin, 1996, pp. 25–40.

[2] S. Albeverio, A.Y. Khrennikov, and V.M. Shelkovich, Theory of p-Adic Distributions: Linear and Nonlinear Models, Cambridge University Press, New York, 2010.

[3] D. Assaf and S. Gadbois, Definition of chaos, Am. Math. Mon. 99(9) (1992), pp. 865–865.

[4] J. Banks, J. Brooks, G. Cairns, G. Davis, and P. Stacey, On Devaney's definition of chaos, Am. Math. Mon. 99(4) (1992), pp. 332–334.
[5] A. Barzanouni, M.S. Divandar, and E. Shah, On properties of expansive group actions, Acta Math. Vietnam. 44(4) (2019), pp. 923–934.
[6] Y.V. Bazaikin, A.S. Galaev, and N.I. Zhukova, Chaos in Cartan foliations, Chaos Interdiscip. J. Nonlinear Sci. 30 (2020), Article ID 103116.
[7] E.V. Bogolepova and N.I. Zhukova, Anosov actions of isometry groups on Lorentzian 2-orbifolds, Lobachevskii Math. J. 42(14) (2021), pp. 3324–3335.
[8] G. Cairns, G. Davis, D. Elton, A. Kolganova, and P. Perversi, Chaotic group actions, Enseign. Math.41 (1995), pp. 123–133.
[9] G. Cairns and A. Kolganova, Chaotic actions of free groups, Nonlinearity 9(4) (1996), pp. 1015–1021.
[10] G. Cairns, A. Kolganova, and A. Nielsen, Topological transitivity and mixing notions for group actions, Rocky Mt. J. Math. 37(2) (2007), pp. 371–397.
[11] F. Chovanec, Cantor sets, Sci. Mil. J. 5(1) (2010), pp. 5–10.
[12] R.L. Devaney, Linked twist mappings are almost Anosov, in In: Nitecki, Z., Robinson, C. (eds) Global Theory of Dynamical Systems. Lecture Notes in Mathematics, 819, Nitecki Z. and Robinson C, ed., Springer, Berlin, Heidelberg, 1980. pp. 121–145.
[13] R.L. Devaney, An Introduction to Chaotic Dynamical Systems, Addisson Wesley, Redwood City, CA, 1986.
[14] M.M. Deza and E. Deza, Encyclopedia of Distances, Springer, Berlin, Heidelberg, 2009.
[15] R. Engelking, General Topology, Mir, Moscow, 1986.
[16] M. Hirasawa and E. Kin, Determination of generalized horseshoe maps inducing all link types, Topol. Appl. 139(1–3) (2004), pp. 261–277.
[17] A. Katok and B. Hasselblatt, Introduction to the Modern Theory of Dynamical Systems, Cambridge University Press, New York, 1997.
[18] A. Kechris, Classical Descriptive Set Theory, Springer-Verlag, New York, 1995.
[19] A.C. Naolekar and P. Sankaran, Chaotic group actions on manifolds, Topol. Appl. 107 (2000), pp. 233–243.
[20] F. Polo, Sensitive dependence on initial conditions and chaotic group actions, Proc. Am. Math. Soc.138(8) (2010), pp. 2815–2826.
[21] E.A. Rogozhina and N.I. Zhukova, Classification of compact Lorentzian 2-orbifolds with noncompact full isometry groups, Sib. Math. J. 53(6) (2012), pp. 1037–1050.
[22] R. Sturman, J.M. Ottino, and S. Wiggins, The Mathematical Foundations of Mixing: The Linked Twist Map as a Paradigm in Applications: Micro to Macro, Fluids to Solids, Cambridge University Press, New York, 2006.
[23] P. Scott, The geometries of 3-manifolds, Bull. Lond. Math. Soc. 15 (1983), pp. 401–487.
[24] N.I. Zhukova, Minimal sets of Cartan foliations, Proc. Steklov Inst. Math. 256 (2007), pp. 105–135.

Remarks on minimal sets on dendrites and finite graphs

Elena Makhrova

ABSTRACT
Let X be a dendrite and let $f : X \to X$ be a continuous map that has an infinite minimal set M in X. In the article we obtain conditions on the structure of a dendrite X and the set M, under which f has an arc horseshoe, and, hence, f has a positive topological entropy. We show that this result is not correct both for continuous maps with an empty set of periodic points, and for continuous maps with a non-empty set of periodic points on finite graphs.

1. Introduction

The topological structure of minimal sets of continuous maps on finite graphs and dendrites is described in [2, 3, 17]. A periodic orbit gives the simplest example of a minimal set. In [10, 11] it is shown that periodic orbits do not determine the dynamics of continuous maps on a dendrite, in contrast with continuous maps on a closed interval (see, e.g. [22]). But currently there are no results on the study of the relationship between the structure of infinite minimal sets and the existence of a horseshoe for dendrite and finite graph maps. In the article we study a relation between the structure of infinite minimal sets and the existence of a horseshoe for continuous maps on dendrites and finite graphs. As a corollary we get a relation between the existence of infinite minimal sets and topological entropy of such maps.

We use $\mathbb{N}$, $\mathbb{C}$ to denote the set of natural numbers and the complex plane, respectively. By $\mathbf{i}$ we denote the imaginary unit on $\mathbb{C}$. Let X be a metric space, a set A be a subset in X. The symbol card (A) stands for the cardinality of A, $\partial(A)$ means the boundary of A.

1.1. Dendrites and finite graphs

By *continuum* we mean a compact connected metric space. Following [11], we define an arc in a continuum X as a connected subset of X, whose closure is homeomorphic to the interval $[0, 1]$ on the real line. An arc containing both end points is said to be closed. A singleton is considered as a degenerate arc. We denote by $[x, y]$ the closed arc connecting points x, y; by (x, y) the set $[x, y]\setminus\{x, y\}$.

A dendrite is a locally connected continuum that does not contain subsets homeomorphic to a circle. *A finite graph* is a continuum that can be presented as the union of finitely many arcs any two of which are either disjoint or intersect each other only in one or both of their endpoints. We note that a finite graph can contain subsets homeomorphic to a circle. A finite graph without subsets homeomorphic to a circle is called *a finite tree*.

Let X be a dendrite or a finite graph with metric d. We use a concept of an *order* of a point $p \in X$ in the sense of Menger-Urysohn as in [11]. We denote by ω the countable ordinal corresponding to the set of positive integers arranged in their natural order.

Definition 1.1 ([16, Ch. 5, §47): , [9, 11]]
Let p be a point in X, $\mathfrak{n} \in \mathbb{N}$. We say that

(1.1.1) the order of p is at most $\mathfrak{n}$ ($\operatorname{ord} p \leq \mathfrak{n}$), if for any $\varepsilon > 0$ there is a $0 < \delta < \varepsilon$ such that the boundary of the δ-neighbourhood of p has cardinality at most $\mathfrak{n}$;

(1.1.2) the order of p is $\mathfrak{n}$ ($\operatorname{ord} p = \mathfrak{n}$), if $\operatorname{ord} p \leq \mathfrak{n}$ and there is no $\mathfrak{m} < \mathfrak{n}$ such that $\operatorname{ord} p \leq \mathfrak{m}$;

(1.1.3) the order of p is ω ($\operatorname{ord} p = \omega$), if there is $\varepsilon_0 > 0$ such that for any $0 < \delta < \varepsilon_0$ the cardinality $\operatorname{card}(\partial U_\delta(p))$ of the boundary of the δ-neighbourhood $U_\delta(p)$ of p is finite, but the set $\{\operatorname{card}(\partial U_\delta(p))\}_{\delta>0}$ is not bounded above (as $\delta \to +0$).

Points of order greater than two (or of order one) in X are called *ramification points* (or *end points*, respectively) of X. We denote by $R(X)$ and $E(X)$ the sets of ramification points and end points of X, respectively.

1.2. Continuous maps

Let $f : X \to X$ be a continuous map of a dendrite X. A point $p \in X$ is said to be *a periodic point of f* provided that there is $n \in \mathbb{N}$ such that $f^n(p) = p$; the least such n is called *period* of p. If $n = 1$, then p is called *a fixed point of f*. A point $p \in X$ is said to be *a recurrent point of f* provided that for each neighbourhood $U(p)$ of a point p there is $n \in \mathbb{N}$ such that $f^n(p) \in U(p)$. *A point p is uniformly recurrent for f* if it is recurrent and for every neighbourhood $U(p)$ of a point p there exists $M \in \mathbb{N}$ such that whenever $f^j(p) \in U(p)$ for $j \geq 0$, then $f^{j+k}(x) \in U(p)$ for some $0 < k \leq M$ (see, e.g. [15]).

We denote by *Fix(f)*, *Per(f)*, *R(f)*, *UR(f)* the sets of fixed points, periodic points, recurrent points, uniformly recurrent points of f, respectively. It follows from definitions that $Fix(f) \subseteq Per(f) \subseteq UR(f) \subseteq R(f)$.

1.3. Minimal sets

Let X be any topological space and let $f : X \to X$ be a continuous map. A nonempty set M in X is called to be *minimal under f* if it is closed, f-invariant and does not contain any proper subset satisfied these conditions. It is known that a compact f-invariant set Y in X contains some minimal sets (see, e.g. [15]).

Definition 1.2: We say that M is *totally minimal under f*, if M is minimal under f^n for all $n \in \mathbb{N}$.

We describe what happens if M is minimal, but not totally minimal (see, e.g. [4, 15, 24]). If M is not minimal for some $n \geq 2$, then there are pairwise disjoint compact subsets $M_i \subset M$, uniquely defined up to the order, with $M = M_0 \cup M_1 \cup \ldots \cup M_{k-1}$, such that $k \geq 2$ is a divisor of n, $f(M_i) = M_{i+1} (mod\, k)$ and M_i is minimal under f^k for each $0 \leq i \leq k-1$. In this case we say that M has a periodic decomposition. A number k is called a length of a periodic decomposition of M. We note that M has at most one periodic decomposition of a length k.

Definition 1.3: We say that a minimal set M is *a relatively totally minimal under f* if there is a periodic decomposition $\{M_0, M_1, \ldots, M_{k-1}\}$ of M such that M_i is a totally minimal set under f^k for every $0 \leq i \leq k-1$.

We note that there are minimal sets that are neither totally minimal under f nor relatively totally minimal under f. Adding machines on the set of all one-sided infinite sequences gives us the fundamental examples of such minimal sets (see, e.g. [4, 8]).

1.4. Horseshoe and topological entropy

We say that a continuous map $f : X \to X$ of a dendrite X has *a horseshoe* if there are disjoint closed subsets A, B in X such that $f(A) \cap f(B) \supset A \cup B$. We denote by (A, B) a horseshoe of f. It is known if f has a horseshoe, topological entropy of f is positive [7]. We say that a map f has an *arc horseshoe*, if f possesses a horseshoe (A, B), where A, B are arcs belonging to the same arc. We note if a dendrite X has an infinite countable set of end points and f has a horseshoe then there is a natural number $n \geq 1$ such that f^n has an arc horseshoe [19].

1.5. Main results

Let $f : X \to X$ be a continuous map of a dendrite X and f has an infinite minimal set M. Theorem 1.4 gives us conditions on the structure of a dendrite X and on a minimal set M under which f has an arc horseshoe and, hence, a positive topological entropy.

Theorem 1.4: *Let X be a dendrite with an infinite countable set of end points, $f : X \to X$ be a continuous map and there exists an infinite minimal set M in X such that M is either totally minimal under f or relatively totally minimal under f. Then there is $n \in \mathbb{N}$ such that f^n has an arc horseshoe and, hence, topological entropy of f is positive.*

Since a finite tree has a finite set of end points then we immediately obtain the next corollary.

Corollary 1.5: *Let X be a finite tree, $f : X \to X$ be a continuous map and there exists an infinite minimal set M in X such that M is either totally minimal under f or relatively totally minimal under f. Then there is $n \in \mathbb{N}$ such that f^n has an arc horseshoe and, hence, topological entropy of f is positive.*

Theorem 1.4 is not correct, if a dendrite X has an uncountable set of end points [14, Example 6]. One of such dendrites is the Gehman dendrite G (see, e.g. [13, 20]). The Gehman dendrite has an infinite countable set of ramification points, whose derivative set (that is the set of all limit points of $E(G)$) is the set of endpoints $E(G)$, which is homeomorphic to the Cantor discontinuum; each ramification point of G has the order three. See different constructions of the Gehman dendrite in [10, 11, 20]. The result of [1] shows that the set of end points of a dendrite X being uncountable is equivalent to the existence of a subcontinuum in X homeomorphic to the Gehman dendrite.

The next statement shows that Theorem 1.4 is correct for dendrites with an uncountable set of end points, but under additional conditions on an infinite minimal set M. To formulate it we need one more notation. Let X be a dendrite. For a minimal set M in X we set $[M] = \bigcup_{x,y\in M}[x, y]$. We note that $[M]$ is the least set containing M. We show that $[M]$ is a closed and connected set in the proof of Theorem 1.4. So $[M]$ is a continuum.

Corollary 1.6: *Let $f : X \to X$ be a continuous map of a dendrite X and there exists an infinite minimal set M in X, which is either totally minimal under f or relatively totally minimal under f. If $[M]$ is a subcontinuum in X with at most countable set of end points then there is $n \in N$ such that f^n has an arc horseshoe in $[M]$ and, hence, topological entropy of f is positive.*

We note that the existence of an infinite minimal set, which is neither totally minimal under f nor relatively totally minimal under f, does not imply the existence of any horseshoe and the positivity of topological entropy of f on dendrites of any structure. We remember the logistic map $f = \lambda x(1 - x)$ on the segment $[0, 1]$, where $\lambda \approx 3,56995\ldots$ is the limiting parameter value of period doubling sequence in f [12]. Adding machine arises here as the restriction of f to an invariant Cantor set that is a minimal set under f (see, e.g. [4, 5], [22, Chapter 1, § 2]). Topological entropy of f is equal to 0 and none of the iteration of f has any horseshoe. In [10, Theorem C] a homeomorphism f on the Gehman dendrite having an uncountable set of end points is constructed. A map f has a minimal set that is neither totally minimal under f nor relatively totally minimal under f. Topological entropy of f equals 0 and f^n does not have any horseshoe for each $n \in \mathbb{N}$.

It is known that an irrational rotation of a circle has a totally minimal set and zero topological entropy. We note also that the set of periodic points of such a map is empty. The next theorem shows that Theorem 1.4 does not correct for continuous maps on finite graphs both with empty set of periodic points and non-empty set of periodic points.

Theorem 1.7: (1.7.1) *There are a finite graph G and a continuous map $f : G \to G$ with an empty set of periodic points such that f has an infinite totally minimal set, a map f^n does not have any horseshoe for each $n \in \mathbb{N}$ and a topological entropy of f equals 0.*

(1.7.2) *There are a finite graph G and a continuous map $f : G \to G$ with a non-empty set of periodic points such that f has an infinite totally minimal set, a map f^n does not have any horseshoe for each $n \in \mathbb{N}$ and a topological entropy of f equals 0.*

2. Auxiliary results

We start with the properties of dendrites.

Lemma 2.1 ([16, Ch. 5, §47): , [9, 20]*]*

Let X be a dendrite. Then

(2.1.1) *for any different points $x, y \in X$ there is a unique arc connecting these points;*
(2.1.2) *the set $R(X)$ of ramification points of a dendrite X is at most countable;*
(2.1.3) *X is a regular continuum, that is, $\operatorname{ord} p \leq \omega$ for any point $p \in X$;*
(2.1.4) *any subcontinuum of a dendrite is a dendrite.*

We also need the next statements.

Theorem 2.2 ([18]): *Let $f : X \to X$ be a continuous map of a dendrite X, Y be a subcontinuum in X. If Y has a countable set of end points and $Y \subset f(Y)$ then $Y \cap Per(f) \neq \emptyset$.*

The following lemma is an adapted version of the classical Birkhoff theorem for this case [6, Charpter 7, §2] (see, also [15]).

Lemma 2.3: *Let $f : X \to X$ be a continuous map of a continuum X and M be a minimal set under f in X. Then*

(2.3.1) *$\overline{\{f^i(z)\}_{i\geq 0}} = M$ for any point $z \in M$;*
(2.3.2) *$z \in UR(f)$ for every point $z \in M$.*

The next fact is a consequence of the result of [21, Lemma 4].

Lemma 2.4 ([21]): *Let X be a dendrite, Y be a subcontinuum in X. Then*

$$\operatorname{card} E(Y) \leq \operatorname{card} E(X).$$

3. Proof of Theorem 1.4 and Corollary 1.6

Proof of Theorem 1.4.: (1) Let M be a totally minimal set. Let $Y = [M] = \bigcup_{x,y\in M}[x, y]$. We show that Y is a closed set.

Suppose on the contrary. Then there exists a point $\alpha \in \overline{Y} \setminus Y$. Since M is a closed set, a point $\alpha \notin M$. Then there exists a δ-neighbourhood $U_\delta(\alpha)$ of a point α such that $U_\delta(\alpha) \cap M = \emptyset$. According to the condition $\alpha \in \overline{Y} \setminus Y$, there is a sequence of points $\{z_j\}_{j\geq 1} \subset Y \cap U_\delta(\alpha)$ such that

$$\lim_{j\to\infty} z_j = \alpha. \tag{1}$$

By the choice of a δ-neighbourhood $U_\delta(\alpha)$, we have $\{z_j\}_{j\geq 1} \cap M = \emptyset$. Therefore, according to definition of Y, for each number $j \in \mathbb{N}$ there are points $x_j, y_j \in M$ such that $z_j \in (x_j, y_j)$. Again by the choice of a δ-neighbourhood $U_\delta(\alpha)$, we obtain $x_j, y_j \notin U_\delta(\alpha)$. But $z_j \in U_\delta(\alpha)$. Therefore, $\partial(U_\delta(\alpha)) \cap [x_j, y_j] \neq \emptyset$ for any number $j \in \mathbb{N}$. We set

$$D = \bigcup_{j\geq 1} (\partial(U_\delta(\alpha)) \cap [x_j, y_j]).$$

We show that D is infinite. Suppose on the contrary. Then D is finite and there exist a sequence of points $\{z_{j_i}\}_{i\geq 1} \subset \{z_j\}_{j\geq 1}$ and different points $u, v \in \bigcup_{j\geq 1}\{x_j, y_j\}$ such that

$\{z_{j_i}\}_{i\geq 1} \subset [u, v]$. According to (1), $\alpha \in [u, v]$. By choice of points u, v and definition of Y, we obtain $[u, v] \subset Y$. Hence, $\alpha \in Y$. It contradicts to the choice of α. Then the set D is not finite.

Since D is infinite then according to Definition 1.1, we have $\mathrm{ord}\,\alpha \neq \mathfrak{n}$, where $\mathfrak{n} \in \mathbb{N}$, and $\mathrm{ord}\,\alpha \neq \omega$, contrary to the statement (2.1.3) of Lemma 2.1. Therefore, our assumption is wrong, and Y is a closed set.

(2) We show that Y is connected.

Let $[x_1, y_1]$, $[x_2, y_2]$ be any two disjoint arcs in Y, where $x_i, y_i \in M$ $(i \in \{1, 2\})$. Then by the statement (2.1.1) of Lemma 2.1, there is an arc $[x_1, x_2] \subset X$. According to definition of Y, $[x_1, x_2] \subset Y$. So Y is a connected set.

(3) It follows from items (1) and (2) that Y is a continuum. By property (2.1.4) of Lemma 2.1, subcontinuum Y is a dendrite. It follows from the definition of Y that

$$E(Y) \subset M. \tag{2}$$

(4) We show that there is a natural number $n \geq 1$ such that f^n has an arc horseshoe in Y.

Since M is a minimal set then $f(M) = M$. Hence, by (2), $Y \subseteq f(Y)$. Since X is a dendrite with a countable set of end points, then according to Lemma 2.4, the set of end points of Y is at most countable. By Theorem 2.2, $Y \cap Per(f) \neq \emptyset$. Let $p \in Y \cap Per(f)$ and m be period of p $(m \geq 1)$. According to (2), $p \notin E(Y)$.

Since M is infinite then it is uncountable. But $E(Y)$ is at most countable, so there is an end point $e \in E(Y)$ such that

$$\mathrm{card}\,(M \cap (p, e]) = c, \tag{3}$$

where c is the continuum cardinality.

Since M is closed and $p \notin M$, then there is a point $a_0 \in (p, e) \cap M$ such that $(p, a_0) \cap M = \emptyset$. Using (3), we have

$$\mathrm{card}\,(M \cap [a_0, e]) = c. \tag{4}$$

Since M is a totally minimal set then by (4), there is a natural number n such that $f^{mn}(a_0) \in (a_0, e]$.

We set $g = f^{mn}$. Then $g([p, a_0]) \supseteq [p, g(a_0)] \supset [p, a_0]$ and, hence,

$$g^i([p, a_0]) \supset [p, a_0] \quad \text{forany } i \in \mathbb{N}. \tag{5}$$

We denote by $X_e(p)$ the component of a point p containing a point e. Since M is a totally minimal set and $p \notin E(Y)$, then there is a natural number s such that $g^s(g(a_0)) \notin X_e(p)$, that is $p \in (a_0, g^{s+1}(a_0))$. We suppose that such s is minimal, that is

$$g^i(g(a_0)) \in X_e(p) \quad \text{for } 1 \leq i \leq s-1. \tag{6}$$

Then $g^s([a_0, g(a_0)] \supseteq [g^s(a_0), g^{s+1}(a_0)]$ and by (6), $[g^s(a_0), g^{s+1}(a_0)] \ni p$. Hence, $g^s([a_0, g(a_0]) \supset [p, g(a_0)]$. So using (5), we obtain

$$g^{s+1}([a_0, g(a_0)]) \supset [p, g(a_0)]. \tag{7}$$

Since $[p, a_0] \subset [p, g(a_0)] \subset g([p, a_0])$, there is a point $a_1 \in (p, a_0)$ such that $g(a_1) = a_0$. Then $[p, a_1] \subset g([p, a_1])$. Repeating these arguments s times, we construct a collection of points $\{a_i\}_{i=1}^{s} \subset (p, a)$ such that $a_i \in (p, a_{i-1})$ and $g(a_i) = a_{i-1}$ for any

$1 \leq i \leq s$. Then

$$g^{s+1}([p, a_s]) \supseteq [p, g(a_0)]. \tag{8}$$

We set $A = [p, a_s]$, $B = [a_0, g(a_0)]$. We note that $A, B \subset [p, g(a_0)]$ and $A \cap B = \emptyset$. According to (7) and (8) a map g^{s+1} has an arc horseshoe A and B.

(5) Let M be a relatively totally minimal set under f. According to Definition 1.3, there is a periodic decomposition $\{M_0, M_1, \ldots, M_{k-1}\}$ of M such that every M_i is a totally minimal under f^k, $0 \leq i \leq k-1$. We set $g = f^k$. Then M_0 is a totally minimal set under g. By item (4), there is $s \in \mathbb{N}$ such that g^s has an arc horseshoe. Then for $n = ks$ a map f^n has an arc horseshoe.

Theorem 1.4 is proved. ■

The correctness of Corollary 1.6 follows from the proof of Theorem 1.4.

Now we construct an example of a continuous map f on a finite tree having a relatively totally minimal set under f.

Example 3.1: We set $X = \bigcup_{j=0}^{n-1} [0, e^{\frac{2\pi j}{n}\mathbf{i}}]$, where $\mathbf{i}$ is an imaginary unit. We note that a continuum X is a finite tree with a unique ramification point 0. We need two auxiliary maps.

We define the first map $\varphi : X \to X$ as follows:

$$\varphi(z) = \begin{cases} 2z, & \text{if } z \in [0, 1/2], \\ 2 - 2z, & \text{if } z \in (1/2, 1], \\ z, & \text{if } z \in X \setminus (0, 1]. \end{cases} \tag{9}$$

We note that $\varphi_{|[0, 1]}$ is the tent map, which has a totally minimal set $M_0 \subset [0, 1]$.

Let $\tau(z) = z \cdot e^{\frac{2\pi}{n}\mathbf{i}}$ for any point $z \in X$. Then for any $1 \leq j \leq n-1$ we have

$$\tau^j([0, 1]) = [0, e^{\frac{2\pi j}{n}\mathbf{i}}], \tag{10}$$

and $\tau^n([0, e^{\frac{2\pi j}{n}\mathbf{i}}]) = [0, e^{\frac{2\pi j}{n}\mathbf{i}}]$ for any $0 \leq j \leq n-1$. We set

$$f = \tau \circ \varphi : X \to X.$$

By (10), $f^j(M_0) \subset [0, e^{\frac{2\pi j}{n}\mathbf{i}}]$ for $1 \leq j \leq n-1$. We set $M_j = f^j(M_0)$, $1 \leq j \leq n-1$. We note that $f^n_{|M_j} = M_j$ and M_j is a totally minimal set under f^n for every $0 \leq j \leq n-1$. We set $M = M_0 \cup M_1 \cup \ldots \cup M_{n-1}$. According to Definition 1.3, f has a relatively totally minimal set M under f, and M has a periodic decomposition of a length n.

At the end of this section we construct an example of a continuous map on a dendrite having a minimal set M such that $[M]$ is a dendrite with an infinite countable set of end points.

Example 3.2: (1) Construction of a dendrite X.

Let $I_n = [\frac{1}{2\cdot 3^{n-1}}; \frac{1}{2\cdot 3^{n-1}} + \frac{\mathrm{i}}{2\cdot 3^{n-1}}]$ be a vertical segment on the complex plane $\mathbb{C}$ for any $n \geq 1$. We set $Y = [0, 1] \cup \bigcup_{n=1}^{\infty} I_n$. We note that Y is a dendrite with a countable set of ramification points $\{\frac{1}{2\cdot 3^{n-1}}\}_{n=1}^{\infty}$. Let

$$X = [-1/4, 0] \cup Y \cup [1, 5/4].$$

Then X is a dendrite, the set $E(X)$ of end points of X is countable.

(2) Construction of a map $f : X \to X$.

We define an auxiliary map $g : [0, 1] \to Y$ as follows:

(a) (1)$g(x) = x$, if $x \in \{0\} \cup [1/2, 1]$;

(b) (2)For any $n \geq 2$ we divide a segment $[\frac{1}{2\cdot 3^{n-1}}, \frac{1}{2\cdot 3^{n-2}}]$ into three parts by points $\frac{2}{3^n}$ and $\frac{1}{3^{n-1}}$. On every segment $[\frac{1}{2\cdot 3^{n-1}}, \frac{2}{3^n}]$, $[\frac{2}{3^n}, \frac{1}{3^{n-1}}]$, $[\frac{1}{3^{n-1}}, \frac{1}{2\cdot 3^{n-2}}]$ we define g as a linear homeomorphism such that $g(\frac{1}{2\cdot 3^{n-1}}) = \frac{1}{2\cdot 3^{n-1}}$, $g(\frac{2}{3^n}) = \frac{1}{2\cdot 3^{n-2}}$, $g(\frac{1}{3^{n-1}}) = \frac{1}{2\cdot 3^{n-2}} + \frac{\mathrm{i}}{2\cdot 3^{n-2}}$ and $g(\frac{1}{2\cdot 3^{n-2}}) = \frac{1}{2\cdot 3^{n-2}}$. Then

$$g\left(\left[\frac{1}{2\cdot 3^{n-1}}, \frac{2}{3^n}\right]\right) = \left[\frac{1}{2\cdot 3^{n-1}}, \frac{1}{2\cdot 3^{n-2}}\right],$$

and

$$g\left(\left[\frac{2}{3^n}, \frac{1}{3^{n-1}}\right]\right) = g\left(\left[\frac{1}{3^{n-1}}, \frac{1}{2\cdot 3^{n-2}}\right]\right) = I_{n-1}. \tag{11}$$

We note that g is a continuous map, $g([0, 1]) = Y$. Let K be the classical Cantor set on $[0, 1]$. We note that $K \cap (0, \frac{1}{3}] \subset \bigcup_{n\geq 2}[\frac{2}{3^n}, \frac{1}{3^{n-1}}]$. Then by (11), $g(K \cap (0, \frac{1}{3}]) \subset \bigcup_{n=1}^{\infty} I_n$. Hence, $g(K) \subset \{0\} \cup \bigcup_{n=1}^{\infty} I_n \cup [\frac{2}{3}, 1]$. We note that $E(Y) \subset g(K)$ and for every $n \in \mathbb{N}$ and any point $x \in g(K) \cap I_n$, $g^{-1}(x) \cap K$ is a singleton. For any point $x \in K \cap [\frac{2}{3}, 1]$, $g^{-1}(x)$ is a singleton.

We also need two auxiliary maps.

Let $\varphi(x) : [0, 1] \to [0, 1]$ be the tent map as in (9). It is known that φ has a totally minimal set; denote it by L.

Let $h : [0, 1] \to [-1/4, 5/4]$ be a linear homeomorphism such that $h(0) = -1/4$, $h(1) = 5/4$, $h(L) = K$.

We set $M = g(K)$. We define a map $\tilde{f} : M \to M$ such that

$$\tilde{f}(x) = g \circ h \circ \varphi \circ h^{-1} \circ (g^{-1}(x) \cap K) \quad \text{foranypoint } x \in M.$$

Since g, h, φ are continuous maps then $\tilde{f}$ is a continuous map. We note that $\tilde{f}(M) = M$. Since $M \subset X$ and M is a closed set then by Urysohn [23], this map can be extended so as to be continuous throughout X. Thus, there exists a continuous map $f : X \to X$ such that $f(x) = \tilde{f}(x)$ for any point $x \in M$. Then f has a minimal set M such that $[M]$ is a dendrite Y with an infinite countable set of end points.

4. Proof of Theorem 1.7

Let $S^1 = \{z \in \mathbb{C} : |z| = 1\}$ be a unit circle. We define an irrational rotation $\varphi : S^1 \to S^1$ as $\varphi(z) = ze^{2\pi \mathrm{i}\theta}$, where θ is any irrational number, $z \in S^1$. Then circle S^1 is a totally minimal set under φ and topological entropy of φ is equal to 0. We note that φ does not have any horseshoe for every $n \in \mathbb{N}$ and $Per(\varphi) = \emptyset$.

Proof of the statement (1.7.1).: Let n be any natural number such that $n \geq 2$. We set $I_j = [e^{\frac{2\pi j}{n}\mathbf{i}}, 2e^{\frac{2\pi j}{n}\mathbf{i}}]$ for any $0 \leq j \leq n-1$. Let

$$G = S^1 \cup \bigcup_{j=0}^{n-1} I_j.$$

Then G is a finite graph with n ramification points $\{e^{\frac{2\pi j}{n}\mathbf{i}}\}_{j=0}^{n-1}$.

We define a map $f : G \to G$ by the following way:

(1) $f(z) = \varphi(z)$, if $z \in S^1$;
(2) for any $0 \leq j \leq n-1, f(z) = \varphi(e^{\frac{2\pi j}{n}\mathbf{i}})$, if $z \in I_j$.

Then f is a continuous map, $f(G) = S^1$ and f has all properties of φ on S^1. ■

Proof of the statement (1.7.2).: Let G be a finite graph from the proof of (1.7.1). We define $f : S^1 \to S^1$ such that $f(z) = \varphi(z)$, if $z \in S^1$. Since for any two points from S^1 there are two different arcs containing these points then we define an arc $[\varphi(e^{\frac{2\pi j}{n}\mathbf{i}}), e^{\frac{2\pi j}{n}\mathbf{i}}]$ as a set $\{z \in S^1 : \frac{2\pi}{n} \leq Arg\, z \leq \frac{2\pi}{n} + \theta\}$.

We divide every segment I_j into two equal segments with a point $\frac{3}{2}e^{\frac{2\pi j}{n}\mathbf{i}}, 0 \leq j \leq n-1$. For any $0 \leq j \leq n-1$ we define linear homeomorphisms

$$f : \left[e^{\frac{2\pi j}{n}\mathbf{i}}, \frac{3}{2}e^{\frac{2\pi j}{n}\mathbf{i}}\right] \to \left[\varphi(e^{\frac{2\pi j}{n}\mathbf{i}}), e^{\frac{2\pi j}{n}\mathbf{i}}\right] \quad \text{and} \quad f : \left[\frac{3}{2}e^{\frac{2\pi j}{n}\mathbf{i}}, 2e^{\frac{2\pi j}{n}\mathbf{i}}\right] \to \left[e^{\frac{2\pi j}{n}\mathbf{i}}, 2e^{\frac{2\pi j}{n}\mathbf{i}}\right]$$

such that $f(e^{\frac{2\pi j}{n}\mathbf{i}}) = \varphi(e^{\frac{2\pi j}{n}\mathbf{i}}), f(\frac{3}{2}e^{\frac{2\pi j}{n}\mathbf{i}}) = e^{\frac{2\pi j}{n}\mathbf{i}}$ and $f(2e^{\frac{2\pi j}{n}\mathbf{i}}) = 2e^{\frac{2\pi j}{n}\mathbf{i}}$.

We note that f is a continuous map, $f(G) = G$, any end point $2e^{\frac{2\pi j}{n}\mathbf{i}}$ in G is a fixed point of f $(0 \leq j \leq n-1)$. Thus, the set $Per(f)$ of periodic points is not-empty, f has an infinite totally minimal set S^1, for any number $n \in \mathbb{N}$ a map f^n does not have any horseshoe and topological entropy of f equals zero.

Theorem 1.7 is proved. ■

Disclosure statement

No potential conflict of interest was reported by the author(s).

Funding Acknowledgement

This work is supported by the grant of the Russian Science Foundation (RSF) (project No 24-21-00242, https://rscf.ru/en/project/24-21-00243/).

References

[1] D. Arévalo, W.J. Charatonic, P.P. Covarrubias, and L. Simón, Dendrites with a closed set of end points, Topol. Appl. 115 (2001), pp. 1–17.

[2] F. Balibrea, R. Hric, and L'. Snoha, Minimal sets on graphs and dendrites, Int. J. Bifurcat. Chaos Appl. Sci. Eng. 13 (2003), pp. 1721–1725.

[3] F. Balibrea, T. Downarowicz, R. Hric, Ľ. Snoha, and V. Špitalský, Almost totally disconnected minimal systems, Ergod. Theory Dyn. Syst. 29 (2009), pp. 737–766.

[4] J. Banks, Regular periodic decompositions for topologically transitive maps, Ergod. Theory Dyn. Syst.17 (1997), pp. 505–529.

[5] Yu.S. Barkovskii and G.M. Levin, A Cantor limit set, Russ. Math. Surv. 35 (1980), pp. 235–236.

[6] G.D. Birkhoff, *Dynamical Systems*, Amer. Math. Soc. Colloq. Publ., Vol. 9, American Mathematical Society, New York, 1927.

[7] A. Blokh and E. Teoh, How little is little enough, Discrete Contin. Dyn. Syst. 9 (2003), pp. 969–978.

[8] L. Block and J. Keesling, A characterization of adding machine maps, Topol. Appl. 140 (2004), pp. 151–161.

[9] J. Charatonik and W. Charatonik, Dendrites, Aportaciones Mat. Comun. 22 (1998), pp. 227–253.

[10] L.S. Efremova and E.N. Makhrova, The dynamics of monotone maps of dendrites, Matem. Sb. 192 (2001), pp. 15–30.

[11] L.S. Efremova and E.N. Makhrova, One-dimensional dynamical systems, Russ. Math. Surv. 76 (2021), pp. 821–881.

[12] M. Feigenbaum, Universal behavior in nonlinear systems, Los Alamos Sci. 1 (1980), pp. 4–27.

[13] H.M. Gehman, Concerning the subsets of a plane continuous curve, Ann. Math. Second Series 27 (1925), pp. 29–46.

[14] Z. Kočan, V. Kornecká-Kurková, and M. Málek, Entropy, horseshoes and homoclinic trajectories on trees, graphs and dendrites, Ergod. Theory Dyn. Syst. 31 (2011), pp. 165–175.

[15] S. Kolyada and Ľ. Snoha, Minimal dynamical systems, Scholarpedia 4 (2009), p. 5803. https://www.scholarpedia.org/article/Minimal_dynamical_systems

[16] K. Kuratowski, *Topology*, Academic Press, New York; PWN, Warsaw, Vol. 2, 1968.

[17] J.-H. Mai, Pointwise-recurrent graph maps, Ergod. Theory Dyn. Syst. 25 (2005), pp. 629–637.

[18] E.N. Makhrova, The structure of dendrites with the periodic point property, Russ. Math. (Iz. VUZ).55 (2011), pp. 33–37.

[19] E.N. Makhrova, Structure of dendrites admitting an existence of arc horseshoe, Russ. Math. (Iz. VUZ) 59 (2015), pp. 52–61.

[20] S. Nadler, Continuum Theory, Marcel Dekker, New York, 1992.

[21] J. Nikiel, A characterization of dendroids with uncountable many end-points in the classical sense, Houst. J. Math. 9 (1983), pp. 421–432.

[22] A.N. Sharkovsky, Yu.L. Maistrenko, and E.Yu. Romanenko, Difference Equations and Their Applications, Naukova Dumka, Kiev, Ukraine, 1986.

[23] P. Urysohn, Über die Mächtigkeit der zusammenhängenden Mengen, Math. Ann. 94 (1925), pp. 262–295.

[24] X. Ye, *D*-function of a minimal set and an extension of Sharkovskiĭ's theorem to minimal sets, Ergod. Theory Dyn. Syst. 12 (1992), pp. 365–376.

Recurrence and nonwandering sets of local dendrite maps

Hafedh Abdelli, Habib Marzougui and Amira Mchaalia

ABSTRACT
Given a local dendrite X with set End(X) of endpoints of X countable and $f : X \to X$ be a continuous map, we let P(f), AP(f), R(f), $\Omega(f)$, EP(f), EAP(f) the sets of periodic points, almost periodic points, recurrent points, nonwandering points, eventually periodic points and eventually almost periodic points of f, respectively. We show that $\mathrm{R}(f) \subset \overline{\mathrm{EAP}(f)}$. On the other hand, we show that $\Omega(f) \subset \overline{\mathrm{EAP}(f)}$, whenever $\mathrm{P}(f) = \emptyset$. As a consequence, we prove that if the set B(X) of branch points is finite, then $\overline{\mathrm{R}(f)} = \overline{\mathrm{AP}(f)}$. We give a counter-example showing that the above results do not hold whenever B(X) is infinite or End(X) is uncountable.

1. Introduction

In the last two decades, there is a growing interest in studying recurrence and the topological structure of nonwandering points in some one-dimensional Peano continua. Examples of continua studied include, graphs and dendrites (see e.g. [10,11,15,16,21,22,27,30,34,35]). Recently local dendrites has given a special attention (see e.g. [1,3,4,28,33]). These form a large class of continua which includes local dendrites. Arai et al. [6], proved that if f is a continuous map of a dendrite X with finitely many branch points into itself and $\Omega(f)$ the set of nonwandering points of f, then $\Omega(f)$ is contained in the closure of the set of eventually periodic points of f (later, this result was extended by Abdelli et al. in [1] to monotone maps on local dendrites having a periodic point and in particular for any monotone dendrite maps). For graph maps f, Rongbao et al. [31] proved that $\Omega(f)$ is contained in the closure of the set ER(f). In the present paper, we extend the above result to local dendrite maps. We prove that if f is a continuous map of a local dendrite X with a countable set of endpoints into itself and if f has no periodic points, then $\Omega(f)$ is contained in the closure of the set of eventually almost periodic points of f (see Theorem 4.1). On the other hand, we prove that if f is a continuous map of a local dendrite X with a countable set of endpoints into itself, then R(f) is contained in the closure of the set of eventually almost periodic points of f (see Theorem 5.1). With regard to almost periodic points, we show that if f is a continuous map of a local dendrite X with finitely many branch points into itself, then $\overline{\mathrm{R}(f)} = \overline{\mathrm{AP}(f)}$ (see Theorem 6.1). Furthermore, we show that if f is a continuous map of a dendrite X into itself with set of branch points of X closed, then $\overline{\mathrm{R}(f)} \subset \overline{\mathrm{EP}(f)}$ (see Theorem 6.4).

2. Definitions and preliminaries

Let X be a compact metric space with metric d and $f : X \longrightarrow X$ a continuous map. (X,f) is called a dynamical system. Let $\mathbb{Z}$, $\mathbb{Z}_+$, $\mathbb{Z}_-$ and $\mathbb{N}$ be the sets of integers, non-negative integers, non-positive integers and positive integers, respectively. For $n \in \mathbb{Z}_+$, denote by f^n the nth iterate of f where $f^0 =$ identity. For any $x \in X$, the subset $O_f(x) = \{f^n(x) : n \in \mathbb{Z}_+\}$ is called the *orbit* of x (under f). A subset $A \subset X$ is called *f-invariant* (resp. strongly f-invariant) if $f(A) \subset A$ (resp. $f(A) = A$), it is further called a *minimal set* (under f) if it is closed, nonempty and does not contain any nonempty f-invariant, closed proper subset of X. For any $x \in X$, the set of limit points of the sequence $(f^n(x))_{n\in\mathbb{N}}$ is called the ω-limit set of x and f, and is denoted by $\omega_f(x)$. We denote by $\mathring{A}$ (resp. $\overline{A}$) the interior (resp. the closure) of A in X and by $\mathrm{diam}(A) = \sup_{(x,y)\in A\times A} d(x,y)$ the diameter of A. The symbol card(A) stands for the cardinality of A. A point $x \in X$ is called:

- *periodic* of period $n \in \mathbb{N}$ if $f^n(x) = x$ and $f^i(x) \neq x$ for $1 \leq i \leq n-1$; if $n = 1$, x is called a *fixed point* of f (i.e. $f(x) = x$),
- *recurrent* if $x \in \omega_f(x)$,
- *almost periodic* if for any neighborhood U of x there is $N \in \mathbb{N}$ such that $\{f^{i+k}(x) : i = 0, 1, \ldots, N\} \cap U \neq \emptyset$, for all $k \in \mathbb{N}$,
- *nonwandering* if for any neighborhood U of x there is $N \in \mathbb{N}$ such that $f^N(U) \cap U \neq \emptyset$.
- *eventually periodic (resp. eventually almost periodic, resp. eventually recurrent)* if there exists $n \in \mathbb{N}$ such that $f^n(x)$ is *periodic (resp. almost periodic, resp. recurrent).*

We denote by P(f), AP(f), R(f) and $\Omega(f)$ the sets of periodic points, almost periodic points, recurrent points and nonwandering points of f, respectively. Also we denote by EP(f), EAP(f) and ER(f) the sets of eventually periodic points, eventually almost periodic points and eventually recurrent points of f, respectively.

Let $h(f)$ denote the topological entropy of f (see [5], for the definition). Now, we recall some basic properties of graphs, dendrites and local dendrites. We use the terminologies from Nadler [29].

A metric space X is said to be :

- a *continuum* if it is compact connected. We called an *arc* each space homeomorphic to the interval $[0, 1]$. If A is an arc in X with endpoints a and b, then A is said to be a free arc in X if the set $A \setminus \{a, b\}$ is open in X.
- a *circle* if it is homeomorphic to the unit circle $\mathbb{S}^1 = \{z \in \mathbb{C} : |z| = 1\}$.
- a *graph* if it is a continuum which can be written as the union of finitely many arcs such that any two of them are either disjoint or intersect only in one or both of their

endpoints. Each of these arcs is called an *edge* of the graph. A point $v \in G$ is called a *branch point* if it admits a neighborhood U in X homeomorphic to the set $\{z \in \mathbb{C} : z^r \in [0, 1]\}$ with the natural topology for some integer $r \geq 3$, with the homeomorphism mapping v to 0. If $r = 1$, then we call v an *endpoint* of X. Denote by $\mathrm{B}(X)$ and $\mathrm{End}(X)$ the sets of branch points and endpoints of X, respectively. An edge is the closure of some connected component of $X \setminus \mathrm{B}(X)$, it is homeomorphic to $[0, 1]$. A subgraph of X is a subset of X which is a graph itself. Every sub-continuum of a graph is a graph [29, Corollary 9.10.1].

- a *dendrite* if it is locally connected continuum containing no homeomorphic copy to a circle. Every sub-continuum of a dendrite is a dendrite [29, Theorem 10.10] and every connected subset of X is arcwise connected [29, Proposition 10.9]. In addition, any two distinct points x, y of a dendrite X can be joined by a unique arc denoted by $[x, y]$ with endpoints x and y. We denote by $\mathrm{End}([x, y]) = \{x, y\}$, $[x, y) = [x, y] \setminus \{y\}$ (resp. $(x, y] = [x, y] \setminus \{x\}$ and $(x, y) = [x, y] \setminus \{x, y\}$). A point $x \in X$ is called an *endpoint* if $X \setminus \{x\}$ is connected. It is called a *branch point* if $D \setminus \{x\}$ has more than two connected components. The number of connected components of $X \setminus \{x\}$ is called the *order* of x. Denote by $\mathrm{End}(X)$ and $\mathrm{B}(X)$ the sets of endpoints, and branch points of X, respectively. A point $x \in X \setminus \mathrm{End}(X)$ is called a *cut point*. The set of cut points of X is dense in X. Following [7, Corollary 3.6], for any dendrite D, we have $\mathrm{B}(D)$ is discrete whenever $\mathrm{End}(X)$ is closed.
- a *local dendrite* if it is a continuum having the property that every of its points has a dendrite neighbourhood. According to [24, Theorem 4], a local dendrite is a locally connected continuum containing only a finite number of circles. As a consequence, every sub-continuum of a local dendrite is a local dendrite [24]. Every graph and every dendrite is a local dendrite. A point $x \in X$ is called a *branch point* of X if there exists a closed neighbourhood D of x which is a dendrite such that x is a branch point of D (i.e. $D \setminus \{x\}$ has more than two connected components). We denoted by $\mathrm{B}(X)$ the set of branch points of X. By [24, Theorem 6, p. 304 and Theorem 7, p. 302], $\mathrm{B}(X)$ is at most countable. A point $e \in X$ is called an *endpoint* of X if it admits a neighbourhood U in X such that U is a dendrite and $U \setminus \{e\}$ is connected. The set of endpoints of X is denoted by $\mathrm{End}(X)$. For any $x \in X$, the cardinal number of connected components of $U \setminus \{x\}$ is called the order of x, where U is any dendrite neighbourhood of x in X.

Let X be a local dendrite. We denote by $\{S_1, \ldots, S_k\}$ all simple closed curves in X. Then we define the *canonical graph* Γ_X as the intersection of all graphs in X containing $\bigcup_{i=1}^k S_i$. Thus Γ_X is a subgraph of X (with $\Gamma_X = \emptyset$ whenever X contains no simple closed curve). An attached dendrite D in X is the closure of a connected components of $X \setminus \Gamma_X$, then D itself is a dendrite. Moreover, it is easy to see that $D \cap \Gamma_X$ is a singleton. The attaching point corresponding to an attached dendrite D is the only point in $D \cap \Gamma_X$. Let D be an attached dendrite and $a, b \in D$ with $a \neq b$, we denote by $[a, b]_D$ the unique arc in D joining a and b (note that it is also the unique arc in X joining a and b). We denote also by $(a, b]_D = [a, b] \setminus \{a\}$, $[a, b)_D = (b, a]_D$ and $(a, b) = [a, b] \setminus \{a, b\}$. We consider $\mathcal{L}$ the class of local dendrites whose canonical graph is nonempty, equivalently the class of local dendrites which are not dendrites. Obviously, if X is a local dendrite which supports a self continuous map without periodic points, then X belongs to $\mathcal{L}$.

A continuous map from a local dendrite (resp. an interval, a topological graph or a dendrite) into itself is called a *local dendrite map* (resp. an *interval map*, a *graph map* or a *dendrite map*).

By *Gehman dendrite*, we mean the unique dendrite (up to homeomorphism) whose set of endpoints is homeomorphic to the Cantor set and whose branch points are all of order 3 [29].

Definition 2.1: Let X be a topological space. A continuous map $r: X \to X$ is called a *retraction* provided that $r \circ r = r$, i.e. r is the identity on its range. A subset A of X for which there is a retraction of X onto A is called a *retract* of X.

For any local dendrite $X \in \mathcal{L}$, Γ_X is retract of X. Indeed, we define the map $r : X \to \Gamma_X$ as follows: $r_{|\Gamma_X} = \mathrm{id}_{\Gamma_X}$ and if $x \in D \setminus \Gamma_X$ where D is an attached dendrite having z as an attaching point, then $r(x) = z$. It is clear that r is a continuous map, so Γ_X is a retract of X.

Given a local dendrite map $f : X \to X$, we associate its *retraction graph map* $g := r \circ f_{|\Gamma_X} : \Gamma_X \to \Gamma_X$. We see that if $x \in \Gamma_X$ with $O_f(x) \subset \Gamma_X$, then $\omega_f(x) = \omega_g(x)$ and $x \in \mathrm{P}(f)$ if and only if $x \in \mathrm{P}(g)$. This follows from the fact that if $O_f(x) \subset \Gamma_X$, then $g^n(x) = f^n(x)$, for any $n \in \mathbb{N}$.

3. Some results

Lemma 3.1: *Let (X,f) be dynamical system and let U be an open connected subset of X. Suppose that there exists $n \in \mathbb{N}$ such that $f^n(U) \cap U \neq \emptyset$. Set $h := f^n$ and $W := \bigcup_{k=0}^{\infty} h^k(U)$. Then the following properties hold:*

1. *W is a sub-connected of X,*
2. *$h(W) \subset W$,*
3. *If $U \cap \overline{E(S,f)} = \emptyset$, then $W \cap E(S,f) = \emptyset$, for any subset S of X, where $E(S,f) = \{x \in X : f^k(x) \in S$, for some $k \in \mathbb{N}\}$.*

Proof: Assertions (1) follows from the fact that the $h^k(U)$, ($k \in \mathbb{N}$), are not pairwise disjoint. Assertion (2) is clear. Let us prove the assertion (3). If $x \in W \cap E(S,f)$, for some $S \subset X$, then there exist $m \in \mathbb{N}$ and $y \in U$ such that $x = f^m(y)$. Since $x \in E(S,f)$, so $y \in E(S,f)$ and thus $U \cap E(S,f) \neq \emptyset$. A contradiction. ■

Lemma 3.2: *Let (X,f) be a dynamical system. Let A be an f-invariant subset of X. Then $\Omega(f) \cap A \subset \Omega(f_{|\overline{A}})$.*

Proof: Lemma 3.2 is done, when $A = \emptyset$. Assume that $A \neq \emptyset$. Let $x \in \Omega(f) \cap A$. Since $x \in \Omega(f)$, there are $(x_k)_{k\in\mathbb{N}} \subset X$ and $(n_k)_{k\in\mathbb{N}} \subset \mathbb{N}$ such that $\lim_{k\to+\infty} x_k = x = \lim_{k\to+\infty} f^{n_k}(x_k)$. As $x \in A$, there exists $N \in \mathbb{N}$ such that $x_k \in A$, for all $k \geq N$. Since A is f-invariant, $f^{n_k}(x_k) = f^{n_k}_{|\overline{A}}(x_k)$, for all $k \geq N$. Thus $x \in \Omega(f_{|\overline{A}})$. ■

Lemma 3.3 ([9, Lemma 4]): *Let (X,f) be a dynamical system and let $x \in X$ such that $\omega_f(x)$ is infinite. If z is an isolated point in $\omega_f(x)$, then it is not a periodic point.*

Lemma 3.4 ([27, Theorem 4.2]): *Let $f : G \longrightarrow G$ be a graph map and $x \in \Omega(f)$. If $x \notin \mathrm{EP}(f)$, then $x \in \Omega(f^n)$, for all $n > 1$.*

Lemma 3.5 ([8]): *Let $(C_n)_{n\geq 1}$ be a sequence of pairwise disjoint connected subsets of a local dendrite (X, d). Then $\lim_{n\to+\infty} \mathrm{diam}(C_n) = 0$.*

Lemma 3.6 ([1, Lemma 6.4]): *Let X be a local dendrite with a countable set of endpoints. Then every sub-local dendrite of X has countable set of endpoints.*

Lemma 3.7: *Let X be a local dendrite, and W be a sub-connected subset of X. Then $\overline{W} \setminus W \subset \mathrm{End}(\overline{W}) \cup \Gamma_{\overline{W}}$.*

Proof: Let $x \in \overline{W} \setminus W$. Assume that $x \notin (\mathrm{End}(\overline{W}) \cup \Gamma_{\overline{W}})$. Let C_1, C_2 be two connected components of $\overline{W}\setminus\{x\}$. As W is dense in $\overline{W}$, then $C_1 \cap W \neq \emptyset \neq C_2 \cap W$. Let $\alpha_1 \in C_1 \cap W$ and $\alpha_2 \in W \cap C_2$. By connectedness of W, there exists an arc I of X with $\gamma(I) = \{\alpha_1, \alpha_2\}$ such that $I \subset W$. Therefore $I \cap (\overline{W} \setminus C_1 \cup C_2) \neq \emptyset$. As $C_1 \cup C_2 \subset \overline{W}\setminus\{x\}$, so $x \in I$ and hence $x \in W$. A contradiction. ■

Lemma 3.8: *Let X be a local dendrite and W be a sub-connected subset of X. Then for any free arc I in $\Gamma_{\overline{W}}$, $\mathrm{card}((\overline{W} \setminus W) \cap I) \leq 3$.*

Proof: Assume on the contrary that $\mathrm{card}((\overline{W} \setminus W) \cap I) \geq 4$, for some free arc I in $\Gamma_{\overline{W}}$. Let $a, b, c, d \in \overline{W} \setminus W \cap I$ be such that $a < b < c < d$ in I. Let $r : \overline{W} \longrightarrow \Gamma_{\overline{W}}$ be the retraction map of $\Gamma_{\overline{W}}$. Set $U_1 = r^{-1}((a, b)_{|I})$ and $U_2 = r^{-1}((c, d)_{|I})$. Since r is continuous and each $(a, b)_I$, $(c, d)_I$ is open subset in $\Gamma_{\overline{W}}$, so U_i is open in $\overline{W}$, for each $i \in \{1, 2\}$. Hence $U_1 \cap W \neq \emptyset$ and $U_2 \cap W \neq \emptyset$. Since U_1 and U_2 are disjoint and W is connected, so W meets the boundary $\partial_Y U_i$ of U_i; $W \cap \partial_{\overline{W}} U_i \neq \emptyset$, $i = 1, 2$. Let $\alpha_i \subset W \cap \partial_{\overline{W}} U_i$, $i = 1, 2$. As $\partial_Y U_1 \subset r^{-1}(\{a, b\})$ and $\partial_Y U_2 \subset r^{-1}(\{c, d\})$, so $\alpha_i \in \overline{W}\setminus\Gamma_W$ and $r(\alpha_1) \neq r(\alpha_2)$. Since W is connected, there exists an arc $J \subset W$ joining α_1 and α_2. This implies that $J \cap \{a, b, c, d\} \neq \emptyset$. A contradiction with $a, b, c, d \notin W$. ■

Proposition 3.9: *Let X be a local dendrite and W a sub-connected of X. Then $\Gamma_{\overline{W}} \setminus W$ is finite.*

Proof: The proposition is done, when $\Gamma_{\overline{W}}$ is empty. So assume that $\Gamma_{\overline{W}}$ is nonempty. We have

$$\Gamma_{\overline{W}} \setminus W = (\overline{W} \setminus W) \cap \Gamma_{\overline{W}} = \bigcup_{i=0}^{k} \left((\overline{W} \setminus W) \cap I_i\right), \quad \text{for some } k \in \mathbb{N},$$

where the finite union is taken over all free arcs of $\Gamma_{\overline{W}}$. Then by Lemma 3.8, $\Gamma_{\overline{W}} \setminus W$ is finite. ■

Lemma 3.10 ([33, Lemma 3.1]): *Let $f : X \to X$ be a local dendrite map and $(x, y) \subset X\setminus(B(X) \cup \overline{P(f)} \cup \Gamma_X)$. If $z \in (x, y) \cap \Omega(f)$, then $O_f(f(z)) \cap (x, y) = \emptyset$. In particular, $\mathrm{card}(O_f(z) \cap (x, y)) \leq 1$.*

Proposition 3.11: *Let X be a local dendrite with $B(X)$ closed and let $f: X \to X$ be a continuous map. Let W be a sub-connected of X verifying:*

(1) *W is f^n-invariant, for some $n \in \mathbb{N}$,*
(2) *$W \cap \mathrm{EP}(f) = \emptyset$.*

Then for any $z \in \Omega(f) \cap W$, we have

$$\omega_{f^n}(z) \subset (B(X) \cap \overline{W}) \cup \Gamma_{\overline{W}} \tag{1}$$

Proof: Assume that there are $z \in \Omega(f) \cap W$ and $x \in \omega_{f^n}(z) \setminus ((B(X) \cap \overline{W}) \cup \Gamma_{\overline{W}})$. Since $B(X)$ is closed in X, so $\overline{W} \setminus ((B(X) \cap \overline{W}) \cup \Gamma_{\overline{W}})$ is an open subset of $\overline{W}$. Moreover from the hypothesis, we have $x \in \overline{W}$. Let C be the connected component of $\overline{W} \setminus ((B(X) \cap \overline{W}) \cup \Gamma_{\overline{W}})$ that contains x. So C is an open free arc of $\overline{W}$, denote it by $C = (a, b)$.

• First we show that $(a, b) \subset W$. Indeed, otherwise, there exists $y_0 \in (a, b) \setminus W$. Hence $y_0 \in \overline{W} \setminus W$. As $y_0 \notin \Gamma_{\overline{W}}$, so by Lemma 3.7, $y_0 \in \mathrm{End}(\overline{W})$. A contradiction.

From the hypothesis, we see that $(a, b) \cap P(f) = \emptyset$ and $\omega_{f^n}(z) \subset \overline{W}$. As $x \in \omega_{f^n}(z) \cap (a, b)$ and $z \notin \mathrm{EP}(f)$, so $O_f(z) \cap (a, b)$ is infinite. On the other hand, there is $p \in \mathbb{N}$ such that $f^p(z) \in (a, b) \cap \Omega(f)$. Since $(a, b) \subset \overline{W} \setminus ((B(X) \cap \overline{W}) \cup \overline{P(f)}) \cup \Gamma_{\overline{W}})$ and $B(\overline{W}) \subset B(X) \cap \overline{W}$, so by Lemma 3.10, $O_f(f^{p+1}p(z)) \cap (a, b) = \emptyset$ and hence $O_f(z) \cap (a, b)$ is finite. A contradiction. ■

4. Nonwandering and eventual almost periodicity

4.1. Case of local dendrite maps without periodic points

The ultimate result of this paragraph is the following theorem.

Theorem 4.1: *Let $f: X \longrightarrow X$ be a local dendrite map with* $\mathrm{P}(f) = \emptyset$ *and* $\mathrm{End}(X)$ *countable. Then* $\Omega(f) \subset \overline{\mathrm{EAP}(f)}$.

Theorem 4.1 does not hold whenever $\mathrm{P}(f) \neq \emptyset$ (see Example 7.1).

We need the following lemma.

Lemma 4.2 ([4, Theorems 3.1 and 4.1]): *Let $f: X \longrightarrow X$ be a local dendrite map with* $\mathrm{P}(f) = \emptyset$ *and* $\mathrm{End}(X)$ *countable. Then Γ_X contains a unique minimal set M of f.*

Proof of Theorem 4.1.: Suppose that there is $z \in \Omega(f) \setminus \overline{\mathrm{EAP}(f)}$. Let C be the connected component of $S^1 \setminus \overline{\mathrm{EAP}(f)}$ containing z. Since $z \in \Omega(f)$, there exists $n \in \mathbb{N}$ such that $f^n(C) \cap C \neq \emptyset$. Set $h := f^n$, $W := \bigcup_{k=0}^{\infty} h^k(C)$ and $g = h_{|\overline{W}}$. By Lemma 3.1, W is connected and h-invariant. So $\overline{W}$ is a sub-local dendrite of X and $(\overline{W}, g)$ is a local dendrite map with $\mathrm{P}(g) = \emptyset$. In particular, $\Gamma_{\overline{W}} \neq \emptyset$. By Lemma 4.2, g admits an infinite minimal set $M \subset \Gamma_{\overline{W}}$. Since $C \cap \mathrm{EAP}(f) = \emptyset$, so by Lemma 3.1, $W \cap \mathrm{EAP}(f) = \emptyset$. Hence $W \cap \mathrm{EAP}(g) = \emptyset$ and in particular, $W \cap M = \emptyset$. Therefore $M \subset \Gamma_{\overline{W}} \setminus W$. By Proposition 3.9, M is finite, a contradiction. This completes the proof. ■

In [4], Abdelli et al. proved that if X is a local dendrite with $\mathrm{End}(X)$ countable and $f: X \longrightarrow X$ is a local dendrite map with $\mathrm{P}(f) = \emptyset$, then $\mathrm{R}(f) = \mathrm{AP}(f)$. In particular, $\mathrm{ER}(f) =$

EAP(f). When we replace "End(X) countable" by "EAP(f) closed", we get the following corollary.

Corollary 4.3: *Let* $f : X \longrightarrow X$ *be a local dendrite map with* $\mathrm{P}(f) = \emptyset$. *If* $\mathrm{EAP}(f)$ *is closed, then* $\mathrm{ER}(f) = \mathrm{EAP}(f)$.

Proof: It is enough to prove that $\mathrm{ER}(f) \subset \mathrm{EAP}(f)$. If $x \in \mathrm{ER}(f)$ and $n \in \mathbb{N}$ such that $f^n(x) \in \mathrm{R}(f)$, then by Theorem 4.1, $f^n(x) \in \overline{\mathrm{EAP}(f)} = \mathrm{EAP}(f)$. Therefore $x \in \mathrm{EAP}(f)$. ■

4.2. Case of local dendrite maps with periodic points

Block et al. [10, Theorem C] proved that if the set of periodic points of a circle map f is nonempty and closed, then $\Omega(f) = \mathrm{P}(f)$. We extend the above result as follows.

Proposition 4.4 (The circle case): *Let* $f : S^1 \longrightarrow S^1$ *be a circle map with* $\mathrm{P}(f) \neq \emptyset$. *Then* $\Omega(f) \subset \overline{\mathrm{EP}(f)}$.

Proof: Assume that there is $z \in \Omega(f) \setminus \overline{\mathrm{EP}(f)}$ and let C be the connected component of $S^1 \setminus \overline{\mathrm{EP}(f)}$ containing z. Then there exists $n \in \mathbb{N}$ such that $f^n(C) \cap C \neq \emptyset$. Set $h := f^n$ and $W := \bigcup_{k=0}^{\infty} h^k(C)$. By Lemma 3.1, W is connected and h-invariant. Moreover, we have $W \cap \mathrm{EP}(f) = \emptyset$. We distinguish two cases:

Case 1. $\overline{W} = S^1$. So $W = S^1 \setminus \{a\}$, for some $a \in S^1$. Hence $\mathrm{P}(f) = \{a\} = \mathrm{Fix}(f)$. By [10, Theorem C], $\Omega(f) = \mathrm{P}(f) = \{a\}$ and thus we get $z = a$, a contradiction.

Case 2. $\overline{W} = [a, b]$, for some $a, b \in S^1$. Since $z \notin \mathrm{EP}(f)$, $O_f(z)$ is infinite and so there exists $n_0 > 1$ such that $y = f^{n_0}(z) \in W = (a, b)$. By Lemma 3.4, $y \in \Omega(f) \cap W = \Omega(h) \cap W$. According to Lemma 3.2, $y \in \Omega(h_{|\overline{W}})$. As $\mathrm{P}(h_{|W}) \subset \{a, b\}$, hence closed, so $\Omega(h_{|\overline{W}}) = \mathrm{P}(h_{|\overline{W}})$ and therefore $z \in \mathrm{EP}(f)$, a contradiction. ■

We recall some definitions and results concerning the structure of ω-limit sets for a graph map, which were obtained by Block in the series of papers ([11,12,14] and [13]). Here we follow the notations from [32] and reformulate his results in a similar way as authors in [32].

Definition 4.5 ([32]): Let $f : G \to G$ be a graph map. A subgraph K of G is called periodic of period k (or k-periodic) if $K, \ldots, f^{k-1}(K)$ are pairwise disjoint and $f^k(K) = K$. If, instead of $f^k(K) = K$, we have only that $f^k(K) \subset K$, the subgraph K is called weakly periodic. Then the set $O_f(K) = \cup_{i=0}^{k} f^i(K)$ is called a cycle of graphs if K is periodic and a weak cycle of graphs if K is weakly periodic.

For an infinite ω-limit set $\omega_f(x)$, we let

$$\mathcal{C}(x) := \{Z : Z \subseteq G \text{ is a cycle of graphs and } \omega_f(x) \subseteq Z\}.$$

Since the whole graph G is weakly 1-periodic, this implies that $\mathcal{C}(x)$ is never empty. The periods of the cycles in $\mathcal{C}(x)$ are either unbounded or bounded.

Definition 4.6: An infinite ω-limit set $\omega_f(x)$ of a graph map f is called a solenoid if the periods of cycles in $\mathcal{C}(x)$ are unbounded.

The following lemma can be derived from [32, Lemma 11].

Lemma 4.7: *If $\omega_f(x)$ is a solenoid, then $\omega_f(x)$ does not contain periodic points.*

If the set of periods of the cycles in $\mathcal{C}(x)$ is bounded, then by [32, Lemma 12], there exists a minimal cycle of graphs $Z \in \mathcal{C}(x)$ (i.e. for any $Y \in \mathcal{C}(x)$, $Z \subset Y$). Denote by

$$E(Z,f) = \{y \in Z : \text{ for any } U \text{ of } y \text{ in } Z, \overline{O_f(U)} = Z\}.$$

Definition 4.8: The set $E := E(Z,f)$ is called *a basic set* if Z contains a periodic point.

From [32], the set E has the following properties:

(1) $\omega_f(x) \subset E$.
(2) E is a maximal ω-limit set.

In particular, if f admits a basic set, then $h(f) > 0$.

Lemma 4.9 ([19, Theorem 8,(iii)]): *Let $f : G \to G$ be a graph map. Then the set of periodic points of f is dense in every basic set of f.*

Proposition 4.10: *Let $f : G \to G$ be a graph map and let $\omega_f(x)$ be an infinite ω-limit set. If $\omega_f(x) \cap P(f) \neq \emptyset$ then $\omega_f(x) \subset \overline{P(f)}$.*

Proof: From Lemma 4.7, $\omega_f(x)$ is not solenoid. So the set of periods of the cycles in $\mathcal{C}(x)$ is bounded. Let $Z \in \mathcal{C}(x)$ be the minimal cycle of graphs. Then $E(Z,f)$ is a basic set (since $\omega_f(x) \cap P(f) \neq \emptyset$). Hence by Lemma 4.9, $\omega_f(x) \subset E(Z,f) \subset \overline{P(f)}$. The proof is complete. ■

In [31], it was proved that if $f : G \to G$ is a graph map, then $\Omega(f) \subset \overline{\mathrm{ER}(f)}$. This result is strengthened as follows.

Theorem 4.11 (The graph case): *Let $f : G \to G$ be a graph map. Then $\Omega(f) \subset \overline{\mathrm{EAP}(f)}$.*

We need the following lemmas.

Lemma 4.12 ([32]): *Let $f : G \to G$ be a graph map. Then $h(f) > 0$ if and only if there exists $x \in G$ such that $\omega_f(x)$ is infinite and $\omega_f(x) \cap \mathrm{P}(f) \neq \emptyset$.*

Lemma 4.13: *Let $f : G \to G$ be a graph map. If $\mathrm{P}(f)$ is closed, then $h(f) = 0$.*

Proof: The proof results from [25, Proposition 4.4]. ■

Lemma 4.14 ([19, Theorem 3.1]): *Let $f : G \to G$ be a graph map. Then $\Omega(f) \subset \overline{\mathrm{ER}(f)}$.*

Proof of Theorem 4.11.: Assume that there is $z \in \Omega(f) \setminus \overline{\mathrm{EAP(f)}} \neq \emptyset$ and let C be the connected component of $G \setminus \overline{\mathrm{EAP}(f)}$ containing z. Then there exists $n \in \mathbb{N}$ such that

$f^n(C) \cap C \neq \emptyset$. Set $h := f^n$ and $W := \bigcup_{k=0}^{\infty} h^k(C)$. By Lemma 3.1, W is h-invariant and connected. Moreover, $W \cap \mathrm{EAP}(f) = \emptyset$ and $O_h(z) \subset W$ is infinite. We denote by $g := h_{|\overline{W}}$.

- $\mathrm{P}(g)$ is closed: Indeed, since $\mathrm{P}(g) \cap W = \emptyset$ and $\partial\overline{W} = \overline{W} \setminus W$ is finite, so $\mathrm{P}(g) \subset \partial\overline{W}$ is finite. Moreover, there exists $p \in \mathbb{N}$ such that $z' = h^p(z) \in W$. By Lemma 3.4, $z' \in \Omega(h)$. As $z' \in W$, so by Lemma 3.2, $z' \in \Omega(g)$. By Lemma 4.14, $z' \in \overline{\mathrm{ER}(g)}$, thus $W \cap \mathrm{ER}(g) \neq \emptyset$.
- $\mathrm{R}(g) \setminus \mathrm{P}(g) \neq \emptyset$: Assume that $\mathrm{R}(g) = \mathrm{P}(g)$, so $\mathrm{ER}(g) = \mathrm{EP}(g)$. Thus $W \cap \mathrm{EP}(g) \neq \emptyset$, a contradiction.

Let $y \in \mathrm{R}(g) \setminus \mathrm{P}(g)$. Then $\omega_g(y)$ is infinite. As $\mathrm{P}(g)$ is closed, then by (Lemmas 4.13 and 4.12), $\omega_g(y) \cap \mathrm{P}(g) = \emptyset$. Therefore, there exists an infinite minimal set $M \subset \omega_g(y)$. Since $\omega_g(y) \subset \overline{W}$ and $M \cap W = \emptyset$, so $M \subset \partial\overline{W}$. Thus M is finite, a contradiction. This finishes the proof of Theorem 4.11. ■

For local dendrite, we put the following question.

Open question. Let X be a local dendrite with End(X) countable and $f : X \longrightarrow X$ be a continuous map with $\mathrm{P}(f) \neq \emptyset$. Is $\Omega(f) \subset \overline{\mathrm{EAP}(f)}$?

We give a partial answer to the above question in the next section.

5. Recurrence and eventual almost periodicity

This section aims to prove the following theorem:

Theorem 5.1: *Let $f : X \to X$ be a local dendrite map with* End(X) *countable. Then* $\mathrm{R}(f) \subset \overline{\mathrm{EAP}(f)}$.

First notice that if $\mathrm{P}(f) = \emptyset$, Theorem 5.1 follows from Theorem 4.1. So in the sequel, we assume that $\mathrm{P}(f) \neq \emptyset$. Next, we give some results, which are needed for the proof of Theorem 5.1.

Lemma 5.2: *Let (X,f) be a dynamical system. Then* $\mathrm{R}(f) \setminus \overline{\mathrm{P}(f)}$ *is f-invariant.*

The proof for this Lemma is almost taken verbatim from [20, page 233]. For convenience of the reader we present here its proof.

Proof: Suppose that there is $x \in \mathrm{R}(f) \setminus \overline{\mathrm{P}(f)}$ such that $f(x) \in \overline{\mathrm{P}(f)}$. Let $\varepsilon > 0$ be such that $B(x,\varepsilon) \cap \mathrm{P}(f) = \emptyset$. As $x \in \mathrm{R}(f)$, there exists $m > 1$ such that $f^m(x) \in B(x,\varepsilon)$. Let $\eta > 0$ be such that $B(f^m(x),\eta) \subset B(x,\varepsilon)$. By continuity of f^{m-1} at $f(x)$, there exists $\delta > 0$ such that $d(z,f(x)) < \delta$ implies that $d(f^{m-1}(z),f^m(x)) < \eta$. There exists $y \in B(f(x),\delta) \cap \mathrm{P}(f)$, so $f^{m-1}(y) \in B(f^m(x),\eta) \cap \mathrm{P}(f) \subset B(x,\varepsilon) \cap \mathrm{P}(f)$. A contradiction. ■

Lemma 5.3: *Given a dynamical system (X,f) and a subset A of X. Let $x \in A \cap (\mathrm{R}(f) \setminus \overline{\mathrm{P}(f)})$. Assume that there exists an open neighbourhood U of x in X such that $U \subset A$. Then the set*

$$S := \{f^n(x) : n \in \{k \in \mathbb{N}, f^k(x) \in U\}\}$$

has no isolated point.

Proof: Denote by $\mathbb{P} := \{k \in \mathbb{N}, f^k(x) \in U\}$. Assume on the contrary that $f^{n_0}(x)$ is an isolated point of S for some $n_0 \in \mathbb{P}$. So there exists a neighbourhood V of $f^{n_0}(x)$ in X such that $V \cap S = \{f^{n_0}(x)\}$. So $(V \cap U) \cap S = \{f^{n_0}(x)\}$. Since $f^{n_0}(x) \in \mathrm{R}(f) \setminus \overline{\mathrm{P}(f)}$ (by Lemma 5.2), so $f^{n_0}(x) \in \omega_f(x) \setminus \mathrm{P}(f)$ and hence there exists $m_0 > n_0$ such that $f^{m_0}(x) = f^{n_0}(x)$. So $f^{n_0}(x) \in \mathrm{P}(f)$. A contradiction. ■

Proposition 5.4: *Let X be a local dendrite with* $\mathrm{End}(X)$ *countable and $f : X \to X$ be a continuous map. Let D be an attached dendrite with z its attaching point. Let $I = [a, b]$ be a sub-arc in D such that $r_I(z) \in \{a, b\}$ and let U be the connected component of $X \setminus \{a, b\}$ containing (a, b). Then the following are equivalent:*

(i) $U \cap \mathrm{R}(f) \neq \emptyset$
(ii) $U \cap \mathrm{P}(f) \neq \emptyset$

Lemma 5.5 ([4, Proposition 3.1]): *Let $f : X \to X$ be a local dendrite map and let D be an attached dendrite with z its attaching point. If there exist $x_0 \in D$ and $e \in \mathrm{End}(D) \setminus \{z\}$ such that $f(x_0) \in [x_0, e]_D$ and $x_0 \in [z, e]_D$, then $\mathrm{Fix}(f) \cap D_{x_0} \neq \emptyset$, where D_{x_0} is the closure of the connected component of $X \setminus \{x_0\}$ that contains $(x_0, e]_D$. In particular if $f(z) \in D$, then f has a fixed point.*

Lemma 5.6 ([20, Lemma 4]): *Let X be a dendrite and let S be a non-empty subset of X without isolated points. If for every arc I in X, the set $I \cap S$ is discrete, then X contains a homeomorphic copy of the Gehman dendrite.*

Notation. Let X be a local dendrite and D be an attached dendrite with z its attaching point. Let $I = [a, b]$ be an arc of D such that $r_I(z) = a$ and U be the connected component of $X \setminus \{a, b\}$ that contains (a, b) and $z' \in (a, b)$. We denote by $D'_{z'}$ the attached dendrite of U such that $D'_{z'} \cap (a, b) = \{z'\}$.

Lemma 5.7: *Under the notations above, suppose that* $\mathrm{End}(X)$ *is countable. If $(D'_{z'} \setminus \{z'\}) \cap \mathrm{R}(f) \neq \emptyset$, then $D'_{z'} \cap \mathrm{P}(f) \neq \emptyset$.*

Proof: Denote by $C = D'_{z'} \setminus \{z'\}$, and let $x \in (D'_{z'} \setminus \{z'\}) \cap \mathrm{R}(f)$ and

$$S := \{f^n(x) : n \in \{k \in \mathbb{N}, f^k(x) \in C\}\}.$$

If $x \in \overline{\mathrm{P}(f)}$, then $D'_{z'} \cap \mathrm{P}(f) \neq \emptyset$. One can assume that $x \in \mathrm{R(f)} \setminus \overline{\mathrm{P}(f)}$. By Lemma 5.3, S has no isolated point. Since $\mathrm{End}(X)$ is countable, so does $\mathrm{End}(D'_{z'})$ (Lemma 3.6). By applying Lemma 5.6, there exists an arc J of $D'_{z'}$ such that $S \cap J$ is not discrete. So $S \cap J$ contains an infinite sub-sequence $(f^{n_k}(x))_k$. So we can find $n_{p_0} < n_{p_1}$ and $e \in \mathrm{End}(D'_{z'}) \setminus \{z\}$ such that $f^{n_{p_0}}(x) \in [z, e]_{D'_{z'}}$ and $f^{n_{p_1}}(x) \in [f^{n_{p_0}}(x), e]_{D'_{z'}}$. Let $\ell = n_{p_1} - n_{p_0}$, then according to Lemma 5.5, $\mathrm{Fix}(f^\ell) \cap D_{f^{n_{p_0}}(x)} \neq \emptyset$, where $D_{f^{n_{p_0}}(x)}$ is the closure of the connected component of $X \setminus f^{n_{p_0}}(x)$ containing $[f^{n_{p_0}}(x), e]_D$. Thus $D_{f^{n_{p_0}}(x)} \subset D'_{z'}$ and so $\mathrm{P}(f) \cap D'_{z'} \neq \emptyset$. This completes the proof. ■

Proof of Proposition 5.4.: It is enough to prove $(i) \Rightarrow (ii)$. Assume on the contrary that $U \cap \mathrm{P}(f) = \emptyset$. By Lemma 5.7, we have for every $z' \in (a, b)$, $(D'_{z'} \setminus \{z'\}) \cap \mathrm{R}(f) = \emptyset$. As $U =$

$\bigcup_{z' \in (a,b)} D'_{z'}$, then $U \cap \mathrm{R}(f) \subset (a, b)$. We let $x \in U \cap R(f) \subset (a, b)$, so there exists $q \in \mathbb{N}$ such that $f^q(x) \in (a, b) \setminus \{x\}$. Without loss of generality, we can assume that $x < f^q(x)$ in (a, b). Define $\xi = f^q$ and $r : X \to I = [a, b]$ be the retraction mapping of I.

Claim. We have $\xi(x) \leq r(\xi^k(x))$, for all $k \in \mathbb{N}$.

Proof of the claim. Suppose that there exists $k_0 \in \mathbb{N}$ such that $\xi(x) > r(\xi^{k_0}(x))$. We may assume that $k_0 = \min\{i \in \mathbb{N} : \xi(x) > r(\xi^i(x))\}$. So $k_0 \geq 2$. Let ψ be the interval map $\psi : I \to I : t \mapsto \psi(t) = r \circ \xi^{k_0-1}(t)$. We have $\psi(x) = r \circ \xi^{k_0-1}(x) \geq \xi(x) > x$ and $\psi(\xi(x)) = r \circ \xi^{k_0}(x) < \xi(x)$. So there exists $x_0 \in]x, \xi(x)[\subset (a, b)$ such that $\psi(x_0) = x_0$. Since $(a, b) \cap \mathrm{P}(f) = \emptyset$, by hypothesis $\xi^{k_0-1}(x_0) = f^{q(k_0-1)}(x_0) \in D'_{x_0} \setminus \{x_0\}$. By Lemma 5.5, $\mathrm{Fix}(f^{q(k_0-1)}) \cap D'_{x_0} \neq \emptyset$, so $\mathrm{P}(f) \cap D'_{x_0} \neq \emptyset$. In particular $U \cap \mathrm{P}(f) \neq \emptyset$, which is a contradiction. This ends the proof of the claim.

From the claim, we have that $\xi(x) \leq r(\xi^k(x))$, for all $k \in \mathbb{N}$. Define $V = r^{-1}(\{t \in I : t < \xi(x)\})$. Then V is an open subset of X containing x (since $r(x) = x < \xi(x)$) and $\xi^k(x) \notin V$, for all $k \in \mathbb{N}$. As $x \in R(f) = R(\xi)$, this leads to a contradiction. ■

Proposition 5.4 fails even for X a graph when $r_I(z) \in (a, b)$ as can be shown by the following example.

Example 5.8: Let us start with the construction of the graph G_1 as a subset of the plane $\mathbb{R}^2$ in the following way: Let $\mathbb{S}^1$ be the unit circle and $I = [a, c]$ be an arc such that $\mathbb{S}^1 \cap I = \{a\}$. Let T be the tree defined by $T = I \cup J \cup K$, where $J = [c, d]$ and $K = [c, e]$ are two arcs such that $I \cap J = I \cap K = \{c\}$. Define the graph $G_1 = \mathbb{S}^1 \cup T$ (see Figure 1).

We define the continuous mapping $f : G_1 \to G_1$ as follows:

- $f_{|\mathbb{S}^1} = R_\alpha$, where R_α is the irrational rotation with angle α.
- $f(I) \subset \mathbb{S}^1$ with $f(c) = a$.
- f sends linearly J (resp. K) to I with $f(d) = f(e) = c$.
- $P(f) = \emptyset$.

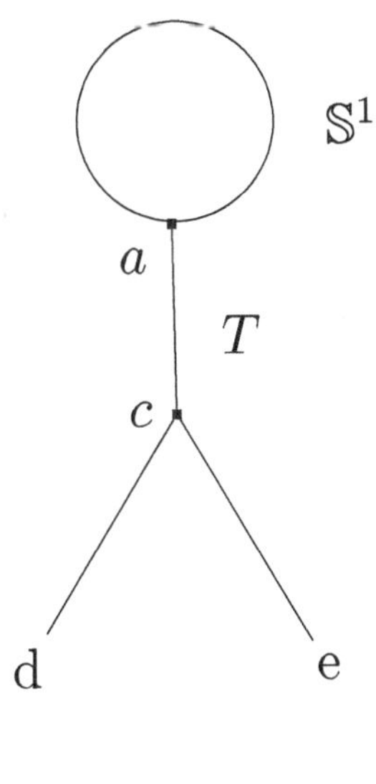

Figure 1. The graph G_1.

Observe that $U = G_1 \setminus \{e, d\}$ is connected which is the connected component U of $G_1 \setminus \{e, d\}$ that contains (e, d). It satisfies $U \cap \mathrm{R}(f) = \mathbb{S}^1$, but $U \cap \mathrm{P}(f) = \emptyset$.

Proposition 5.9: *Let $f : X \to X$ be a local dendrite map and let $g : \Gamma_X \to \Gamma_X$ be its associated retraction graph map. Let $x \in P(g) \setminus P(f)$ with period $n \in \mathbb{N}$ under g. Then there exists an open neighbourhood U of x in Γ_X such that $g^n(U) = \{x\}$.*

Proof: Since $x \in \mathrm{P}(g) \setminus \mathrm{P}(f)$, there exists $k \in \mathbb{N}$ such that $f^k(x) \notin \Gamma_X$. Denote by $p := \min\{i \in \mathbb{N}, f^i(x) \notin \Gamma_X\}$, so $f^{p-i}(x) = g^{p-i}(x)$, for any $i \in \{1, \ldots, p\}$. Let D be the attached dendrite in X containing $f^p(x)$ with attaching point z, so $\{z\} = D \cap \Gamma_X$ and $z = r(f^p(x))$. Let W be an open neighbourhood of $f^p(x)$ in X. By continuity of f, there exists an open neighbourhood V of $f^{p-1}(x)$ in Γ_X (with respect to induced topology) such that $f(V) \subset W$. As $V \subset \Gamma_X$, so by continuity of g^{p-1}, there exists an open neighbourhood U of x in Γ_X such that $g^{p-1}(U) \subset V$. Then $f(g^{p-1}(U)) \subset f(V) \subset W$. It follows that

$$r \circ f(g^{p-1}(U)) \subset r \circ f(V) \subset r(W) = \{z\},$$

and thus $g^p(U) = \{z\}$. Then $g^p(U) = \{g^p(x)\}$. Moreover, we have $p \leq n$; indeed, otherwise, $f^n(x) \in \Gamma_X$ and so $x \in P(f)$, which is a contradiction. Now as $g^n(x) = x$, so $g^n(U) = \{x\}$. ■

Corollary 5.10: *Let $f : X \to X$ be a local dendrite map and $g : \Gamma_X \to \Gamma_X$ be its associated retraction graph map. Assume that $\omega_g(x) \cap \left(\mathrm{P}(g) \setminus \mathrm{P}(f)\right) \neq \emptyset$, for some $x \in \Gamma_X$. Then $\omega_g(x)$ is finite.*

Proof: By hypothesis, there exists $y \in \omega_g(x) \cap \left(\mathrm{P}(g) \setminus \mathrm{P}(f)\right)$. Let n be the period of y under g. By Proposition 5.9, there exists an open neighbourhood U of y in Γ_X such that $g^n(U) = \{y\}$. As $y \in \omega_g(x)$, there exists $p \in \mathbb{N}$ such that $g^p(x) \in U$ and so $g^{n+p}(x) = y$. Therefore, $\omega_g(x)$ is finite. ■

Lemma 5.11 ([26, Theorem 3.5]): *Let X be a dendrite with End(X) countable and $f : X \to X$ be a continuous map. Then $\overline{\mathrm{R}(f)} = \overline{\mathrm{P}(f)}$.*

Now we are ready to prove our main result of this section.

Proof of Theorem 5.1.: Arguing by contradiction, assume that there is $z \in \mathrm{R}(f) \setminus \overline{\mathrm{EAP}(f)}$. Let C be the connected component of $X \setminus \overline{\mathrm{EAP}(f)}$ containing z. Since $z \in \mathrm{R}(f)$, there exists $n \in \mathbb{N}$ such that $f^n(C) \cap C \neq \emptyset$. Set $h = f^n$, $W = \bigcup_{k=0}^{\infty} h^k(C)$ and $g = h_{|\overline{W}}$. By Lemma 3.1, W is connected and g-invariant, moreover $W \cap \mathrm{EAP}(f) = \emptyset$. Observe that $\overline{W}$ is a sub-local dendrite of X and $g : \overline{W} \to \overline{W}|$ is a local dendrite map. By Lemma 3.6, the set End$(\overline{W})$ is countable. We distinguish two cases.

Case 1. $\Gamma_{\overline{W}} = \emptyset$. So $\overline{W}$ is a dendrite with countable set of endpoints. By Lemma 5.11, $\overline{\mathrm{R}(g)} = \overline{\mathrm{P}(g)}$. Hence $z \in \overline{\mathrm{P}(g)} \subset \overline{\mathrm{EAP}(f)}$. A contradiction.

Case 2. $\Gamma_{\overline{W}} \neq \emptyset$. In this case, we have the following:

Claim. $O_g(z) \subset \Gamma_{\overline{W}}$.

Proof: Suppose that there exists $p \in \mathbb{Z}_+$ such that $t := g^p(z) \in \overline{W} \setminus \Gamma_{\overline{W}}$. Let C' be the connected component of $\overline{W} \setminus \Gamma_{\overline{W}}$ containing t. By Lemma 5.2, $t \in \mathrm{R}(g) \setminus \overline{\mathrm{P}(g)}$. So there exists $\varepsilon > 0$ such that $B(t,\varepsilon) \subset C'$ and $B(t,\varepsilon) \cap \overline{\mathrm{P}(g)} = \emptyset$. Let D be a dendrite neighbourhood of t such that $D \subset B(t, \frac{\varepsilon}{2})$. Let U be an open neighbourhood of t such that $U \subset D$. By Lemma 5.3, the set $S := \{g^n(t) : n \in \{k \in \mathbb{N}, g^k(t) \in U\}\}$ has no isolated point in D. According to Lemma 5.6, there exists an arc $I = [a,b] \subset D$ such that $S \cap I$ have an accumulation point (since End(D) is countable (Lemma 3.6)). So there exists a sub-sequence $(n_k)_{k\in\mathbb{N}}$ of positive integers such that $g^{n_k}(t) \in (a,b) \cap U$, for all $k \in \mathbb{N}$. We let $y_k = g^{n_k}(t)$. By taking a sub-sequence, we can assume that $y_{k+1} \in (y_k, y_{k+2}) \subset I$, for all $k \in \mathbb{N}$. Let $(U_k)_{k\in\mathbb{N}}$ be the sequence of connected components of $\overline{W} \setminus \{y_{2k-1}, y_{2k+1}\}$ containing the open arc (y_{2k-1}, y_{2k+1}). It is easy to see that the U_k, $(k \in \mathbb{N})$ are mutually disjoint. By Lemma 3.5, there exists $N \in \mathbb{N}$ such that $\mathrm{diam}(U_i) < \frac{\varepsilon}{2}$, for all $i \geq N$. Set $p \geq N$ such that $\pi(z) \in \{y_{2p-1}, y_{2p+1}\}$, where $\pi : \overline{W} \to [y_{2p-1}, y_{2p+1}]$ is the retraction map of the arc $[y_{2p-1}, y_{2p+1}]$ and z is the attached point associated to the connected component C. Since $\mathrm{diam}(U_p) < \frac{\varepsilon}{2}$, so $U_p \subset B(t,\varepsilon)$: indeed, if $x \in U_p$, then

$$d(x,t) \leq d(x, y_{2p}) + d(y_{2p}, t) \leq \frac{\varepsilon}{2} + \frac{\varepsilon}{2} = \varepsilon$$

(since $y_{2p} \in U_p \cap D \subset U_p \cap B(t, \frac{\varepsilon}{2})$) and so $x \in B(t,\varepsilon)$. According to Proposition 5.4, $U_p \cap \mathrm{P}(g) \neq \emptyset$ and hence $B(t,\varepsilon) \cap \mathrm{P}(g) \neq \emptyset$, which is a contradiction. This finishes the proof of the claim. ■

Now we complete the proof of Theorem 5.1 by distinguishing two sub-cases:

Case 2.1. $\omega_g(z) \cap \mathrm{P}(g) = \emptyset$. In this case $\omega_g(z)$ contains an infinite minimal set M. So $M \subset \Gamma_{\overline{W}} \setminus W$. This contradicts Proposition 3.9.

Case 2.2. $\omega_g(z) \cap \mathrm{P}(g) \neq \emptyset$. Let $\gamma = r \circ g_{|\Gamma_{\overline{W}}} : \Gamma_{\overline{W}} \to \Gamma_{\overline{W}}$ be the canonical graph map associated to g. By the claim above, $O_g(z) \subset \Gamma_{\overline{W}}$ and therefore $\omega_g(z) = \omega_\gamma(z)$. So $\omega_\gamma(z) \cap \mathrm{P}(\gamma) \neq \emptyset$ (since $\omega_g(z) \subset \Gamma_{\overline{W}}$). Set E be the maximal ω-limit that contains $\omega_\gamma(z)$. Thus E is a basic set for γ. Since $z \in \mathrm{R}(\gamma) \setminus \mathrm{P}(\gamma)$, $\omega_\gamma(z)$ is infinite. By Corollary 5.10, we have $E \cap (\mathrm{P}(\gamma) \setminus \mathrm{P}(g)) = \emptyset$. In particular, $\mathrm{P}(\gamma_{|E}) \subset \mathrm{P}(g) = \emptyset$. Since $E = \overline{\mathrm{P}(\gamma_{|E})}$ Lemma 4.9, $E \subset \overline{\mathrm{P}(g)}$. Therefore, $z \in \omega_\gamma(z) \subset E \subset \overline{\mathrm{P}(g)}$, which is a contradiction, since $z \in \mathrm{R}(g) \setminus \overline{\mathrm{EAP}(g)} \subset \mathrm{R}(g) \setminus \overline{\mathrm{P}(g)}$. This ends the proof of Theorem 5.1. ■

Theorem 5.1 is not true in general for maps on local dendrite with uncountable set of endpoints (see Example 7.1).

Corollary 5.12: *Let X be a local dendrite with End(X) countable and $f : X \to X$ be a continuous map. If* EAP(f) *is closed, then* ER(f) = EAP(f).

6. Recurrence, almost periodicity and eventual periodicity

Hattab [17, Lemma 3.1] proved that if $f : G \to G$ is a graph map, then $\overline{\mathrm{R}(f)} = \overline{\mathrm{AP}(f)}$. This also holds for a dendrite map $f : X \to X$ if and only if End(X) countable (see [2]). We extend this property to local dendrite with finitely many branch points.

Theorem 6.1: *Let X be a local dendrite with* B(X) *finite and $f : X \to X$ be a continuous map. Then* $\overline{\mathrm{R}(f)} = \overline{\mathrm{AP}(f)}$.

The following lemma is a direct consequence of [33, lemma 3.4].

Lemma 6.2: *Let $f : X \to X$ be a local dendrite map and $I := [a, b]$ be a free arc such that $I \cap \mathrm{P}(f) = \emptyset$. If $\mathrm{int}_X(I) \cap \mathrm{R}(f) \neq \emptyset$, then there exists $n \in \mathbb{N}$ such that $f^n(a) \in (a, b]$.*

Proof of Theorem 6.1.: It suffices to show that $\mathrm{R}(f) \subset \overline{\mathrm{AP}(f)}$. Let $x \in \mathrm{R}(f) \setminus \overline{\mathrm{AP}(f)}$ and let V be an open neighbourhood of x such that $V \cap \mathrm{B}(X) \subset \{x\}$ and $V \cap \mathrm{AP}(f) = \emptyset$. Since $x \in \mathrm{R}(f)$, there exists $n \in \mathbb{N}$ such that $f^n(x) \in V \setminus \{x\}$. Let consider C the connected component of $V \setminus \{x\}$ containing $f^n(x)$, thus C is an open arc of X. According to Theorem 5.1, there exists $t \in \mathrm{EAP}(f) \cap C \setminus \{f^n(x)\}$ (since $f^n(x) \in \mathrm{R}(f)$). Let $p \in \mathbb{N}$ such that $f^p(t) \in \mathrm{AP}(f)$. Since $C \cap \mathrm{AP}(f) = \emptyset$, so $O_f(f^p(t)) \cap C = \emptyset$. As $f^n(x) \in \mathrm{R}(f^p)$ and $C \cap \mathrm{P}(f^p) = \emptyset$, then by Lemma 6.2, there exists $k \in \mathbb{N}$ such that $f^{kp}(t) \in C$. A contradiction. ■

Remark 6.3: Theorem 6.1 fails whenever $B(X)$ is infinite (see Example 7.1).

Theorem 6.4: *Let $f : X \longrightarrow X$ be a dendrite map. If $\mathrm{B}(X)$ is closed, then $\mathrm{R}(f) \subset \overline{\mathrm{EP}(f)}$.*

Proof: Suppose that there is $z \in \mathrm{R}(f) \setminus \overline{\mathrm{EP}(f)}$. Let C be the connected component of $X \setminus \overline{\mathrm{EP}(f)}$ containing z. Since $z \in \mathrm{R}(f)$, there exists $n \in \mathbb{N}$ such that $f^n(C) \cap C \neq \emptyset$. Set $h = f^n$, $W = \bigcup_{k=0}^{\infty} h^k(C)$ and $g = h_{|\overline{W}}$. By Lemma 3.1, W is connected and g-invariant, moreover $W \cap \mathrm{EP}(f) = \emptyset$. So $\overline{W}$ is a sub-dendrite of X and $(\overline{W}, g)$ is a dendrite map. By Proposition 3.11, $\omega_g(z) \subset \mathrm{B}(X) \cap \overline{W}$ (since $\Gamma_{\overline{W}} = \emptyset$). As $z \in \mathrm{R}(g) \setminus \overline{\mathrm{EP}(g)}$, $\omega_g(z)$ is infinite. As the set $\mathrm{B}(X)$ is at most countable, then so does $\omega_g(z)$. It follows that $\omega_g(z)$ has an isolated point a and hence $a \notin P(g)$ (by Lemma 3.3). Since $z \in \omega_g(z)$, there is an infinitely many $k \in \mathbb{N}$ such that $g^k(z) = a$, so $a \in \mathrm{P}(g)$, a contradiction. The proof is complete. ■

Remark 6.5: (1) Theorem 6.4 fails whenever $\mathrm{B}(X)$ is not closed. Indeed, consider the Gehman dendrite map $f : X \to X$ constructed in [18, Example 1.6]. In that example, $\mathrm{B}(X)$ is not closed, $\mathrm{R}(f) = \mathrm{End}(X) \cup \{p\}$, but $\mathrm{EP}(f) = \{p\}$.

(2) Theorem 6.4 fails whenever X is a local dendrite with $\mathrm{B}(X)$ closed, as shown by the following example of graph maps.

Example 6.6: Let G_1 be the graph given in Example 5.8. There exists a continuous mapping $g : G_1 \to G_1$ with $\mathrm{P}(g) \neq \emptyset$ such that $\mathrm{R}(g) \not\subset \overline{\mathrm{EP}(g)}$.

Proof: Let $b \in (a, c)$ and set $I = [a, b]$, $J = [b, c]$, $K = [c, d]$ and $L = [c, e]$. We define the continuous mapping $g : G_2 \to G_2$ as follows:

- $g_{|\mathbb{S}^1} = R_\alpha$, where R_α is the irrational rotation with angle α.
- $g(I) \subset \mathbb{S}^1$.
- g sends linearly $J = [b, c]$ to $[a, c]$ with $g(c) = c$.
- g sends linearly $K \cup L$ to J.
- $\mathrm{P}(g) \cap T = \{c\}$.

We see that $\mathrm{R}(g) = \mathbb{S}^1 \cup \{c\}$ and $\mathrm{EP}(g) = \{c\}$. Therefore, $\mathrm{R}(g) \not\subset \overline{\mathrm{EP}(g)}$. ■

7. A counter-example

In this section, we give an example showing that Theorems 4.1, 5.1, 6.1 and 6.4 are not true in general when we remove in each one, one of its hypothesis.

Example 7.1: There exist a local dendrite X with End(X) uncountable and B(X) infinite and non closed, and a continuous map $f : X \longrightarrow X$ with $\mathrm{P}(f) \neq \emptyset$ satisfying $\mathrm{R}(f) \not\subseteq \overline{\mathrm{EAP}(f)}$.

Example 7.1 shows that Theorem 5.1 fails whenever End(X) is uncountable.

Proof: We consider the local dendrite X in Example [4, Example 5.1] called the Gehman man local dendrite X. For the sake and completeness, we recall its construction.

- Let recall the construction of the Gehman dendrite G (see e.g. [23]). Let $B_0 = [p, p_0]$, $B_1 = [p, p_1]$ be two arcs in $\mathbb{R}^2$ of length 1 intersecting at the point p, and for every $n \in \mathbb{N}$, $B_{i_1 i_2 \ldots i_n} = [p_{i_1 i_2 \ldots i_{n-1}}, p_{i_1 i_2 \ldots i_n}]$ is an arc in $\mathbb{R}^2$ of length $\frac{1}{n}$, where every i_k is either 0 or 1 (see Figure 3). The Gehman dendrite is $G = \overline{\bigcup_{n \in \mathbb{N}} X_n}$, where $X_n := \bigcup_{1 \leq k \leq n} \bigcup_{i_1, i_2, \ldots i_k \in \{0,1\}} B_{i_1 i_2, \ldots i_k}$ is a tree (i.e. a dendrite with finitely many endpoints). The endpoint set End(G) is homeomorphic to a Cantor set. It can be written as $\mathrm{End}(G) = \{p_\sigma : \sigma(k) \in \{0, 1\}\}$, where $p_\sigma = \lim_{n \to +\infty} p_{i_1, i_2, \ldots, i_n}$ whenever $\sigma = (i_k)_{k \geq 1}$. It is plain that branching points in G can be arranged in such a way that each of the following intervals contains branching points (or point p) exactly at endpoints. Let $a \in \mathbb{S}^1$ and $[a, p]$ be an arc such that $G \cap ([a, p] \cup \mathbb{S}^1) = \{p\}$ for some p of order 2 in G. Let $D = [a, p] \cup G$. We define X as follows: $X = \mathbb{S}^1 \cup D$ (see Figure 2). Obviously X is a local dendrite with B(X) infinite and End(X) = End(G) is uncountable.
- Now to construct the map f on X, we need the following proposition.

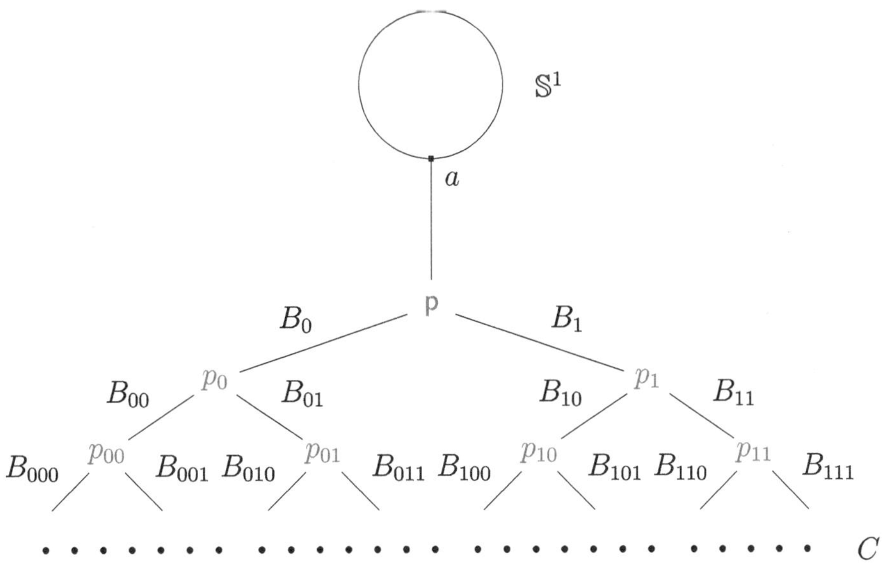

Figure 2. The Gehman man local dendrite [4].

Proposition 7.2 ([2, Proposition 3.3]): *There exists a homeomorphism $\varphi : C \longrightarrow C$ of the Cantor set C to itself satisfying the following properties:*

(a) *There is a unique fixed point ω, and the full orbit of any other point is dense in C.*
(b) $\mathrm{AP}(\varphi) = \{\omega\}$ *and* $\overline{\mathrm{R}(\varphi)} = C$.

We can assume that $\mathrm{End}(X) = C$ (up to homeomorphism).

Assume that the unit circle $\mathbb{S}^1$ oriented counterclockwise. Let $b, c \in \mathbb{S}^1 \backslash \{a\}$ with $b \neq c$. Consider the map $g : \mathbb{S}^1 \longrightarrow \mathbb{S}^1$ defined by the following way:

$$\begin{cases} g([b,c]_{|\mathbb{S}^1}) = [b,c]_{|\mathbb{S}^1},\ g_{|[b,c]} \text{ is a homeomorphism;} \\ g([c,b]_{|\mathbb{S}^1}) = [c,b]_{|\mathbb{S}^1},\ g_{|[c,b]} \text{ is a homeomorphism;} \\ \mathrm{P}(g) = \mathrm{Fix}(g) = \{b,c\}. \end{cases}$$

We see that g is a homeomorphism circle map. From [3, Lemma 2.4], we have $\mathrm{R}(g) = \mathrm{P}(g) = \{b, c\}$. Now we define the map f as follows:

$$\begin{cases} f(p) = a; \\ f_{|\mathbb{S}^1} = g; \\ f([p,a]) = [a, g(a)]_{|\mathbb{S}^1},\ f_{|[p,a]} \text{ is a homeomorphism;} \\ f([p_0,p]) = f([p_1,p]) = [p,a],\ f_{|[p_0,p]} \text{ is a homeomorphism;} \\ f(p_{i_1 i_2 \ldots i_n}) = p_{i_2 i_3 \ldots i_n}; \\ f_{|\mathrm{End}(X)} = \varphi. \end{cases}$$

It is easy to see that f is continuous and that satisfies:

(1) $\overline{\mathrm{P}(f)} = \mathrm{AP}(f) = \{b, c, \omega\}$.
(2) $\overline{\mathrm{R}(f)} = \{b, c\} \cup \mathrm{End}(X)$.
(3) For any $x \in X \backslash (\mathbb{S}^1 \cup \mathrm{End}(X))$, there exists $N_x \in \mathbb{N}$ such that for any $n \geq N_x, f^n(x) \in \mathbb{S}^1 \backslash \{b, c\}$.

We shall show that $\mathrm{EAP}(f) = \mathrm{AP}(f) = \{b, c, \omega\}$. First we have that $\mathrm{EAP}(f) \cap X \setminus (\mathbb{S}^1 \cup \mathrm{End}(X)) = \emptyset$. Indeed, suppose that there is $x \in \mathrm{EAP}(f) \cap X \setminus (\mathbb{S}^1 \cup \mathrm{End}(X))$. By (3), there exists $N_x \in \mathbb{N}$ such that for any $n \geq N_x$, $f^n(x) \in \mathbb{S}^1 \backslash \{b, c\}$. Also there exists $p \in \mathbb{N}$ such that $f^p(x) \in \mathrm{AP}(f)$. By (1), $f^p(x) \in \{b, c, \omega\}$. Set $n = \max\{N_x, p\}$. Then $f^n(x) \in \mathrm{AP}(f) \cap \mathbb{S}^1 \setminus \{b, c\} = \{\omega\}$, this contradicts the construction of f. Subsequently, $\mathrm{EAP}(f) \subset \mathbb{S}^1 \cup \mathrm{End}(X)$. Therefore,

$$\mathrm{EAP}(f) = \mathrm{EAP}(f_{|\mathbb{S}^1}) \cup \mathrm{EAP}(f_{|\mathrm{End}(X)}) = \mathrm{EAP}(g) \cup \mathrm{EAP}(\varphi) = \{b, c, \omega\}.$$

We conclude that $\mathrm{EAP}(f) = \mathrm{AP}(f) = \{b, c, \omega\}$. It is obvious that $\mathrm{R}(f) \not\subseteq \overline{\mathrm{EAP}(f)}$. ■

As a consequence:

Corollary 7.3: *The local dendrite map of Example 7.1 satisfies:*

(i) $\Omega(f) \not\subseteq \overline{\mathrm{EAP}(f)}$.

(ii) $R(f) \not\subseteq \overline{EP(f)}$.
(iii) $\overline{R(f)} \neq \overline{AP(f)}$.

From Corollary 7.3, we have the following remarks.

- Assertion (i) shows that Theorem 4.1 fails whenever $P(f) \neq \emptyset$.
- Assertion (ii) shows that Theorem 6.4 fails whenever $B(X)$ is not closed.
- Assertion (iii) shows that Theorem 6.1 fails whenever $B(X)$ is infinite.

Acknowledgments

This work was supported by the research unit: "Dynamical systems and their applications", (UR17ES21), Ministry of Higher Education and Scientific Research, Faculty of Science of Bizerte, Bizerte, Tunisia.

Disclosure statement

No potential conflict of interest was reported by the author(s).

Funding

This work was supported by the research unit: "Dynamical systems and their applications", (UR17ES21), Ministry of Higher Education and Scientific Research, Faculty of Science of Bizerte, Bizerte, Tunisia.

References

[1] H. Abdelli, H. Abouda, and H. Marzougui, *Nonwandering points of monotone local dendrite maps revisited*, Topol. Appl. 250 (2018), pp. 61–73.
[2] H. Abdelli and H. Marzougui, *Recurrence and periodicity on dendrites*, Chaos Solitons Fractals 87 (2016), pp. 17–18.
[3] H. Abdelli and H. Marzougui, *Invariant sets for monotone local dendrite maps*, Internat. J. Bifur. Chaos Appl. Sci. Engrg. 26 (2016), pp. 1650150.
[4] H. Abdelli, I. Naghmouchi, and H.E. Rezgui, *Local dendrite maps without periodic points*, Topol. Appl. 305 (2022), pp. 107901.
[5] R.L. Adler, A.G. Konheim, and M.H. McAndrew, *Topological entropy*, Trans. Amer. Math. Soc. 114 (1965), pp. 309–319.
[6] T. Arai, N. Chinen, and T. Suda, *Continuous maps of dendrites with finitely many branch points and nonwandering sets*, Topology Proc. 26 (2002), pp. 29–36.
[7] D. Arevalo, W.J. Charatonik, P.P. Covarrubias, and L. Simon, *Dendrites with a closed set of endpoints*, Topology Appl. 115 (2001), pp. 1–17.
[8] G. Askri and I. Naghmouchi, *Topological size of scrambled sets for local dendrite maps*, Topology Appl. 164 (2014), pp. 95–104.
[9] L.S. Block and W.A. Coppel, Dynamics in One Dimension, Lecture Notes in Math, Vol. 1513, Springer-Verlag, Berlin, 1992.
[10] L. Block, E. Coven, I. Mulvey, and Z. Nitecki, *Homoclinic and non-wandering points for maps of the circle*, Ergodic Theory Dynam. Syst. 3 (1983), pp. 521–532.
[11] A.M. Blokh, *Dynamical systems on one-dimensional branched manifolds I (Russian)*, Teor. Funktsiĭ Funktsional. Anal. I Prilozhen. 46 (1986), pp. 8–18. English translation in J. Soviet Math. 48(5) (1990), pp. 500–508.

[12] A.M. Blokh, *Dynamical systems on one-dimensional branched manifolds II (Russian)*, Teor. Funktsiĭ Funktsional. Anal. I Prilozhen. No. 47 (1987), pp. 67–77. English translation in J. Soviet Math. 48(6) (1990), pp. 668–674.

[13] A.M. Blokh, *Dynamical systems on one-dimensional branched manifolds III (Russian)*, Teor. Funktsiĭ Funktsional. Anal. I Prilozhen. 48 (1987), pp. 32–46. English translation in J. Soviet Math. 49(2) (1990), pp. 875–883.

[14] A.M. Blokh, *On transitive mappings of one-dimensional branched manifolds, Differential-difference equations and problems of mathematical physics (Russian)*, Akad. Nauk Ukrain. SSR, Inst. Mat., Kiev. Vol. 131, 1984, pp. 3–9.

[15] L.S. Efremova and E.N. Makhrova, *The dynamics of monotone maps of dendrites*, Sb. Math. 192(6) (2001), pp. 807–821.

[16] L.S. Efremova and E.N. Makhrova, *On the center of continuous maps of dendrites*, J. Differ. Equ. Appl. 9 (2003), pp. 381–392.

[17] H. Hattab, *Relatively recurrent for graph map*, Proc. Amer. Math. Soc. 139 (2011), pp. 2087–2092.

[18] H. Hosaka and H. Kato, *Continuous maps of trees and nonwandering sets*, Topology Appl. 81 (1997), pp. 35–46.

[19] R. Hric and M. Malek, *Omega limit sets and distributional chaos on graphs*, Topology Appl. 153 (2006), pp. 2469–2475.

[20] A. Illanes, *A characterization of dendrites with the periodic-recurrent property*, Topology Proc. 23 (1998), pp. 221–235.

[21] H. Kato, *The depth of centres of maps on dendrites*, J. Austral. Math. Soc. Ser. A. 64(1) (1998), pp. 44–53.

[22] Z. Kočan, V. Kurková, and M. Málek, *On the centre and the set of ω-limit points of continuous maps on dendrites*, Topology Appl. 156(18) (2009), pp. 2923–2931.

[23] Z. Kočan, V. Kurková, and M. Málek, *Counter examples of continuous maps on dendrites*, J. Difference Equ. Appl. 22 (2016), pp. 253–271.

[24] K. Kuratowski, Topology, Vol. 2, Academic Press, New-York, 1968.

[25] J. Li, P. Oprocha, Y. Yang, and T. Zeng, *On dynamics of graph maps with zero topological entropy*, Nonlinearity 30 (2017), pp. 4260–4276.

[26] J.H. Mai and E. Shi, $\overline{R} = \overline{P}$ *for maps of dendrites X with* $\mathrm{Card}(\mathrm{End}(X)) < c$, Internat. J. Bifur. Chaos Appl. Sci. Engrg. 19 (2009), pp. 1391–1396.

[27] J. Mai, J.G. Zhang, and T. Sun, *Recurrent points and non-wandering points of graph maps*, J. Math. Anal. Appl. 383 (2011), pp. 553–559.

[28] E.N. Makhrova and K.S. Vaniukova, *On the set of non-wandering points of monotone maps on local dendrites*, NOMA'15 International Workshop on Nonlinear Maps and Applications, J. Phys. Conf. Ser. Vol. 692, 2016, p. 012012.

[29] S.B. Nadler, Continuum Theory: An Introduction, Monographs and Textbooks in Pure and Applied Mathematics, Vol. 158, Marcel Dekker Inc., New York, 1992.

[30] I. Naghmouchi, *Dynamical properties of monotone dendrite maps*, Topology Appl. 159(1) (2012), pp. 144–149.

[31] G. Rongbao, S. Taixiang, and Z. Tingting, *Non-wandering set of a continuous graph map*, Appl. Math. J. Chinese Univ. Ser. B. 18 (2003), pp. 477–481.

[32] S. Ruette and L. Snoha, *For graph maps, one scrambled pair implies li-Yorke chaos*, Proc. Amer. Math. Soc. 142 (2014), pp. 2087–2100.

[33] G. Su, T. Sun, F. Zeng, and B. Qin, *The depths and the attracting centres for continuous maps on local dendrites with the number of branch points being finite*, Topology Appl. 272 (2020), pp. 107067.

[34] T. Sun and T. Xi, *The centre and the depth of the centre for continuous maps on dendrites with finite branch points*, Qual. Theory Dyn. Syst. 16(3) (2017), pp. 697–702.

[35] X. Ye, *Non-wandering points and the depth of a graph map*, J. Austral. Math. Soc. Ser. A. 69(2) (2000), pp. 143–152.

Diffeomorphisms with infinitely many Smale horseshoes

Xu Zhang and Guanrong Chen

ABSTRACT
A class of planar diffeomorphims is formulated, with infinitely many coexisting Smale horseshoes, where the Lebesgue measure of the parameters with such strange dynamics is infinite. On each horseshoe, there exists a uniformly hyperbolic invariant set, on which the map is topologically conjugate to the two-sided full-shift on two symbols. Moreover, the topological entropy is infinite in certain parameter regions.

1. Introduction

The existence of an attractor, especially a chaotic attractor, is an important topic to study in the theory of differentiable dynamical systems.

Let M be a closed manifold, which is compact without boundaries, and $\mathrm{Diff}^r(M)$ be the set of rth differentiable diffeomorphisms with $r \in \mathbb{N} \cup \{\infty\}$. For any positive integer $r > 1$, it was shown by Newhouse [30] that there exists an open set $\mathcal{U} \subset \mathrm{Diff}^r(M)$ with $\dim(M) = 2$, where the set of diffeomorphisms exhibiting a homoclinic tangency is dense. This result shows that Axiom-A (or hyperbolic) diffeomorphisms cannot be dense in the space of surface diffeomorphisms. Further, for dissipative diffeomorphisms, it was proved in [31] the existence of a residual (Baire's second cathegory) subset $\mathcal{R}$ of $\mathcal{U}$ with $\dim(M) \geq 2$, such that each diffeomorphism in $\mathcal{R}$ has infinitely many sinks. Later, it was verified in [32] that such open sets actually can be found arbitrarily near a diffeomorphism on M with $\dim(M) \geq 2$, and each diffeomorphism in the open sets has a homoclinic tangency. The existence of C^2-diffeomorphism with infinitely many Axiom-A strange attractors on a two-dimensional manifold was proved by Gambaudo and Tresser [16]. The existence of a dense subset of $\mathrm{Diff}^\infty(M)$-diffeomorphisms with the coexistence of infinitely many Hénon-like strange attractors (or repellers) on a two-dimensional manifold was proved by Colli [12]. Later, these results were generalized to higher-dimensional manifolds [27]. Some numerical experiments also reveal the existence of infinitely many strange attractors, e.g. two-dimensional maps with trigonometric functions [26, 37], memristor-type chaotic systems [29, 38], and some others [28].

Differing from the coexistence of infinitely many attractors or repellers observed by Newhouse [30–32], a global description of dynamical systems is the Palis conjecture (finitude of attractors) [10, 33]: There exists a C^r, $r \geq 1$, dense subset of dynamical systems on any compact manifold such that each diffeomorphism in this subset has a finite number of attractors, whose basins cover almost all of the points in the manifold in the Lebesgue measure. This conjecture has been confirmed in the context of unimodal maps on intervals. At present, this conjecture seems to be true for surface diffeomorphisms. The main difficulty for understanding this kind of systems is due to the Newhouse observations mentioned above. Another important related open question is whether Palis' conjecture may hold for maps with parameters within a positive Lebesgue-measure region in the parameter space, or this conjecture holds for generic parametrized families of diffeomorphisms [9, 21, 34].

One of the most well-known planar diffeomorphism is the Hénon map [22], which has a simple expression but very complicated dynamics. Friedland and Milnor [15] showed that any planar polynomial map with polynomial inverse from the real or complex plane to itself is either conjugate to a composition of generalized Hénon maps, or is dynamically trivial. The study of the existence of Smale horseshoes and hyperbolic invariant sets for Hénon map is also an important and interesting problem. Devaney and Nitecki [13] investigated the quadratic Hénon map $H(x, y) = (1 - ax^2 + y, bx)$, and proved that for a fixed b and a sufficiently large a, there is a horseshoe and a hyperbolic invariant set on which it is topologically conjugate to the two-sided full-shift on two symbols. Dullin and Meiss [14] obtained similar results for the generalized cubic Hénon map. By using the techniques of complex dynamics, Bedford and Smillie [4, 5] verified that the Hénon map $L(x, y) = (a - x^2 - by, x)$ with $|b| < 0.08$, for any fixed b and sufficiently large a, has a hyperbolic horseshoe, where the formulae for L and H are topologically conjugate. Cao et al. [11] applied the methods of real dynamical systems to investigate the Hénon-like families of diffeomorphisms on $\mathbb{R}^2$, which are small C^2-perturbations of the Hénon map in the form of H, and proved that for sufficiently large a, the nonwandering set is uniformly hyperbolic. Hoensch [24] considered an abstract class of diffeomorphisms, which could be applied to study the orientation-preserving Hénon maps with small Jacobian determinants, and proved the existence of hyperbolic invariant sets for the maps. The existence of Smale horseshoes and uniformly hyperbolic invariant sets were proved by Zhang [36] for generalized Hénon maps. Some three-dimensional Hénon maps were studied by Gonchenko et al. [17–20]. The works of Benedicks, Carleson, and others [6–8] have shown the existence of chaotic attractors, in the sense of having positive Lyapunov exponents and invariant measures, in some parameter regions of positive measures. Some complex maps such as the Hénon map were also studied by many, including Bedford, Smillie, etc. [1–3] (see also references therein).

Inspired by the aforementioned studies and results, the dynamical behaviour of the following planar diffeomorphism is considered in this article:

$$F : (x, y) \in \mathbb{R}^2 \longrightarrow (y, p(y) - \delta x) \in \mathbb{R}^2, \tag{1}$$

where

$$p(y) = \rho g(y) = \rho y^m \sin(y), \tag{2}$$

in which ρ and δ are nonzero real parameters, and m is a positive integer. Motivated in particular by the works of Devaney and Nitecki [13], as well as many others, on the Hénon maps, the parameter regions for the map (1) are analyzed.

It will be shown that the map F has infinitely many Smale horseshoes and infinitely many uniformly hyperbolic invariant sets, and moreover on each invariant set, F is topologically conjugate to the two-sided full-shift on two symbols.

Theorem 1.1: *For the map (1) with $m \geq 2$, consider the following parameter regions:*

(i) $\rho > 0$ *and* $\delta > 0$;
(ii) $\rho > 0$ *and* $-1 < \delta < 0$;
(iii) $\rho < 0$ *and* $\delta < -1$.

Then, given any (ρ, δ) from the domains above, there exist infinitely many coexisting disjoint Smale horseshoes, and on each horseshoe, there is a uniformly hyperbolic invariant set on which the map is topologically conjugate to the two-sided full-shift on two symbols. Furthermore, for Case (i), the topological entropy is infinite.

Remark 1.1: The map (1) has two parameters $(\rho, \delta) \in \mathbb{R}^2$. So, the parameter space is $\mathbb{R}^2$. The parameter region for Case (i) is $(0, \infty) \times (0, \infty)$; the parameter region for Case (ii) is $(0, \infty) \times (-1, 0)$; and the parameter region for Case (iii) is $(-\infty, 0) \times (-\infty, -1)$. Hence, the measures of the parameter regions in $\mathbb{R}^2$ for these cases are infinite.

The diffeomorphims in this article have strange dynamics:

- This type of maps is different from the generalized Hénon map $F(x, y) = (y, p(y) - \delta x)$, where $p(y)$ is a polynomial, and the generalized Hénon map has only a finite number of strange attractors or Smale horseshoes.
- The existence of maps with infinite topological entropy means that this kind of diffeomorphisms is different from a class of diffeomorphisms constructed by using the topological method, where one can glue some standard Smale horseshoes (two-fold horseshoe) together, such that the coexistence of infinitely many horseshoes can be found on the plane, since the topological entropy might be only $\log 2$ [35]. Moreover, for a C^1-differentiable map defined on a compact Riemannian manifold, it is known that there is a finite upper bound for the topological entropy; namely, the topological entropy is finite [25].
- The Lebesgue measure of the parameter regions with strange dynamics is infinite. In [12, 31], the diffeomorphisms with infinitely many sinks for Hénon-like attractors is verified to exist in dense subsets, where the measure of dense set in the parameter spaces is unknown (the measure of the dense set might be zero or positive, for example, the measure of the rational number and the irrational number on the real line).

Therefore, this type of maps are valuable models for understanding the global dynamics of planar diffeomorphisms.

The rest of the paper is organized as follows. Section 2 presents some basic concepts about hyperbolicity and symbolic dynamics. In Section 3, the proof of Theorem 1.1 is provided, where the coexistence of infinitely many Smale horseshoes and hyperbolic invariant sets is proved, showing also that on each horseshoe there is an invariant set on which the general map is topologically conjugate to the two-sided full-shift on two symbols.

2. Preliminaries

In this section, the definition of uniformly hyperbolic invariant set and some basic concepts about symbolic dynamics (see, e.g. [35]) are introduced.

Definition 2.1: An invariant set Λ has a hyperbolic structure for a diffeomorphism f on a manifold M if: (i) at each point p in Λ the tangent space to M splits into a direct sum of $\mathbf{E}_p^u$ and $\mathbf{E}_p^s$, $T_p = \mathbf{E}_p^u \oplus \mathbf{E}_p^s$, (ii) the splitting is invariant under the derivative map in the sense that $Df_p(\mathbf{E}_p^u) = \mathbf{E}_{f(p)}^u$ and $Df_p(\mathbf{E}_p^s) = \mathbf{E}_{f(p)}^s$, (iii) $\mathbf{E}_p^u$ and $\mathbf{E}_p^s$ vary continuously with p, and (iv) there exist $0 < \mu < 1$ and $C \geq 1$ such that, for all $n \geq 0$,

$$|Df_p^n \mathbf{v}^s| \leq C\mu^n |\mathbf{v}^s|, \quad \forall\, \mathbf{v}^s \in \mathbf{E}_p^s,$$
$$|Df_p^{-n} \mathbf{v}^u| \leq C\mu^n |\mathbf{v}^u|, \quad \forall\, \mathbf{v}^u \in \mathbf{E}_p^u.$$

If the invariant set Λ has a hyperbolic structure for f, it is called a uniformly hyperbolic invariant set.

Let $S_0 := \{1, 2, \ldots, m\}$, $m \geq 2$, and let

$$\sum\nolimits_m := \{\alpha = (\cdots, a_{-2}, a_{-1}, a_0, a_1, a_2, \ldots) : a_i \in S_0,\ i \in \mathbb{Z}\}$$

be the two-sided sequence space. The distance on $\sum_m$ is defined by $d(\alpha, \beta) = \sum_{i=-\infty}^{\infty} \frac{d(a_i, b_i)}{2^{|i|}}$, for $\alpha = (\ldots, a_{-2}, a_{-1}, a_0, a_1, a_2, \ldots)$, $\beta = (\ldots, b_{-2}, b_{-1}, b_0, b_1, b_2, \ldots) \in \sum_m$, where $d(a_i, b_i) = 1$ if $a_i \neq b_i$, and $d(a_i, b_i) = 0$ if $a_i = b_i$, $i \in \mathbb{Z}$. Then, $(\sum_m, d)$ is a complete compact metric space and also a Cantor set.

Moreover, $(\sum_m, \sigma)$ is said to be the two-sided symbolic dynamical system on m symbols, or simply a two-sided full-shift on m symbols, where the shift map $\sigma : \sum_m \to \sum_m$ is defined by $\sigma(\alpha) = (\ldots, b_{-2}, b_{-1}, b_0, b_1, b_2, \ldots)$ for any $\alpha = (\ldots, a_{-2}, a_{-1}, a_0, a_1, a_2, \ldots) \in \sum_m$, where $b_i = a_{i+1}$ for any $i \in \mathbb{Z}$.

3. Existence of Smale horseshoes

In this section, the statements of Theorem 1.1 are verified. The whole arguments are divided into three parts. First, the invariant sets are introduced, and the topological semi-conjugacy between the maps on the invariant sets and the symbolic dynamics is obtained (see Lemmas 3.1–3.4). Second, the uniform hyperbolicity of the invariant sets are shown and the topological conjugacy is established (see Lemmas 3.5–3.10). Finally, the proof of Theorem 1.1 is given and the statement of the infinite entropy is verified (see Lemma 3.11).

For the map (1), it is clear that

$$F^{-1}(x,y) = (\delta^{-1}(p(x) - y), x),$$

and

$$DF = \begin{pmatrix} 0 & 1 \\ -\delta & p'(y) \end{pmatrix}, \quad DF^{-1} = \begin{pmatrix} \delta^{-1}p'(x) & -\delta^{-1} \\ 1 & 0 \end{pmatrix}. \tag{3}$$

For any $i_0 \in \mathbb{N}$, set

$$\alpha_{i_0} := 2i_0\pi \quad \text{and} \quad \alpha_{i_0+1} := (2i_0+1)\pi.$$

Denote

$$S_{i_0} := [\alpha_{i_0}, \alpha_{i_0+1}] \times [\alpha_{i_0}, \alpha_{i_0+1}]$$

and

$$\Lambda_{i_0} := \bigcap_{j=-\infty}^{+\infty} F^j(S_{i_0}).$$

Then, it follows directly from the definition that $\Lambda_{i_0} \subset S_{i_0}$ and Λ_{i_0} is an invariant set of F.

Set

$$a := (\alpha_{i_0}, \alpha_{i_0}), \quad b := (\alpha_{i_0+1}, \alpha_{i_0}), \quad c := (\alpha_{i_0+1}, \alpha_{i_0+1}), \quad d := (\alpha_{i_0}, \alpha_{i_0+1}).$$

By direct calculation, one has

$$F(a) = (\alpha_{i_0}, -\delta\alpha_{i_0}), \quad F(b) = (\alpha_{i_0}, -\delta\alpha_{i_0+1}),$$
$$F(c) = (\alpha_{i_0+1}, -\delta\alpha_{i_0+1}), \quad F(d) = (\alpha_{i_0+1}, -\delta\alpha_{i_0});$$

and

$$F^{-1}(a) = (-\delta^{-1}\alpha_{i_0}, \alpha_{i_0}), \quad F^{-1}(b) = (-\delta^{-1}\alpha_{i_0}, \alpha_{i_0+1}),$$
$$F^{-1}(c) = (-\delta^{-1}\alpha_{i_0+1}, \alpha_{i_0+1}), \quad F^{-1}(d) = (-\delta^{-1}\alpha_{i_0+1}, \alpha_{i_0}).$$

In the following discussions, fix δ and a constant $\lambda > \max\{|\delta|^{1/2}, 1\}$.

Denote

$$M_0 := \max\{\lambda + |\delta|, \ 1 + \lambda|\delta|\},$$

and set

$$[a,b] := [\alpha_{i_0}, \alpha_{i_0+1}] \times \{\alpha_{i_0}\}, \quad [b,c] := \{\alpha_{i_0+1}\} \times [\alpha_{i_0}, \alpha_{i_0+1}],$$
$$[d,c] := [\alpha_{i_0}, \alpha_{i_0+1}] \times \{\alpha_{i_0+1}\}, \quad [a,d] := \{\alpha_{i_0}\} \times [\alpha_{i_0}, \alpha_{i_0+1}].$$

Consider three different situations:

(a) $\rho > 0$ and $\delta > 0$;
(b) $\rho > 0$ and $\frac{\alpha_{i_0}}{\alpha_{i_0+1}} > -\delta > 0$;
(c) $\rho < 0$ and $\delta < -\frac{\alpha_{i_0+1}}{\alpha_{i_0}}$.

These three different cases will be considered throughout this section.

Now, recall that $\text{arccot} : (-\infty, \infty) \to (0, \pi)$ is the inverse function of *cot*, and is a decreasing function.

Lemma 3.1: *For any fixed $i_0 \in \mathbb{N}$, there is a unique point $y^*_{i_0} \in (2i_0\pi + \frac{\pi}{2}, (2i_0 + 1)\pi)$ such that*

- $g'(y^*_{i_0}) = 0;$
- $g'(y) > 0$ *for any* $y \in (2i_0\pi, y^*_{i_0});$
- $g'(y) < 0$ *for any* $y \in (y^*_{i_0}, (2i_0 + 1)\pi).$

In Case (a), *there exist $y^L_{i_0} \in (\alpha_{i_0}, y^*_{i_0})$ and $y^R_{i_0} \in (y^*_{i_0}, \alpha_{i_0+1})$ such that*

$$\left|\frac{g'(y)}{g(y)}\right| > \frac{M_0}{(1+\delta)\alpha_{i_0}}, \quad y \in (\alpha_{i_0}, y^L_{i_0}] \cup [y^R_{i_0}, \alpha_{i_0+1}), \tag{4}$$

where

$$y^L_{i_0} = 2i_0\pi + \text{arccot}\left(\frac{M_0}{(1+\delta)\alpha_{i_0}} - \frac{m}{\alpha_{i_0+1}}\right) \tag{5}$$

and

$$y^R_{i_0} = 2i_0\pi + \text{arccot}\left(-\frac{M_0}{(1+\delta)\alpha_{i_0}} - \frac{m}{\alpha_{i_0+1}}\right). \tag{6}$$

In Case (b), *there exist $y^L_{i_0} \in (\alpha_{i_0}, y^*_{i_0})$ and $y^R_{i_0} \in (y^*_{i_0}, \alpha_{i_0+1})$ such that*

$$\left|\frac{g'(y)}{g(y)}\right| > \frac{M_0}{\alpha_{i_0} + \delta\alpha_{i_0+1}}, \quad y \in (\alpha_{i_0}, y^L_{i_0}] \cup [y^R_{i_0}, \alpha_{i_0+1}), \tag{7}$$

where

$$y^L_{i_0} = 2i_0\pi + \text{arccot}\left(\frac{M_0}{\alpha_{i_0} + \delta\alpha_{i_0+1}} - \frac{m}{\alpha_{i_0+1}}\right) \tag{8}$$

and

$$y^R_{i_0} = 2i_0\pi + \text{arccot}\left(-\frac{M_0}{\alpha_{i_0} + \delta\alpha_{i_0+1}} - \frac{m}{\alpha_{i_0+1}}\right). \tag{9}$$

In Case (c), *there exist $y^L_{i_0} \in (\alpha_{i_0}, y^*_{i_0})$ and $y^R_{i_0} \in (y^*_{i_0}, \alpha_{i_0+1})$ such that*

$$\left|\frac{g'(y)}{g(y)}\right| > \frac{M_0}{|\alpha_{i_0+1} + \delta\alpha_{i_0}|}, \quad y \in (\alpha_{i_0}, y^L_{i_0}] \cup [y^R_{i_0}, \alpha_{i_0+1}), \tag{10}$$

where

$$y^L_{i_0} = 2i_0\pi + \text{arccot}\left(\frac{M_0}{|\alpha_{i_0+1} + \delta\alpha_{i_0}|} - \frac{m}{\alpha_{i_0+1}}\right) \tag{11}$$

and

$$y^R_{i_0} = 2i_0\pi + \text{arccot}\left(-\frac{M_0}{|\alpha_{i_0+1} + \delta\alpha_{i_0}|} - \frac{m}{\alpha_{i_0+1}}\right). \tag{12}$$

Proof: For any $y \in (2i_0\pi, (2i_0+1)\pi)$, one has $g(y) > 0$. By direct computation, one obtains

$$\frac{g'(y)}{g(y)} = \frac{my^{m-1}\sin(y) + y^m\cos(y)}{y^m\sin(y)} = \frac{m}{y} + \cot(y), \tag{13}$$

$$\frac{m}{\alpha_{i_0+1}} \le \frac{m}{y} \le \frac{m}{\alpha_{i_0}}, \tag{14}$$

and

$$\lim_{y \to 2i_0\pi+} \cot(y) = +\infty, \qquad \lim_{y \to (2i_0+1)\pi-} \cot(y) = -\infty.$$

So, there is a unique point $y^*_{i_0} \in (2i_0\pi, (2i_0+1)\pi)$ such that $g'(y^*_{i_0}) = 0$, $g'(y) > 0$ for any $y \in (2i_0\pi, y^*_{i_0})$, and $g'(y) < 0$ for any $y \in (y^*_{i_0}, (2i_0+1)\pi)$. It is clear that $y^*_{i_0} > 2i_0\pi + \frac{\pi}{2}$.

Now, (4) is verified. By (13) and (14), one has, for $y \in (\alpha_{i_0}, y^L_{i_0}]$,

$$\frac{m}{y} + \cot(y) \ge \frac{m}{\alpha_{i_0+1}} + \cot(y) \ge \frac{M_0}{(1+\delta)\alpha_{i_0}};$$

and for $y \in [y^R_{i_0}, \alpha_{i_0+1})$,

$$\frac{m}{y} + \cot(y) \le \frac{m}{\alpha_{i_0}} + \cot(y) \le -\frac{M_0}{(1+\delta)\alpha_{i_0}}.$$

Similarly, (7) and (10) can be verified. ■

Lemma 3.2: *Fixed δ. One has*

- *For Case (a), if*

$$\rho > \frac{(1+\delta)\alpha_{i_0+1}}{(\alpha_{i_0} + \pi/2)^m}, \tag{15}$$

then $F(S_{i_0}) \cap S_{i_0}$ and $F^{-1}(S_{i_0}) \cap S_{i_0}$ have two connected components, respectively.

- *For Case (b), if*

$$\rho > \frac{\alpha_{i_0+1} + \delta\alpha_{i_0}}{(\alpha_{i_0} + \pi/2)^m}, \tag{16}$$

then $F(S_{i_0}) \cap S_{i_0}$ and $F^{-1}(S_{i_0}) \cap S_{i_0}$ have two connected components, respectively.

- *For Case (c), if*

$$\rho < \frac{\alpha_{i_0} + \delta\alpha_{i_0+1}}{(\alpha_{i_0} + \pi/2)^m}, \tag{17}$$

then $F(S_{i_0}) \cap S_{i_0}$ and $F^{-1}(S_{i_0}) \cap S_{i_0}$ have two connected components, respectively.

Remark 3.1: Figures 1–6 illustrate the images of S_{i_0} under F or F^{-1} for Cases (a)–(c).

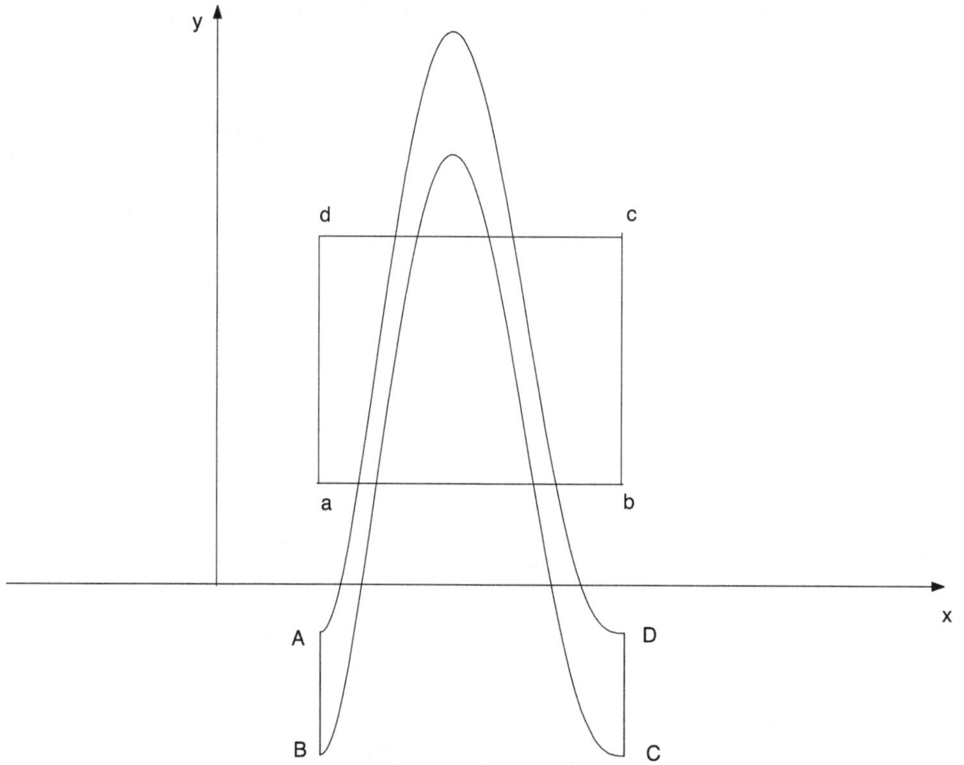

Figure 1. The illustration diagram for the image of S_{i_0} under F in Case (a) of Lemma 3.2, where $a := (\alpha_{i_0}, \alpha_{i_0})$, $b := (\alpha_{i_0+1}, \alpha_{i_0})$, $c := (\alpha_{i_0+1}, \alpha_{i_0+1})$, $d := (\alpha_{i_0}, \alpha_{i_0+1})$, and $A = F(a)$, $B = F(b)$, $C = F(c)$, $D = F(d)$.

Proof: It is easy to verify that, in Cases (a)–(c),

$$F([a,b]) \cap S_{i_0} = \emptyset, \quad F([d,c]) \cap S_{i_0} = \emptyset, \quad F^{-1}([a,d]) \cap S_{i_0} = \emptyset,$$
$$F^{-1}([b,c]) \cap S_{i_0} = \emptyset.$$

Consider the second component of F. The topological conditions for the existence of the horseshoes are derived as follows.

Case (a): If

$$\rho > \frac{(1+\delta)\alpha_{i_0+1}}{(\alpha_{i_0} + \pi/2)^m} = (1+\delta) \cdot \frac{\alpha_{i_0} + \pi}{(\alpha_{i_0} + \pi/2)^m},$$

then

$$p(\alpha_{i_0} + \pi/2) - \delta\alpha_{i_0+1} = \rho(\alpha_{i_0} + \pi/2)^m - \delta\alpha_{i_0+1} > \alpha_{i_0+1}.$$

This, together with Lemma 3.1, implies that $F(S_{i_0}) \cap S_{i_0}$ has two connected components.

If

$$\rho > \frac{(1+\delta)\alpha_{i_0+1}}{(\alpha_{i_0} + \pi/2)^m},$$

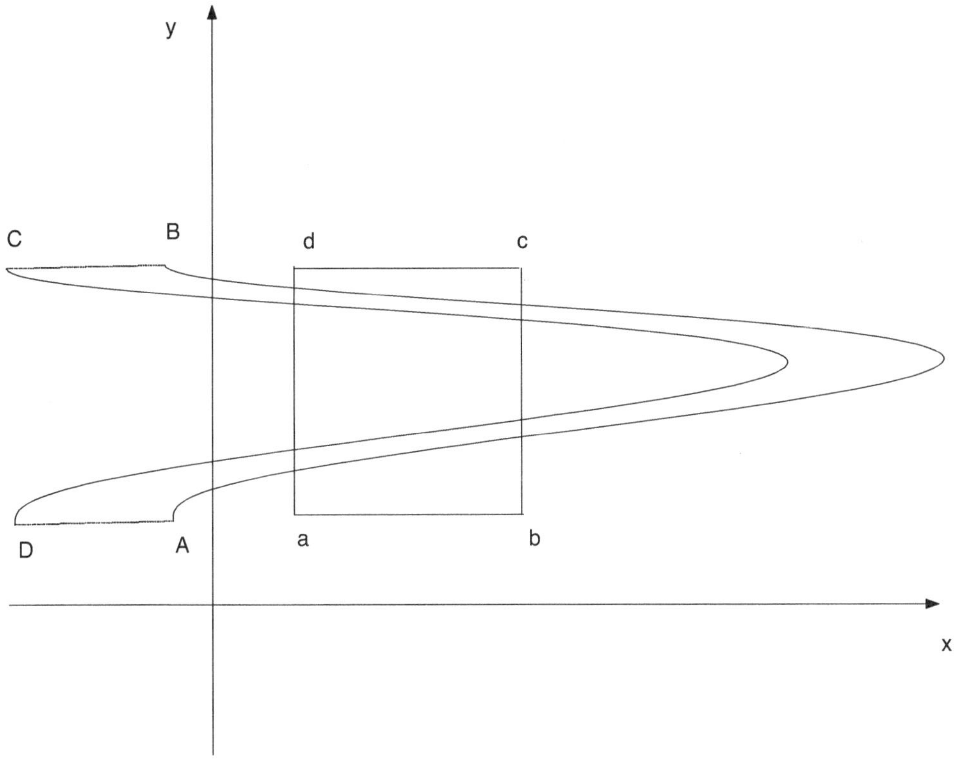

Figure 2. The illustration diagram for the image of S_{i_0} under F^{-1} in Case (a) of Lemma 3.2, where $a := (\alpha_{i_0}, \alpha_{i_0})$, $b := (\alpha_{i_0+1}, \alpha_{i_0})$, $c := (\alpha_{i_0+1}, \alpha_{i_0+1})$, $d := (\alpha_{i_0}, \alpha_{i_0+1})$, and $A = F^{-1}(a)$, $B = F^{-1}(b)$, $C = F^{-1}(c), D = F^{-1}(d)$.

then

$$\delta^{-1}(p(\alpha_{i_0} + \pi/2) - \alpha_{i_0+1}) = \delta^{-1}(\rho(\alpha_{i_0} + \pi/2)^m - \alpha_{i_0+1}) > \alpha_{i_0+1}.$$

This, together with Lemma 3.1, implies that $F^{-1}(S_{i_0}) \cap S_{i_0}$ has two connected components.

Case (b): Note that

$$\alpha_{i_0+1} + \delta\alpha_{i_0} > \alpha_{i_0} + \delta\alpha_{i_0+1} > 0.$$

If

$$\rho > \frac{\alpha_{i_0+1} + \delta\alpha_{i_0}}{(\alpha_{i_0} + \pi/2)^m},$$

then

$$p(\alpha_{i_0} + \pi/2) - \delta\alpha_{i_0} = \rho(\alpha_{i_0} + \pi/2)^m - \delta\alpha_{i_0} > \alpha_{i_0+1}$$

and

$$\delta^{-1}(p(\alpha_{i_0} + \pi/2) - \alpha_{i_0+1}) = \delta^{-1}(\rho(\alpha_{i_0} + \pi/2)^m - \alpha_{i_0+1}) < \alpha_{i_0}.$$

This, together with Lemma 3.1, implies that $F(S_{i_0}) \cap S_{i_0}$ and $F^{-1}(S_{i_0}) \cap S_{i_0}$ have two connected components, respectively.

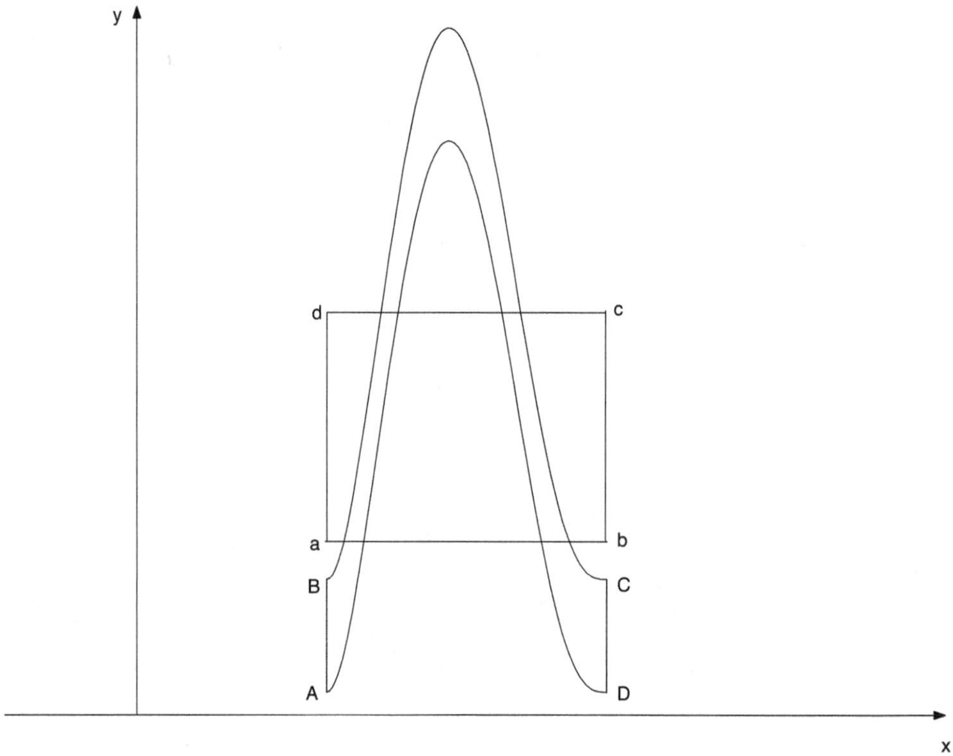

Figure 3. The illustration diagram for the image of S_{i_0} under F in Case (b) of Lemma 3.2, where $a := (\alpha_{i_0}, \alpha_{i_0})$, $b := (\alpha_{i_0+1}, \alpha_{i_0})$, $c := (\alpha_{i_0+1}, \alpha_{i_0+1})$, $d := (\alpha_{i_0}, \alpha_{i_0+1})$, and $A = F(a)$, $B = F(b)$, $C = F(c)$, $D = F(d)$.

Case (c): Note that

$$\alpha_{i_0} + \delta\alpha_{i_0+1} < \alpha_{i_0+1} + \delta\alpha_{i_0} < 0.$$

If

$$\rho < \frac{\alpha_{i_0} + \delta\alpha_{i_0+1}}{(\alpha_{i_0} + \pi/2)^m},$$

then

$$p(\alpha_{i_0} + \pi/2) - \delta\alpha_{i_0+1} = \rho(\alpha_{i_0} + \pi/2)^m - \delta\alpha_{i_0+1} < \alpha_{i_0}$$

and

$$\delta^{-1}(p(\alpha_{i_0} + \pi/2) - \alpha_{i_0}) = \delta^{-1}(\rho(\alpha_{i_0} + \pi/2)^m - \alpha_{i_0}) > \alpha_{i_0+1}.$$

This, together with Lemma 3.1, implies that $F(S_{i_0}) \cap S_{i_0}$ and $F^{-1}(S_{i_0}) \cap S_{i_0}$ have two connected components, respectively. ■

Lemma 3.3: *Set*

$$m_{i_0} := \min\{g(y_{i_0}^L), g(y_{i_0}^R)\}.$$

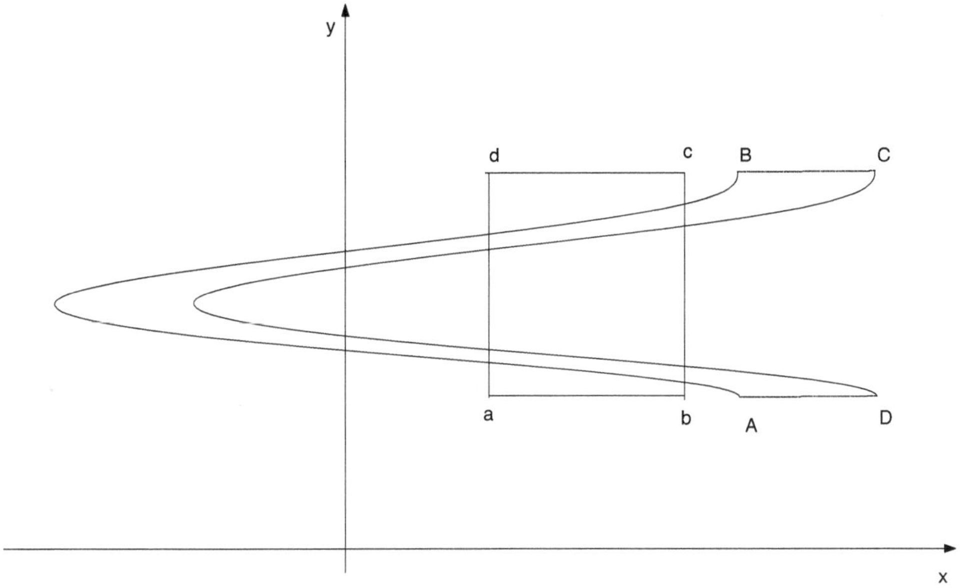

Figure 4. The illustration diagram for the image of S_{i_0} under F^{-1} in Case (b) of Lemma 3.2, where $a := (\alpha_{i_0}, \alpha_{i_0})$, $b := (\alpha_{i_0+1}, \alpha_{i_0})$, $c := (\alpha_{i_0+1}, \alpha_{i_0+1})$, $d := (\alpha_{i_0}, \alpha_{i_0+1})$, and $A = F^{-1}(a)$, $B = F^{-1}(b)$, $C = F^{-1}(c)$, $D = F^{-1}(d)$.

Assume that

(a)′ for Case (a),

$$\rho > \frac{(1+\delta)\alpha_{i_0+1}}{m_{i_0}}; \tag{18}$$

(b)′ for Case (b),

$$\rho > \frac{\alpha_{i_0+1} + \delta\alpha_{i_0}}{m_{i_0}}; \tag{19}$$

(c)′ for Case (c),

$$|\rho| > \frac{|\alpha_{i_0} + \delta\alpha_{i_0+1}|}{m_{i_0}}. \tag{20}$$

Then,

(1) *for any* $z = (x, y) \in S_{i_0} \cap F^{-1}(S_{i_0})$, *one has*

$$|p'(y)| > M_0;$$

(2) *for any* $z = (x, y) \in S_{i_0} \cap F(S_{i_0})$, *one has*

$$|p'(x)| > M_0.$$

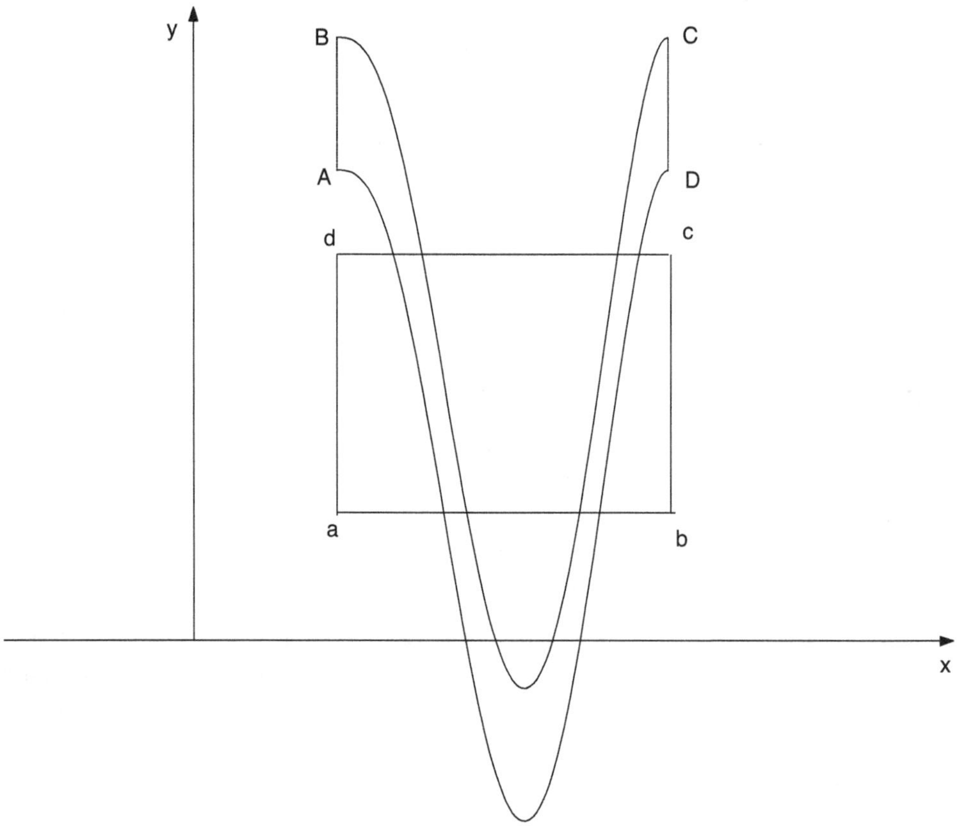

Figure 5. The illustration diagram for the image of S_{i_0} under F in Case (c) of Lemma 3.2, where $a := (\alpha_{i_0}, \alpha_{i_0})$, $b := (\alpha_{i_0+1}, \alpha_{i_0})$, $c := (\alpha_{i_0+1}, \alpha_{i_0+1})$, $d := (\alpha_{i_0}, \alpha_{i_0+1})$, and $A = F(a)$, $B = F(b)$, $C = F(c)$, $D = F(d)$.

Proof: In Case (a), it can be verified by direct calculation that, for any $z = (x, y) \in S_{i_0} \cap F^{-1}(S_{i_0})$, one has

$$(1+\delta)\alpha_{i_0} \le p(y) \le (1+\delta)\alpha_{i_0+1};$$

for any $z = (x, y) \in S_{i_0} \cap F(S_{i_0})$, one has

$$(1+\delta)\alpha_{i_0} \le p(x) \le (1+\delta)\alpha_{i_0+1}.$$

By (18), for any $z = (x, y) \in S_{i_0} \cap F^{-1}(S_{i_0})$, one has $y \in (\alpha_{i_0},\ y^L_{i_0}] \cup [y^R_{i_0},\ \alpha_{i_0+1})$; for any $z = (x, y) \in S_{i_0} \cap F(S_{i_0})$, one has $x \in (\alpha_{i_0},\ y^L_{i_0}] \cup [y^R_{i_0},\ \alpha_{i_0+1})$, where $y^L_{i_0}$ is defined in (5), and $y^R_{i_0}$ is specified in (6).

By (4), one has

$$\left|\frac{p'(y)}{p(y)}\right| = \left|\frac{g'(y)}{g(y)}\right| > \frac{M_0}{(1+\delta)\alpha_{i_0}}, \quad y \in (\alpha_{i_0}, y^L_{i_0}] \cup [y^R_{i_0}, \alpha_{i_0+1}).$$

In Case (b), for any $z = (x, y) \in S_{i_0} \cap F^{-1}(S_{i_0})$, one has

$$\alpha_{i_0} + \delta\alpha_{i_0+1} \le p(y) \le \alpha_{i_0+1} + \delta\alpha_{i_0};$$

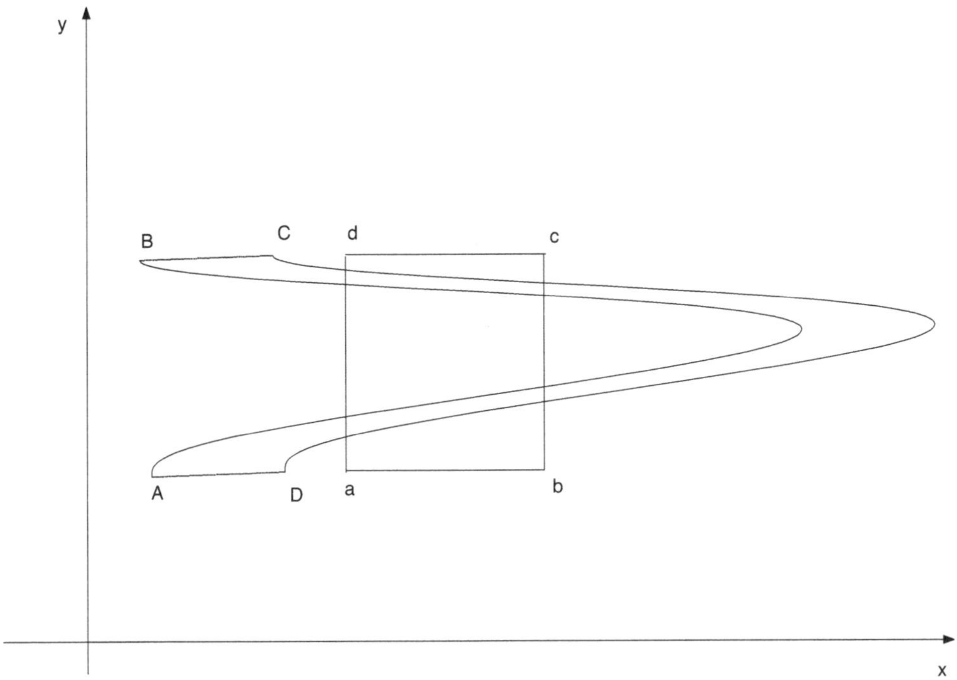

Figure 6. The illustration diagram for the image of S_{i_0} under F^{-1} in Case (c) of Lemma 3.2, where $a := (\alpha_{i_0}, \alpha_{i_0})$, $b := (\alpha_{i_0+1}, \alpha_{i_0})$, $c := (\alpha_{i_0+1}, \alpha_{i_0+1})$, $d := (\alpha_{i_0}, \alpha_{i_0+1})$, and $A = F^{-1}(a)$, $B = F^{-1}(b)$, $C = F^{-1}(c)$, $D = F^{-1}(d)$.

for any $z = (x, y) \in S_{i_0} \cap F(S_{i_0})$, one has

$$\alpha_{i_0} + \delta\alpha_{i_0+1} \leq p(x) \leq \alpha_{i_0+1} + \delta\alpha_{i_0}.$$

By (20), for any $z = (x, y) \in S_{i_0} \cap F^{-1}(S_{i_0})$, one has $y \in (\alpha_{i_0},\ y^L_{i0}] \cup [y^R_{i0},\ \alpha_{i_0+1})$; for any $z = (x, y) \in S_{i_0} \cap F(S_{i_0})$, one has $x \in (\alpha_{i_0},\ y^L_{i0}] \cup [y^R_{i0},\ \alpha_{i_0+1})$, where $y^L_{i_0}$ is defined in (8), and $y^R_{i_0}$ is specified in (9).

By (7), one has

$$\left|\frac{p'(y)}{p(y)}\right| = \left|\frac{g'(y)}{g(y)}\right| > \frac{M_0}{\alpha_{i_0} + \delta\alpha_{i_0+1}}, \quad y \in (\alpha_{i_0}, y^L_{i0}] \cup [y^R_{i0}, \alpha_{i_0+1}).$$

Similarly, in Case (c), for any $z = (x, y) \in S_{i_0} \cap F^{-1}(S_{i_0})$, one has

$$|\alpha_{i_0+1} + \delta\alpha_{i_0}| \leq |p(y)| \leq |\alpha_{i_0} + \delta\alpha_{i_0+1}|;$$

for any $z = (x, y) \in S_{i_0} \cap F(S_{i_0})$, one has

$$|\alpha_{i_0+1} + \delta\alpha_{i_0}| \leq |p(x)| \leq |\alpha_{i_0} + \delta\alpha_{i_0+1}|.$$

By (19), for any $z = (x, y) \in S_{i_0} \cap F^{-1}(S_{i_0})$, one has $y \in (\alpha_{i_0},\ y^L_{i0}] \cup [y^R_{i0},\ \alpha_{i_0+1})$; for any $z = (x, y) \in S_{i_0} \cap F(S_{i_0})$, one has $x \in (\alpha_{i_0},\ y^L_{i0}] \cup [y^R_{i0},\ \alpha_{i_0+1})$, where y^L_{i0} is defined in (11), and y^R_{i0} is specified in (12).

By (10), one has

$$\left|\frac{p'(y)}{p(y)}\right| = \left|\frac{g'(y)}{g(y)}\right| > \frac{M_0}{|\alpha_{i_0+1} + \delta\alpha_{i_0}|}, \quad y \in (\alpha_{i_0}, y^L_{i_0}] \cup [y^R_{i_0}, \alpha_{i_0+1}).$$

Combining the above gives the conclusion that in Cases (a)–(c), for any $z = (x, y) \in S_{i_0} \cap F^{-1}(S_{i_0})$, one has

$$|p'(y)| > M_0;$$

and for any $z = (x, y) \in S_{i_0} \cap F(S_{i_0})$, one has

$$|p'(x)| > M_0.$$

The proof is complete. ■

Lemma 3.4: *Assume that*

- *for Case (a),*

$$\rho > \max\left\{\frac{(1+\delta)\alpha_{i_0+1}}{(\alpha_{i_0} + \pi/2)^m}, \frac{(1+\delta)\alpha_{i_0+1}}{m_{i_0}}\right\}; \tag{21}$$

- *for Case (b),*

$$\rho > \max\left\{\frac{\alpha_{i_0+1} + \delta\alpha_{i_0}}{(\alpha_{i_0} + \pi/2)^m}, \frac{\alpha_{i_0+1} + \delta\alpha_{i_0}}{m_{i_0}}\right\}; \tag{22}$$

- *for Case (c),*

$$|\rho| > \max\left\{\frac{|\alpha_{i_0} + \delta\alpha_{i_0+1}|}{(\alpha_{i_0} + \pi/2)^m}, \frac{|\alpha_{i_0} + \delta\alpha_{i_0+1}|}{m_{i_0}}\right\}. \tag{23}$$

Then, $F : \Lambda_{i_0} \to \Lambda_{i_0}$ is topologically semi-conjugate to $\sigma : \sum_2 \to \sum_2$ in the above three cases.

Proof: By induction, it can be shown that $F : \Lambda_{i_0} \to \Lambda_{i_0}$ is topologically semi-conjugate to $\sigma : \sum_2 \to \sum_2$ for any a satisfying (21)–(23) in Cases (a)–(c), respectively. For more details on the topological semi-conjugacy, one can refer to the arguments used in the proof of Theorem 4.1 in [35, Page 253]. ■

Now, the uniformly hyperbolicity of the invariant sets are shown and the topological conjugate is established.

For the given constant $\lambda > 1$ and any point $z \in \Lambda$, define the cones $C^u(z)$ and $C^s(z)$ as follows:

$$C^u(z) = \{(\xi, \eta) \in T_z\mathbb{R}^2 : |\xi| \leq \lambda^{-1}|\eta|\},$$
$$C^s(z) = \{(\xi, \eta) \in T_z\mathbb{R}^2 : |\eta| \leq \lambda^{-1}|\xi|\}.$$

For any vector $(\xi, \eta) \in T_z\mathbb{R}^2$, the norm $|(\xi, \eta)|_* = \max\{|\xi|, |\eta|\}$ is used, which measures the largest component of the vector. For more on the cones, it is referred to [23].

Lemma 3.5: *For the given λ and ρ satisfying (21)–(23) in Cases* (a)–(c) *respectively, the following statements are true:*

(1) *for all $z \in S_{i_0} \cap F^{-1}(S_{i_0})$ and $v \in C^u(z)$, one has that $DF_z v \in C^u(F(z))$ and $|DF_z v|_* \geq \lambda |v|_*$;*
(2) *for all $z \in S_{i_0} \cap F(S_{i_0})$ and $v \in C^s(z)$, one has that $DF_z^{-1} v \in C^s(F^{-1}(z))$ and $|DF_z^{-1} v|_* \geq \lambda |v|_*$.*

Proof: This can be verified by Lemma 3.3 and direct computations with the expressions of DF and DF^{-1} in (3). For more details, one is referred to the arguments in [35, Page 282, Lemma 4.3]. ∎

For each $z \in \Lambda_{i_0}$, it follows from Lemma 3.5 that the set

$$\bigcap_{j=0}^{n} DF^{j}_{F^{-j}(z)} C^u(F^{-j}(z))$$

is a nested set of the cones at z. So, the infinite intersection is a nonempty cone, that is,

$$\mathbf{E}^u_z := \bigcap_{j=0}^{+\infty} DF^{j}_{F^{-j}(z)} C^u(F^{-j}(z)) \neq \emptyset.$$

Similarly, one has

$$\mathbf{E}^s_z := \bigcap_{j=0}^{-\infty} DF^{j}_{F^{-j}(z)} C^s(F^{-j}(z)) \neq \emptyset.$$

Lemma 3.6: *For any $z = (x, y) \in \Lambda_{i_0}$, $\mathbf{E}^u_z$ and $\mathbf{E}^s_z$ are lines.*

Proof: By contradiction argument, suppose that $\mathbf{E}^u_z$ is not a line. Then, there exist at least two lines. It suffices to show that the angle between these two lines is zero. For more details, one is referred to the arguments in [35, Pages 283–284]. ∎

By Lemmas 3.5 and 3.6, one has the following result.

Lemma 3.7: *For any ρ satisfying (21)–(23) in Cases* (a)–(c) *respectively, the connected components of Λ_{i_0} are points.*

Proof: It suffices to show that the radius of any connected subset of Λ_{i_0} shrinks to zero by Lemma 3.3. For more details, one is referred to the arguments in [35, Page 285]. ∎

Lemma 3.8: *For any ρ satisfying (21)–(23) in Cases* (a)–(c) *respectively, Λ_{i_0} is a hyperbolic invariant set for F.*

Proof: This is derived by Lemmas 3.5–3.7. ∎

Lemma 3.9: *For any ρ satisfying (21)–(23) in Cases* (a)–(c) *respectively, $F : \Lambda_{i_0} \to \Lambda_{i_0}$ is topologically conjugate to $\sigma : \sum_2 \to \sum_2$.*

Proof: This can be verified by a method similar to that used in Theorem 4.1 of [35, Page 253]. ■

Lemma 3.10: *Consider three different situations:*

(a) $\rho > 0$ *and* $\delta > 0$;
(b) $\rho > 0$ *and* $\frac{\alpha_{i_0}}{\alpha_{i_0+1}} > -\delta > 0$;
(c) $\rho < 0$ *and* $\delta < -\frac{\alpha_{i_0+1}}{\alpha_{i_0}}$.

Then, for any fixed δ and sufficiently large $|\rho|$, F has a Smale horseshoe and $\Lambda_{i_0} \subset S_{i_0}$ is a uniformly hyperbolic invariant set, such that $F : \Lambda_{i_0} \to \Lambda_{i_0}$ is topologically conjugate to $\sigma : \sum_2 \to \sum_2$.

Remark 3.2: For a fixed δ, the estimates of ρ are given in (21)–(23).

Proof: Lemma 3.10 can be completed by using Lemmas 3.2–3.9. ■

Now, it is ready to present the proof of Theorem 1.1.

Proof of Theorem 1.1.: The key is to show that the terms on the right-hand side of the equalities (21)–(23) approach zero as $i_0 \to +\infty$, that is,

- for Case (i),

$$\lim_{i_0 \to +\infty} \max \left\{ \frac{(1+\delta)\alpha_{i_0+1}}{(\alpha_{i_0} + \pi/2)^m}, \frac{(1+\delta)\alpha_{i_0+1}}{m_{i_0}} \right\} = 0;$$

- for Case (ii),

$$\lim_{i_0 \to +\infty} \max \left\{ \frac{\alpha_{i_0+1} + \delta\alpha_{i_0}}{(\alpha_{i_0} + \pi/2)^m}, \frac{\alpha_{i_0+1} + \delta\alpha_{i_0}}{m_{i_0}} \right\} = 0;$$

- for Case (iii),

$$\lim_{i_0 \to +\infty} \max \left\{ \frac{|\alpha_{i_0} + \delta\alpha_{i_0+1}|}{(\alpha_{i_0} + \pi/2)^m}, \frac{|\alpha_{i_0} + \delta\alpha_{i_0+1}|}{m_{i_0}} \right\} = 0.$$

It suffices to show Case (i). Other situations can be proved similarly. By (5) and direct computation, one has

$$\lim_{i_0 \to +\infty} \left(\frac{M_0}{(1+\delta)\alpha_{i_0}} - \frac{m}{\alpha_{i_0+1}} \right) = 0.$$

So,

$$\lim_{i_0 \to +\infty} \operatorname{arccot}\left(\frac{M_0}{(1+\delta)\alpha_{i_0}} - \frac{m}{\alpha_{i_0+1}}\right) = \frac{\pi}{2}.$$

Thus,

$$g(y_{i_0}^L) \approx \left(2i_0\pi + \frac{\pi}{2}\right)^m \quad \text{as } i_0 \to +\infty.$$

Similarly, one has

$$g(y_{i_0}^R) \approx \left(2i_0\pi + \frac{\pi}{2}\right)^m \quad \text{as } i_0 \to +\infty.$$

So,

$$\lim_{i_0 \to +\infty} \frac{(1+\delta)\alpha_{i_0+1}}{m_{i_0}} = 0$$

By (15), one has

$$\lim_{i_0 \to +\infty} \frac{(1+\delta)\alpha_{i_0+1}}{(\alpha_{i_0} + \pi/2)^m} = 0.$$

Therefore, for any fixed δ and ρ, there is a positive number N such that, for any $i_0 > N$, one has

$$\max\left\{\frac{(1+\delta)\alpha_{i_0+1}}{(\alpha_{i_0} + \pi/2)^m}, \frac{(1+\delta)\alpha_{i_0+1}}{m_{i_0}}\right\} < \rho.$$

To this end, by (a) of Lemma 3.10, for any $i_0 > N$, S_{i_0} corresponds to a Smale horseshoe and Λ_{i_0} corresponds to a uniformly hyperbolic invariant set, on which the map is topologically conjugate to the two-sided full-shift on two symbols. This proves Case (i).

For the other two cases, similar arguments apply. ■

Lemma 3.11: *For Case (i) in Theorem 1.1, the topological entropy is infinite.*

Proof: For the given m and ρ, and for any positive integer K, there is a large positive integer i_0 such that

$$(2i_0\pi)^m > (1+\delta)(2(i_0 + K - 1)\pi + \pi).$$

Consider the square region $[2i_0\pi, 2(i_0 + K - 1)\pi + \pi] \times [2i_0\pi, 2(i_0 + K - 1)\pi + \pi]$. The map restricted to this square is a $2K$-fold Smale horseshoe. By a simple topological argument, it can be shown that there is an invariant set on which the map is topologically semi-conjugate with the full-shift on $2K$ symbols. For this full-shift on $2K$ symbols, the transition matrix is

$$\begin{pmatrix} 1 & 1 & \cdots & 1 \\ 1 & 1 & \cdots & 1 \\ \vdots & \vdots & \ddots & \vdots \\ 1 & 1 & \cdots & 1 \end{pmatrix}_{2K \times 2K}.$$

The topological entropy of this full-shift on $2K$ symbols is known to be $\log(2K)$ [35].

Letting K go to infinity shows that the topological entropy is infinite. ■

Acknowledgments

The authors thank the reviewers for their helpful comments and suggestions.

Disclosure statement

No potential conflict of interest was reported by the author(s).

Funding

This research is supported by the Natural Science Foundation of Shandong Province, China (Grant No. ZR2023MA047).

References

[1] E. Bedford and J. Smillie, Polynomial diffeomorphisms of $\mathbb{C}^2$. VI: Connectivity of *J*, Ann. Math. 148 (1998), pp. 695–735.
[2] E. Bedford and J. Smillie, Polynomial diffeomorphisms of $\mathbb{C}^2$. VII: Hyperbolicity and external rays, Ann. Sci. Ecole Norm. Sup. 4 Série 32 (1999), pp. 455–497.
[3] E. Bedford and J. Smillie, Polynomial diffeomorphisms of $\mathbb{C}^2$. VIII: Quasiexpansion, Amer. J. Math. 124 (2002), pp. 221–271.
[4] E. Bedford and J. Smillie, Real polynomial diffeomorphisms with maximal entropy: I. Tangencies, Ann. Math. 160 (2004), pp. 1–26.
[5] E. Bedford and J. Smillie, Real polynomial diffeomorphisms with maximal entropy: II. Small Jacobian, Ergod. Theory Dynam. Sys. 26 (2005), pp. 1259–1283.
[6] M. Benedicks and L. Carleson, The dynamics of the Hénon map, Ann. Math. 133 (1991), pp. 73–169.
[7] M. Benedicks and M. Viana, Solution of the basin problem for Hénon-like attractors, Invent. Math.143 (2001), pp. 375–434.
[8] M. Benedicks and L.S. Young, Sinai-Bowen-Ruelle measure for certain Hénon maps, Invent. Math.112 (1993), pp. 541–576.
[9] P. Berger, Generic family with robustly infinitely many sinks, Invent. Math. 205 (2016), pp. 121–172.
[10] C. Bonatti, L. Díaz, and M. Viana, *Dynamics Beyond Uniform Hyperbolicity*, *Encyclopaedia of Mathematics* Vol. 102, Springer, Berlin Heidelberg, 2005.
[11] Y. Cao, S. Luzzato, and I. Rios, The boundary of hyperbolicity for Hénon-like families, Ergod. Theory Dynam. Syst. 28 (2008), pp. 1049–1080.
[12] E. Colli, Infinitely many coexisting strange attractors, Ann. Inst. Henri Poincaré 15 (1998), pp. 539–579.
[13] R.L. Devaney and Z. Nitecki, Shift automorphisms in the Hénon mapping, Commun. Math. Phys.67 (1979), pp. 137–146.
[14] H.R. Dullin and J.D. Meiss, Generalized Hénon maps: The cubic diffeomorphisms of the plane, Physica D 143 (2000), pp. 262–289.
[15] S. Friedland and J. Milnor, Dynamical properties of plane polynomial automorphisms, Ergod. Theory Dynam. Sys. 9 (1989), pp. 67–99.
[16] J. Gambaudo and C. Tresser, Diffeomorphisms with infinitely many strange attractors, J. Complexity 6 (1990), pp. 409–416.
[17] S.V. Gonchenko and M.C. Li, Shilnikov's cross-map method and hyperbolic dynamics of three-dimensional Hénon-like maps, Regul. Chaot. Dyn. 15 (2010), pp. 165–184.
[18] S.V. Gonchenko, I.I. Ovsyannikov, C. Simó, and D. Turaev, Three-dimensional Hénon-like maps and wild Lorenz-like attractors, Int. J. Bifurcat. Chaos 15 (2005), pp. 3493–3508.
[19] S.V. Gonchenko, J.D. Meiss, and I.I. Ovsyannikov, Chaotic dynamics of three-dimensional Hénon maps that originate from a homoclinic bifurcation, Regul. Chaot. Dyn. 11 (2006), pp. 191–212.

[20] S.V. Gonchenko, L.P. Shilnikov, and D.V. Turaev, On global bifurcations in three-dimensional diffeomorphisms leading to wild Lorenz-like attractors, Regul. Chaot. Dyn. 14 (2009), pp. 137–147.
[21] A. Gorodetski and V. Kaloshin, How often surface diffeomorphisms have infinitely many sinks and hyperbolicity of periodic points near a homoclinic tangency, Adv. Math. 208 (2007), pp. 710–797.
[22] M. Hénon, A two-dimensional mapping with a strange attractor, Commun. Math. Phys. 50 (1976), pp. 69–77.
[23] M.W. Hirsch and C. Pugh, *Global analysis*, in *Stable Manifolds and Hyperbolic Sets*, Proc. Sympos. Pure Math. Amer. Math. Soc. Vol. XIV, Providence, RI, 1970, pp. 133–163.
[24] U.A. Hoensch, Some hyperbolicity results for Hénon-like diffeomorphisms, Nonlinearity 21 (2008), pp. 587–611.
[25] A. Katok and B. Hasselblatt, *Introduction to the Modern Theory of Dynamical Systems*, Cambridge University Press, UK, 1995.
[26] S. Kong, C. Li, H. Jiang, Q. Lai, and X. Jiang, A 2D hyperchaotic map with conditional symmetry and attractor growth, Chaos 31 (2021), Article ID 043121.
[27] B. Leal, High dimension diffeomorphisms exhibiting infinitely many strange attractors, Ann. Inst. Henri Poincaré 25 (2008), pp. 587–607.
[28] Y. Li, C. Li, S. Zhang, G. Chen, and Z. Zeng, A self-reproduction hyperchaotic map with compound lattice dynamics, IEEE Trans. Ind. Electron. 69 (2022), pp. 10564–10572.
[29] Y. Li, C. Li, Y. Zhao, and S. Liu, Memristor-type chaotic mapping, Chaos 32 (2022), Article ID 021104.
[30] S.E. Newhouse, Non-density of axiom A on S^2, Proc. A.M.S. Symp. Pure Math. 14 (1970), pp. 191–202.
[31] S.E. Newhouse, Diffeomorphisms with infinitely many sinks, Topology 12 (1974), pp. 9–18.
[32] S.E. Newhouse, The abundance of wild hyperbolic sets and non-smooth stable sets for diffeomorphisms, Publ. Math. Inst. Hautes Etudes Sci. 50 (1979), pp. 101–151.
[33] J. Palis, A global view of dynamics and a conjecture on the denseness of finitude of attractors, Astérisque 261 (2000), pp. 335–347.
[34] J. Palis and M. Viana, High dimension diffeomorphisms displaying infinitely many periodic attractors, Ann. Math. 140 (1994), pp. 207–250.
[35] C. Robinson, Dynamical Systems: Stability, Symbolic Dynamics and Chaos, CRC Press, Florida, 1999.
[36] X. Zhang, Hyperbolic invariant sets of the real generalized Hénon maps, Chaos Solit. Fractals 43 (2010), pp. 31–41.
[37] L. Zhang, Y. Liu, Z. Wei, H. Jiang, and Q. Bi, A novel class of two-dimensional chaotic maps with infinitely many coexisting attractors, Chin. Phys. B 29 (2020), Article ID 060501.
[38] X. Zhang, C. Li, E. Dong, Y. Zhao, and Z. Liu, A conservative memristive system with amplitude control and offset boosting, Int. J. Bifurcat. Chaos 32 (2022), Article ID 2250057.

Index

Note: **Bold** page numbers refer to tables; *italic* page numbers refer to figures and page numbers followed by "n" denote endnotes.

Kolibri 360 GmbH
Ferdinand-Jühlke-Straße 7
99095 Erfurt, Deutschland
produktsicherheit@kolibri360.de